Properties and Performance of Concrete Materials and Structures

Properties and Performance of Concrete Materials and Structures

Editors

Piotr Smarzewski
Adam Stolarski

MDPI • Basel • Beijing • Wuhan • Barcelona • Belgrade • Manchester • Tokyo • Cluj • Tianjin

Editors
Piotr Smarzewski
Military University of
Technology in Warsaw
Poland

Adam Stolarski
Military University of
Technology
Poland

Editorial Office
MDPI
St. Alban-Anlage 66
4052 Basel, Switzerland

This is a reprint of articles from the Special Issue published online in the open access journal *Crystals* (ISSN 2073-4352) (available at: https://www.mdpi.com/journal/crystals/special_issues/Concrete_Performance).

For citation purposes, cite each article independently as indicated on the article page online and as indicated below:

LastName, A.A.; LastName, B.B.; LastName, C.C. Article Title. *Journal Name* **Year**, *Volume Number*, Page Range.

ISBN 978-3-0365-5601-7 (Hbk)
ISBN 978-3-0365-5602-4 (PDF)

© 2022 by the authors. Articles in this book are Open Access and distributed under the Creative Commons Attribution (CC BY) license, which allows users to download, copy and build upon published articles, as long as the author and publisher are properly credited, which ensures maximum dissemination and a wider impact of our publications.

The book as a whole is distributed by MDPI under the terms and conditions of the Creative Commons license CC BY-NC-ND.

Contents

About the Editors vii

Preface to "Properties and Performance of Concrete Materials and Structures" ix

Piotr Smarzewski and Adam Stolarski
Properties and Performance of Concrete Materials and Structures
Reprinted from: *Crystals* 2022, 12, 1193, doi:10.3390/cryst12091193 1

Di Qin, Yidan Hu and Xuemei Li
Waste Glass Utilization in Cement-Based Materials for Sustainable Construction: A Review
Reprinted from: *Crystals* 2021, 11, 710, doi:10.3390/cryst11060710 7

Shahriar Shahbazpanahi, Moslem Khalili Tajara, Rabar H. Faraj and Amir Mosavi
Studying the C–H Crystals and Mechanical Properties of Sustainable Concrete Containing Recycled Coarse Aggregate with Used Nano-Silica
Reprinted from: *Crystals* 2021, 11, 122, doi:10.3390/cryst11020122 41

Sayed Mohamad Soleimani, Abdel Rahman Alaqqad, Adel Jumaah, Naser Mohammad and Alanoud Faheiman
Incorporation of Recycled Tire Products in Pavement-Grade Concrete: An Experimental Study
Reprinted from: *Crystals* 2021, 11, 161, doi:10.3390/ cryst11020161 57

Shiming Liu, Miaomiao Zhu, Xinxin Ding, Zhiguo Ren, Shunbo Zhao, Mingshuang Zhao and Juntao Dang
High-Durability Concrete with Supplementary Cementitious Admixtures Used in Corrosive Environments
Reprinted from: *Crystals* 2021, 11, 196, doi:10.3390/cryst11020196 77

Xinxin Ding, Changyong Li, Minglei Zhao, Jie Li, Haibin Geng and Lei Lian
Tensile Behavior of Self-Compacting Steel Fiber Reinforced Concrete Evaluated by Different Test Methods
Reprinted from: *Crystals* 2021, 11, 251, doi:10.3390/cryst11030251 91

Wei Xia, Jinyu Xu and Liangxue Nie
Research on the Mechanical Performance of Carbon NanofiberReinforced Concrete under Impact Load Based on Fractal Theory
Reprinted from: *Crystals* 2021, 11, 387, doi:10.3390/cryst11040387 103

David O. Nduka, Babatunde J. Olawuyi, Olabosipo I. Fagbenle and Belén G. Fonteboa
Effect of $K_yAl_4(Si_{8-y})O_{20}(OH)_4$ Calcined Based-Clay on the Microstructure and Mechanical Performances of High-Performance Concrete
Reprinted from: *Crystals* 2021, 11, 1152, doi:10.3390/cryst11101152 119

Mukhtar Oluwaseun Azeez and Ahmed Abd El Fattah
Service Life Modeling of Concrete with SCMs Using Effective Diffusion Coefficient and a New Binding Model
Reprinted from: *Crystals* 2020, 10, 967, doi:10.3390/cryst10110967 141

Petr V. Sivtsev and Piotr Smarzewski
Hardening Parameter Homogenization for J2 Flow with Isotropic Hardening of Steel Fiber-Reinforced Concrete Composites
Reprinted from: *Crystals* 2021, 11, 776, doi:10.3390/cryst11070776 161

Hisham Alabduljabbar, Rayed Alyousef, Hossein Mohammadhosseini and Tim Topper
Bond Behavior of Cleaned Corroded Lap Spliced Beams Repaired with Carbon Fiber Reinforced Polymer Sheets and Partial Depth Repairs
Reprinted from: *Crystals* **2020**, *10*, 1014, doi:10.3390/cryst10111014 . . . . **187**

Pavlina Mateckova, Vlastimil Bilek and Oldrich Sucharda
Comparative Study of High-Performance Concrete Characteristics and Loading Test of Pretensioned Experimental Beams
Reprinted from: *Crystals* **2021**, *11*, 427, doi:10.3390/cryst11040427 . . . . **201**

Mostafa Moghadasi, Soheil Taeepoor, Seyed Saeid Rahimian Koloor and Michal Petrů
The Effect of Lateral Load Type on Shear Lag of Concrete Tubular Structures with Different Plan Geometries
Reprinted from: *Crystals* **2020**, *10*, 897, doi:10.3390/cryst10100897 . . . . **219**

About the Editors

Piotr Smarzewski

Piotr Smarzewski is an Associate Professor of Civil Engineering at Faculty of Civil Engineering and Geodesy, Military University of Technology in Warsaw, Poland. He obtained his M.Sc. degree, with a specialty in Building and Engineering Structures, from the Lublin University of Technology (LUT) in 1997. Following two years of structural engineering practice, from 1997 to 1999, he has devoted his career to teaching and research since 1999. He obtained his Ph.D. degree in Civil Engineering in 2008 from the LUT, and then his postdoctoral degree (habilitation) in Engineering and Technical Sciences in the discipline of Civil Engineering and Transport from the Poznan University of Technology in 2021. Piotr Smarzewski's research studies have led to more than 80 publications in technical journals and conference proceedings. He is the author or co-author of two textbooks, twelve chapters in handbooks, and editor or co-editor of two books. He is an Editorial Board Member of two journals indexed in the JCR list, Guest Co-Editor of two Special Issues in the JCR journals, and Review Board Member of the MDPI and Frontiers Publishing Houses.

Adam Stolarski

Adam Stolarski is a full professor of Civil Engineering at Faculty of Civil Engineering and Geodesy, Military University of Technology, Warsaw, Poland, where he received his Ph.D. in 1983, D.Sc. in 1990, and Professor title in 2005. He is the author and co-author of over 160 scientific publications and 15 monographic studies, including "Nonlinear analysis of impulse-loaded reinforced concrete bar structures", "Modeling of the static and dynamic behavior of inelastic reinforced concrete deep beams". He is the co-author of two patents concerning "The system of flat meshes with a truss system of bars for concrete reinforcement" and "Spatial reinforcement of concrete" granted by Polish Patent Office. He was the supervisor of 7 doctoral dissertations. In the years 2005–2013, he was the deputy dean, and in the years 2016–2019, the dean of the Faculty of Civil Engineering and Geodesy at Military University of Technology.

Preface to "Properties and Performance of Concrete Materials and Structures"

The collection of papers published in this Special Issue introduces the current research on the properties and performance of concrete materials and structures.

The objective of the Special Issue is to publish scientists and engineers from around the world, presenting the latest developments in the fields of concrete engineering, mechanics and computation. The papers published in this Special Issue present a discussion of the modern and future trends in experimental, analytical and numerical investigations on concrete materials and structures. This Special Issue also provides an opportunity to disseminate interdisciplinary knowledge among researchers and civil engineers from around the world.

The main topics covered in the accepted papers include the following:

1. Mechanical properties, durability and microstructure of cement-based materials containing different waste materials.

2. Mechanical and microstructural properties, and a durability assessment of high-performance concrete.

3. Mechanical properties of self-compacting steel-fiber-reinforced concrete.

4. Dynamic properties of carbon-nanofiber-reinforced concrete.

5. Models and numerical simulations for concrete-containing supplementary cementitious materials and steel fibers.

6. Application of carbon-fiber-reinforced polymer materials as partial-depth concrete beam repair.

7. Behavior and application of pre-stressed high-performance concrete bridge beams.

8. Analytical and numerical models for framed tube structures.

Particular thanks are due to all authors of the published papers. Grateful thanks are also extended to peer reviewers for their significant help in reviewing the papers and sharing their extensive knowledge and experience in their comments and concerns. The tireless efforts of the management and staff at MDPI in their editorial support in the implementation of this project are highly appreciated.

We sincerely hope that the papers published in this Special Issue will contribute to the development of science and technology in the field of concrete materials and structures.

Piotr Smarzewski and Adam Stolarski
Editors

Editorial

Properties and Performance of Concrete Materials and Structures

Piotr Smarzewski [1,*] and Adam Stolarski [2,*]

1 Department of Structural Engineering, Faculty of Civil Engineering and Architecture, Lublin University of Technology, Nadbystrzycka 40, 20-618 Lublin, Poland
2 Faculty of Civil Engineering and Geodesy, Military University of Technology, 2 Gen. Sylwestra Kaliskiego, 00-908 Warsaw, Poland
* Correspondence: p.smarzewski@pollub.pl (P.S.); adam.stolarski@wat.edu.pl (A.S.); Tel.: +48-698695284 (P.S.)

1. Introduction

Concrete is one of the ancient and most widely used construction material. In recent decades, numerous advances and developments being made in the field of concrete, which were implemented in practical applications. Nowadays, the term modern concrete refers to concrete with good workability, high fracture toughness, high mechanical strength and chemical durability. Structural elements from such materials extends the frontiers of the design and enables the implementation of outstand, durable, ecological and safe structures of the highest quality.

The aim of this Special Issue is to publish current research on concrete composites based on Portland cement or other blended cements and binders containing inclusions of waste materials, special aggregates, e.g., from recycling and/or fibers. The Special Issue focuses on presenting the results of research on the properties and performance of concrete composites, novel experimental techniques, analytical methods, modelling, design, production and practical applications of these materials, and studies regarding the behavior of structural components, in situ performance, renovation, durability and sustainability of structures made of these composites. The next section provides a brief summary of each papers published.

2. Content of Special Issue

Qin et al. [1] performed a scientometric review on the utilization of waste glass (WG) in cement-based materials (CBM) along with an extensive discussion. The article uses scientometric analysis and a comprehensive manual review. Among other things, the scientometric analysis was conducted to establish the current research trends, to identify the publication fields, the sources with the most publications, the most quoted articles and authors, and the countries with a significant contribution to the field of WG utilization in CBMs for sustainable construction. In addition, the sustainable aspects of WG utilization in construction materials were reviewed, as well as the effect of WG on the workability, compressive strength, splitting tensile strength, flexural strength, microstructure and durability of CBMs was evaluated. The scientometric analysis exposed a remarkable increase in the number of publications on this topic in the last 5 years. It was observed that the largest number of documents have been published in Journal of Cleaner Production, Construction and Building Materials, and Resources, Conservation, and Recycling. Moreover, India, China and the United Kingdom contributed the most documents in the current research field. It was reported that WG can be used in CBMs as aggregate replacement and cement replacement, thus protecting natural resources, solving waste management problems, reducing CO_2 emissions by reducing cement demand, protecting the environment from toxic chemicals, and producing cost-effective CBMs. It was found that generally finer glass particles increased, while coarser WG particles decreased the mechanical properties of

Citation: Smarzewski, P.; Stolarski, A. Properties and Performance of Concrete Materials and Structures. Crystals 2022, 12, 1193. https://doi.org/10.3390/cryst12091193

Received: 19 August 2022
Accepted: 21 August 2022
Published: 25 August 2022

Publisher's Note: MDPI stays neutral with regard to jurisdictional claims in published maps and institutional affiliations.

Copyright: © 2022 by the authors. Licensee MDPI, Basel, Switzerland. This article is an open access article distributed under the terms and conditions of the Creative Commons Attribution (CC BY) license (https:// creativecommons.org/licenses/by/ 4.0/).

CBMs. Increasing these properties is possible by replacing WG up to 25% of the cement or up to 20% of natural aggregate. The addition of WG can help improve the microstructure and reduce the permeability of CBMs, thus enhancing their durability. On the other hand, WG can reduce the resistance to carbonation. It was suggested that the amount, size, and type of WGs used in CBMs are adequate to achieve the appropriate mechanical properties and durability dependent on the anticipated applications. As a result of the discussion, it was identified, inter alia, that it is a necessity to explore the influence of WG on the rheological properties of CBM in terms of amount, type, particle size, and morphology of WG particles, or to determine the effect of the content and particle size of WG on the durability of CBMs at different w/b ratios.

Shahbazpanahi et al. [2] investigated the mechanical properties and microstructure of sustainable concrete produced by replacing the natural coarse aggregate (NCA) with recycled coarse aggregate containing used nano-silica (RCA-UNS). In the first group, specimens from the control normal concrete were studied. In the second group, specimens with 30%, 40% and 50% of NCA replacement by coarse aggregate obtained from crushed normal concrete from the first group and 0.5% addition of nano-silica were performed. In the third group, specimens with 30%, 40% and 50% of NCA replacement by RCA-UNS obtained from 90-day crushed specimens from the second group were made. Water absorption, fresh concrete slump, and compressive strength were determined and compared through Fourier transform infrared spectroscopy (FT-IR), X-ray diffraction (XRD), and scanning electron microscopy (SEM) tests. The results showed that the water absorption of the RCA-UNS specimens decreased compared to the control specimens. Moreover, the results of the 28-day compressive strength test showed that the compressive strength in the third group increased by 12.8%, 10.9% and 10% after replacing 30%, 40% and 50% of NAC by RCA-NS in compared to the control specimens. The SEM results displayed that the 30% RCA-UNS specimens produced additional C–S–H, and the XRD and FT-IR graphs illustrated that in the RCA-UNS specimens more C–H crystals were consumed and converted to C–S–H. 30% replacement of NAC by RCA-UNS was found to be the best replacement for the production of sustainable concrete.

Soleimani et al. [3] performed a study the mechanical properties of green concrete containing recycled tire by-products for pavements. All tire by-products were tested individually and hybrid to investigate the concrete assets and determine their effect on the reference mixture. Eleven concrete mixtures were produced with different doses of shredded rubber (SR) or crumbed rubber (CR) or steel fibers (StF) from tire recycling, as well as twelve hybrid concrete mixtures containing different doses of various tire by-products were developed. Then, the impact of waste on the slump, compressive strength, splitting tensile strength, and modulus of rupture of the concrete were evaluated. The incorporation of SR to the reference concrete mixture had a significant impact on the 7 and 28-day compressive and splitting tensile strength. The results showed that the inclusion of CR had a detrimental effect on all the concrete properties tested, but the splitting tensile strength and modulus of rupture were the most evident. On the other hand, the introduction of 0.1% and 0.2% StF resulted in an increase in the 7-day compressive strength and the modulus of rupture. It was found that the incorporation of 5% rubber products and 0.1% steel fibers caused at least the maintenance of the reference mixture properties. The conclusions of this study showed that it is possible to hybridize all recycled tire materials to produce feasible pavement-grade concrete suited for hot weather conditions.

Liu et al. [4] examined the resistance to chloride penetration, sulfate attack and frost of high-performance concrete (HPC). For this purpose, fifteen concretes with a different water-binder ratio with changes in the content of fly ash (FA), silica fume (SF), comminuted granulated blast furnace slag (GGBS), and admixture of sulfate corrosion resistance (AS) were designed. The compressive strength, the total electric flux of chloride permeability, the sulfate resistance coefficient and the freeze-thaw indexes of HPC were determined. The results showed that the compressive strength and the durability of HPC depends on the chemical composition, fineness and pozzolanic activity of the supplementary cementitious admixtures. GGBS had a negative effect on the HPC properties. On the other hand, SF and FA presented beneficial effects on concrete, also when used in conjunction with GGBS. The AS also improved the compressive strength, the resistance to chloride penetration, and the sulfate corrosion resistance of HPC. The concretes were characterized by compressive strength ranging from 70 MPa to 113 MPa, except for the HPCs admixed with GGBS or GGBS + FA. All HPCs were the highest grade over F400 for the frost resistance, with a relative dynamic modulus of elasticity no less than 60% and a weight loss rate of no larger than 5%. The concretes with admixtures of 7% FA, 8% SF, and 8% GGBS or 7% FA, 8% SF, 8% GGBS, and 10–12% AS with a water to binder ratio of 0.29, a total binder of 500 kg/m^3, and the compressive strength of about 100 MPa presented the highest grades of resistance to chloride penetration, sulfate corrosion, and frost.

Ding et al. [5] assessed the tensile strength of self-compacting steel fiber reinforced concrete (SFRC). Seven groups of self-compacting SFRC with steel fibers with the hooked ends with a length of 25.1 mm, 29.8 mm and 34.8 mm were prepared with a volume fraction ranging from 0.4% to 1.4%. The axial tensile tests and the splitting tensile tests were carried out. The results showed that the axial tensile strength was higher than the splitting tensile strength. Moreover, it was noted that the axial tensile work and toughness were not related to the length of the steel fiber. Additionally, the equations for the prediction of tensile strength of self-compacting SFRC were proposed taking into account the effects of fiber distribution, fiber ratio, and volume fraction.

Xia et al. [6] investigated the dynamic compressive strength, impact toughness, and fragmentation size distribution law of the plain concrete and the carbon nanofiber reinforced concrete with 0.1%, 0.2%, 0.3%, and 0.5% volume content. Tests were performed under impact load by using the Φ100 mm split-Hopkinson pressure bar. The influence of the strain rate and the dosage of carbon nanofibers (CNF) on the dynamic mechanical performance of concrete was analyzed. It was reported that the dynamic compressive strength and the impact toughness increased with the improvement of the strain rate level at the same fiber content. On the other hand, at the same strain rate, the impact toughness increased with the increase in the fiber dosage, but the dynamic compressive strength initially increased and then decreased. It was also observed that the higher the strain rate level was, the higher the number of crushed concrete fragments with the lower the size, and the larger the fractal particle dimension were obtained. It was found that the optimal dosage of CNF in order to improve the dynamic compressive strength of concrete is 0.3%.

Nduka et al. [7] carried out studies to determine the potential use of meta-illite calcined clay (MCC) as a supplementary cementitious material (SCM) in a binary Portland cement for the production of high-performance concrete (HPC). Quantitative analyses of the chemical composition, mineral phases, morphology, calcination efficiency, and physical properties were performed using X-ray fluorescence (XRF), scanning electron microscopy/energy dispersive X-rays (SEM/EDX), X-ray diffraction (XRD), Fourier transform infrared/attenuated total reflection (FTIR/ATR), thermogravimetric analysis (TGA), laser particle sizing and Brunauer-Emmett-Teller nitrogen absorption method (BET) to obtain the properties of the cementitious materials. Moreover, the influence of MCC on the workability, compressive strength, splitting tensile strength, and flexural strength as well as HPC microstructure were determined. The XRF results displayed that the MCC had a high useful oxides content. In turn, the XRD results showed that MCC was predominantly an illite-based clay mineral calcined, as revealed by TGA, at a maximum temperature of 650 °C. Furthermore, the addition of MCC at a 5–15% cement replacement increased the HPC slump flow. The inclusion of MCC with a 10% cement replacement best improved the porosity of the HPC resulting in increased mechanical properties. It was recommended that the addition of MCC within 10% cement replacement should be adopted for low w/b Class I HPC without detrimental effects on the mechanical and microstructural properties of concrete.

Oluwaseun Azeez and Abd El Fattah [8] developed a new model to predict the effective diffusivity of concrete taking into account the effects of binding, age, temperature, carbonation, and free chloride. A new algorithm was developed to determine the corrosion initiation time and to predict the concentration of free chloride at various depths in the concrete containing supplementary cementitious materials (SCM). The transport model uses the calibrated effective diffusion by considering the environmental impacts and experimental data of the binding capacity of concrete. Different mixtures of ordinary Portland cement and SCM were tested to determine the experimental binding capacity values used in the algorithm. Chloride profiles were measured on concrete blocks exposed to daily seawater, as well as exposed to harsh weather conditions for two years at the east coast of Saudi Arabia. Linear polarization and chloride profiling assessed the performance of the concrete mixtures in the corrosion environmental. The results generated by the model were compared with the performance of concrete blocks. Statistical analysis proved good accuracy of the model using experimental data from the binding capacity. The proposed transport model was evidenced to be effective in predicting free chloride profiles using the effective diffusion and binding capacity.

Sivtsev and Smarzewski [9] performed numerical modeling of the stress-strain state of steel fiber reinforced concrete using the method of numerical homogenization. In this paper, the description of the anisotropic nature of hardening of the composite material and the numerical homogenization for the J2 flow with isotropic hardening was proposed. The model problem was the deformation of the composite material with a periodic arrangement of inclusions in the form of steel fibers, assuming purely elastic properties for the fibers. Numerical homogenization of the elasticity and plasticity parameters were performed on the representative element. The calculated effective parameters were used to solve the problem on a coarse mesh. The accuracy of the application of the computational algorithm was checked on model problems in comparison with the hardening parameters of the base composite material. In accordance with the obtained results, the proposed model of homogenization of the hardening coefficient demonstrated satisfactory results when one of the components of the strain tensor was prevailed. However, when the components of the strain tensor had comparable values, the error values were already higher. This was due to the anisotropy of the plastic flow, which cannot be fully accounted for in the simple numerical change in the hardening coefficient. It was found that both approximations work quite accurately at small plastic deformations.

Alabduljabbar et al. [10] examined the bond behavior of a cleaned corroded reinforcing bar repaired with a partial depth concrete repair or a partial depth concrete repair using carbon fiber reinforced polymer (CFRP) sheets. Twelve lap splice beams were tested under static loading. The experiment variable was the repair method, i.e., a partial depth repair with pre-packaged self-consolidating concrete (SCC) in six lap splice beams, and additional confinement with CFRP sheets in other six beams. The test results of the repaired lap splice beams were compared with the results for a monolithic lap splice beam. The study showed that the average bond strength increased with increasing the reinforcing bar mass loss for all bonded lengths. The partial depth SCC repairing of the beams improved the average bond strength compared to the monolithic beams. For the lap splice beams repaired with a partial depth, higher concrete strength was obtained than for the monolithic beams. In addition, the beams confined with CFRP sheets displayed an increase in bond strength by 34–49%, an increase in the equivalent slip by 56–260%, a higher maximum load by 49% and a higher corresponding deflection by 191% compared to the unconfined beams.

Mateckova et al. [11] presented two variants of high-performance concrete (HPC), which were developed from the modification of ordinary concrete used for the production of pretensioned bridge beams. Both variants were produced in industrial conditions with commonly used raw materials. The basic mechanical properties of HPC and the resistance to chloride penetration were tested and compared. Moreover, the tests of model experimental pretensioned beams with a length of 7 m prepared of normal strength concrete and one variant of HPC were carried out, the load-deformation relationships were determined, and the calculation method of the load capacity were verified. The tests of pretensioned beams indicates the convenience of the calculation model of the ultimate bending moment capacity for structural beams prepared of concrete with compressive strength exceeding the validity limit of the design code. The research also proved the increase in HPC resistance to chloride penetration compared to ordinary concrete.

Moghadasi et al. [12] studied the effect of height, plan geometry, and lateral load type on shear lag behavior of framed tube structures. The possible relation between the shear lag and the type of lateral load acting on structural systems was investigated. Twelve models with four different heights and three different plan geometry contrary to three different lateral load types were considered. Various plan geometry including rectangular, triangular and hexagon was modeled and subjected to the wind and earthquake load. It was observed that all types of structures subjected to the wind load had a higher value of shear lag factor in comparison with structures subjected to the static and dynamic earthquake loads. In addition, hexagon shaped plan structures had the most reasonable behavior versus lateral loads. In particular, the average of shear lag factors in the three types of analyses were about 25–29% less in the hexagon shaped plan structures compared to the control rectangular shaped plan structure.

The above-mentioned papers can significantly contribute to the development of advanced concrete materials and structures.

Author Contributions: Conceptualization, P.S. and A.S.; writing—original draft preparation, P.S. and A.S.; writing—review and editing, P.S. and A.S. All authors have read and agreed to the published version of the manuscript.

Funding: This research received no external funding.

Acknowledgments: We thank all the authors and peer reviewers for their valuable contributions to this Special Issue. We also thank the management and staff of MDPI for editorial support in the implementation of this project.

Conflicts of Interest: The authors declare no conflict of interest.

References

1. Qin, D.; Hu, Y.; Li, X. Waste Glass Utilization in Cement-Based Materials for Sustainable Construction: A Review. *Crystals* **2021**, *11*, 710. [CrossRef]
2. Shahbazpanahi, S.; Tajara, M.K.; Faraj, R.H.; Mosavi, A. Studying the C–H Crystals and Mechanical Properties of Sustainable Concrete Containing Recycled Coarse Aggregate with Used Nano-Silica. *Crystals* **2021**, *11*, 122. [CrossRef]
3. Soleimani, S.M.; Alaqqad, A.R.; Jumaah, A.; Mohammad, N.; Faheiman, A. Incorporation of Recycled Tire Products in Pavement-Grade Concrete: An Experimental Study. *Crystals* **2021**, *11*, 161. [CrossRef]
4. Liu, S.; Zhu, M.; Ding, X.; Ren, Z.; Zhao, S.; Zhao, M.; Dang, J. High-Durability Concrete with Supplementary Cementitious Admixtures Used in Corrosive Environments. *Crystals* **2021**, *11*, 196. [CrossRef]
5. Ding, X.; Li, C.; Zhao, M.; Li, J.; Geng, H.; Lian, L. Tensile Behavior of Self-Compacting Steel Fiber Reinforced Concrete Evaluated by Different Test Methods. *Crystals* **2021**, *11*, 251. [CrossRef]
6. Xia, W.; Xu, J.; Nie, L. Research on the Mechanical Performance of Carbon Nanofiber Reinforced Concrete under Impact Load Based on Fractal Theory. *Crystals* **2021**, *11*, 387. [CrossRef]
7. Nduka, D.O.; Olawuyi, B.J.; Fagbenle, O.I.; Fonteboa, B.G. Effect of $K_yAl_4(Si_{8-y})O_{20}(OH)_4$ Calcined Based-Clay on the Microstructure and Mechanical Performances of High-Performance Concrete. *Crystals* **2021**, *11*, 1152. [CrossRef]
8. Oluwaseun Azeez, M.; Abd El Fattah, A. Service Life Modeling of Concrete with SCMs Using Effective Diffusion Coefficient and a New Binding Model. *Crystals* **2020**, *10*, 967. [CrossRef]
9. Sivtsev, P.V.; Smarzewski, P. Hardening Parameter Homogenization for J2 Flow with Isotropic Hardening of Steel Fiber-Reinforced Concrete Composites. *Crystals* **2021**, *11*, 776. [CrossRef]
10. Alabduljabbar, H.; Alyousef, R.; Mohammadhosseini, H.; Topper, T. Bond Behavior of Cleaned Corroded Lap Spliced Beams Repaired with Carbon Fiber Reinforced Polymer Sheets and Partial Depth Repairs. *Crystals* **2020**, *10*, 1014. [CrossRef]
11. Mateckova, P.; Bilek, V.; Sucharda, O. Comparative Study of High-Performance Concrete Characteristics and Loading Test of Pretensioned Experimental Beams. *Crystals* **2021**, *11*, 427. [CrossRef]
12. Moghadasi, M.; Taeepoor, S.; Rahimian Koloor, S.S.; Petrů, M. The Effect of Lateral Load Type on Shear Lag of Concrete Tubular Structures with Different Plan Geometries. *Crystals* **2020**, *10*, 897. [CrossRef]

Review

Waste Glass Utilization in Cement-Based Materials for Sustainable Construction: A Review

Di Qin [1], Yidan Hu [1] and Xuemei Li [2,*]

[1] Changchun Institute of Technology, Changchun 130021, China; qindi@ccit.edu.cn (D.Q.); huyidan@ccit.edu.cn (Y.H.)
[2] Changchun Sci-Tech University, Changchun 130600, China
* Correspondence: 100464@cstu.edu.cn

Abstract: The construction industry has a significant environmental impact, contributing considerably to CO_2 emissions, natural resource depletion, and energy consumption. The construction industry is currently trending towards using alternative construction materials in place of natural materials and cement, thereby reducing the environmental impact and promoting sustainability. Two approaches have been used in this review: scientometric analysis and a comprehensive manual review on the waste glass (WG) utilization in cement-based materials (CBMs) as a sustainable approach. Scientometric analysis is conducted to find out the current research trend from available bibliometric data and to identify the relevant publication fields, sources with the most publications, the most frequently used keywords, the most cited articles and authors, and the countries that have made the most significant contribution to the field of WG utilization in CBMs. The effect of WG on the mechanical properties of CBMs was found to be inconsistent in the literature. The inconsistent effects of WG impede its acceptance in the construction sector. This study intends to shed light on the arguments and tries to explain the opposing perspectives. This article summarizes the findings of various research groups and recommends new viewpoints based on the assessment of fundamental processes. The effect of utilizing WG on fresh and hardened properties of CBMs, including workability, compressive strength, split-tensile strength, and flexural strength, are reviewed. Furthermore, the microstructure and durability of composites containing WG are investigated. Different limitations associated with WG use in CBMs and their possible solution are reported. This study will assist researchers in identifying gaps in the present research. Additionally, the scientometric review will enable researchers from diverse regions to exchange novel ideas and technologies, collaborate on research, and form joint ventures.

Keywords: cement-based materials; waste glass; sustainable construction; mechanical properties; durability

Citation: Qin, D.; Hu, Y.; Li, X. Waste Glass Utilization in Cement-Based Materials for Sustainable Construction: A Review. *Crystals* **2021**, *11*, 710. https://doi.org/10.3390/cryst11060710

Academic Editors: Piotr Smarzewski and Adam Stolarski

Received: 8 June 2021
Accepted: 17 June 2021
Published: 21 June 2021

Publisher's Note: MDPI stays neutral with regard to jurisdictional claims in published maps and institutional affiliations.

Copyright: © 2021 by the authors. Licensee MDPI, Basel, Switzerland. This article is an open access article distributed under the terms and conditions of the Creative Commons Attribution (CC BY) license (https://creativecommons.org/licenses/by/4.0/).

1. Introduction

Sustainability trends have accelerated recently as a result of resource constraints, resulting in an increased number of emerging issues from managerial, strategic, and operational perspectives. Additionally, the construction sector significantly contributes to society's requirements by improving people's quality of life [1–3]. Despite this, this industry generates between 45 and 65% of waste disposed of in landfills, accounting for 35% of global CO_2 emissions. Additionally, the construction industry and its related activities generate substantial amounts of harmful emissions, accounting for nearly 30% of global greenhouse gas emissions resulting from construction operations, with transportation and processing of construction materials accounting for 18% of these emissions [4]. The value of sustainability research in the field of civil construction is self-evident. In this regard, firms are increasingly aware that ensuring a competitive advantage is contingent on more than just client satisfaction based on low costs and the quality of the product or service provided. Clients expect businesses to be ethical, environmentally conscious, and socially responsible [5]. As a result, sustainability in construction must be seriously considered. Researchers have concentrated on a variety of factors in order to achieve construction sustainability. One of them is the utilization of waste materials in construction materials.

Several processes, including mining, manufacturing, agricultural production, electricity generation, iron and steel metallurgy, and electronic goods, generate massive amounts of solid waste. Many hazardous solid wastes are flammable, chemically reactive, incendiary, corrosive, and infectious, and their discharge and disposal have resulted in large financial losses [6,7]. As a result, solid waste recycling or reuse in construction materials would be preferable [8]. Cement-based materials (CBMs) are extensively used construction materials [9–12]. Researchers have used various approaches to enhance the performance of CBMs [13–16]. Waste materials can be used in concrete as an aggregate substitute [17–22], as fibers for reinforcement [23–28], as well as a cement substitute [29–35] to improve CBMs' performance. As a result of the reduced cement consumption, natural resources can be conserved, and CO_2 emissions can be reduced. Furthermore, it has been discovered that incorporating waste materials into CBMs improves their mechanical performance [36–38]. Each year, tens of millions of tons of waste glass (WG) are produced in the United States [39], with a huge quantity of WG being disposed of in landfills. Although many cities are producing more WG at a growing rate, landfill space is becoming scarce, particularly in major cities. Glass is chemically stable as compared to certain forms of solid waste, such as plastic and wood. Glass that has been buried in soil for a long time is non-biodegradable [40]. Furthermore, certain glass, for example, cathode ray tube (CRT) glass, holds toxic elements such as mercury, cadmium, lead, and beryllium, polluting subsurface soil and water [41]. China produces CRT glass at a rate of more than 43 million tons per year [42], posing a major environmental threat and endangering public health. Glass production requires a considerable amount of energy as silica must be melted at an elevated temperature for hours [43]. For container glass, the temperature is kept at 1500 °C for 24 and 72 h for plate glass [44]. Producing 1 kg of plate glass utilizes approximately 17 MJ (mega-joule) of fossil fuel energy and emits approximately 0.6 kg of CO_2 [45]. In Europe, annual energy consumption for glass manufacturing exceeds 350 PJ (peta-joule), accounting for approximately 20% of overall industrial energy utilization [45]. Thus, recycling WG effectively is generating increasing interest worldwide. Recycling glass for use in the manufacture of glass products is a popular method of reusing WG. However, recycling is a complicated process. To produce glass plates and containers, WG ought to be cleaned, sorted, and melted [46]. Another way to recycle WG is to use it to make construction materials. WG can be crushed and mixed with cement and aggregates in CBMs to partially replace cement/aggregates [21,47–51]. The use of WG in CBMs has several benefits. First, the WG utilized in CBMs does not need to be melted, so minimizing the energy requirement. Second, WG management is significantly simplified. For example, glass cleaning and sorting are not required. Third, WG consumption will be high due to the extensive use of CBMs in construction. Fourth, toxic elements in glass can be locked and solidified within CBMs. Present research indicates that recycling WG in CBMs is a better approach [47,48]. Thus, the use of WG in construction materials as an aggregate replacement will contribute to save natural resources and solve waste management problems. Its use as a cement replacement will help to reduce the cement demand and ultimately decrease CO_2 emission.

As a result, there is a need to investigate sustainability in construction, which has been discussed in this paper. Manual reviews are insufficient for building a deep and consistent relationship between different literature sections. Scientific mapping and network visualization of bibliographic coupling, co-citations, and co-occurrence are currently among the most difficult parts of modern research. A scientometric review can deal with large amounts of data without adding to the complexity of responding to the fundamental limitations of earlier manual evaluations. To solve the fundamental drawbacks of traditional reviews, scientometric analysis is used in conjunction with traditional reviews in this study. More specifically, a thorough examination of the keywords co-occurrence, co-authorship by researchers, bibliographic coupling of sources, articles, and countries actively contributing to the subject of the utilization of WG in concrete for sustainable construction is conducted. The effect of WG on the mechanical properties (MPs), microstructure, and durability of CBMs are reviewed. Various limitations associated with WG utilization are identified, and their possible solution is detailed. In the current study, scientometric analysis is used in conjunction with an in-depth discussion to achieve the following objectives: (1) To identify the most relevant publishing field, publication sources, most frequently used keywords, most cited authors and works, and nations with the greatest influence in the field of WG utilization for sustainability in construction. (2) To examine the current level of research and its focus on a variety of elements throughout the last two decades. (3) To identify research gaps so that future research can be directed in the proper direction.

2. Research Significance

In recent decades, a substantial study has been done to explore the elements that contribute to sustainability in construction, and some useful results have been reached. Review studies were also conducted; however, these were primarily manual reviews. This study is based on scientometric analysis, as well as an in-depth discussion on WG utilization in CBMs as a sustainable approach. The reason for selecting WG from the various waste materials is because it is non-biodegradable and contains toxic chemicals that have a greater detrimental impact on the environment if disposed of in landfills. Researchers from various locations may benefit from the graphical depiction based on a scientometric evaluation in forging research alliances, forming joint ventures, and sharing breakthrough technologies and ideas as a result of this research. Furthermore, the advantages associated with WG utilization in construction are described. Specifically, the impact of WG on the mechanical performance of CBMs and its sustainability aspects are reviewed. Different restrictions related to the use of WG are also discussed, as well as possible remedies. Finally, possible future studies are recommended.

3. Methodology

This study employed two approaches: a scientometric analysis review [52–55] and an in-depth discussion on WG utilization in construction materials. The primary reason for instituting a scientometric review method is that subjective analyses of civil engineering studies by researchers have been shown to be prone to error. Scientometrics, by itself, provides a more rational and less skewed outcome, as it is not biased by any individual's perspective [56–58]. This study examines and articulates findings spanning two decades. Maps and connections between bibliometric data are used to quantify research progress in this report, resulting in a quantitative assessment.

Numerous publications have been written in the area under study, and it is critical to locate the most accurate database. According to Aghaei et al. [59], the two most effective, comprehensive, and objective databases for conducting literature searches are Web of Science and Scopus. Scopus has a broader coverage and more up-to-date bibliometric data than Web of Science [59–61]. Scopus was used to compile the bibliometric data for the current analysis on the utilization of WG for sustainable construction. The searched keyword in Scopus was "waste glass in concrete", which resulted in 1488 documents. Options for data refinement were used to weed out irrelevant publications. From the "document type", only "article" and "review" were selected. The "source type" field was kept to "journal", while the "language" field was "English". The "publication year" was limited from 2001 to 2021. The "subject area" was kept "Engineering, Material Science, and Environmental Science". After applying these limiting filters, the resulting documents were 737. Researchers from a variety of fields have previously conducted research using similar methods [62–64]. Scientometric reviews make use of science visualization, a technique developed by scholars for analyzing bibliometric data for a variety of purposes [65]. It describes the difficulties researchers face when conducting manual reviews and also establishes a connection between sources, keywords, authors, articles, and countries within a particular research area [66]. Scopus data were saved in the Comma Separated Values (CSV) format for subsequent analysis with an appropriate software tool. The science mapping and visualization were created using the software tool VOSviewer (version: 1.6.16). VOSviewer is a freely available visualization application that is widely used in a variety of fields and comes highly recommended in the literature [67–71]. Hence, the VOSviewer was used to accomplish the current study's objectives. The analysis was conducted in VOSviewer, with the "type of data" set to "create a map from bibliographic data" and the "data source" set to "read data from bibliographic database files". The CSV file downloaded from Scopus was imported into VOSviewer and analyzed in a few simple steps while ensuring data consistency and reliability. As part of the science mapping review, the sources of documents, keyword co-occurrence, citation network, co-authorship, articles, bibliometric overlapping, and country contributions were analyzed. Additionally, the cumulative number of citations to articles was tallied. Additionally, the connections between authors, publications, and countries were charted. Maps were used to visualize various parameters, their relationships, and co-occurrence, while tables were used to summarize their quantitative values. Additionally, the keywords were thoroughly reviewed and summarized in the discussion section in order to develop the major research themes. The sequence of the scientometric analysis is depicted in Figure 1.

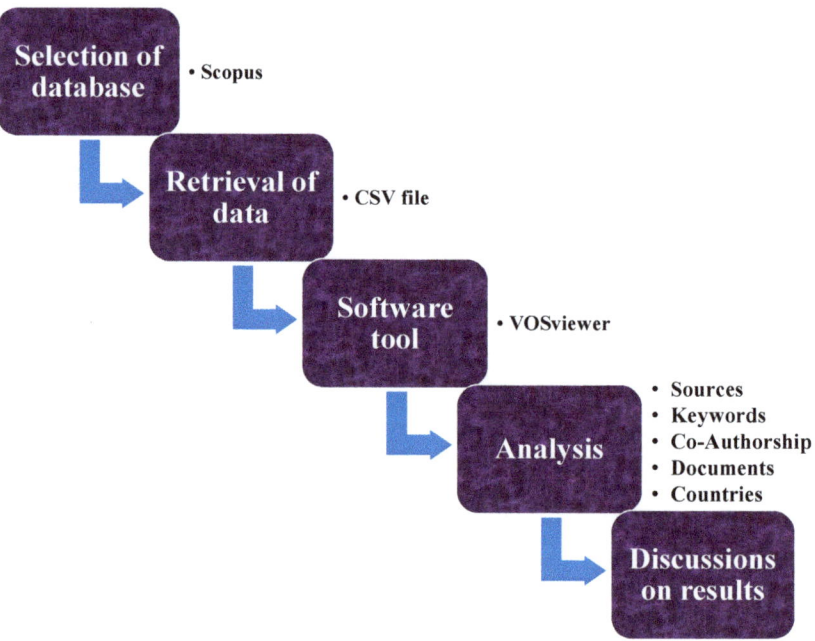

Figure 1. Sequence of scientometric analysis. CSV: comma separated values.

4. Results and Discussion on Scientometric Analysis

4.1. Subject Area and Annual Publication Pattern of Articles

The Scopus analyzer was used to search the Scopus database in order to determine the most significant research areas. The analysis revealed that the top three fields based on the number of documents were determined to be Engineering, Materials Science, and Environmental Science, containing 34.7%, 21.9%, and 16.0% of the total documents, respectively, as illustrated in Figure 2. These fields account for around 72.6% of the total number of documents searched in the Scopus database. Both journal articles and review articles were compared for the overall documents. Journal articles contributed 90.9%, and review articles contributed 9.1%, respectively. Figure 3 depicts the annual publication pattern in the current study field from 2001 to 2021. A gradual increase in the number of publications on the utilization of WG in concrete has been observed up to 2016. However, a remarkable increase was observed in the last 5 years. It is fascinating to discover that scholars are now focusing their studies on sustainable construction methods.

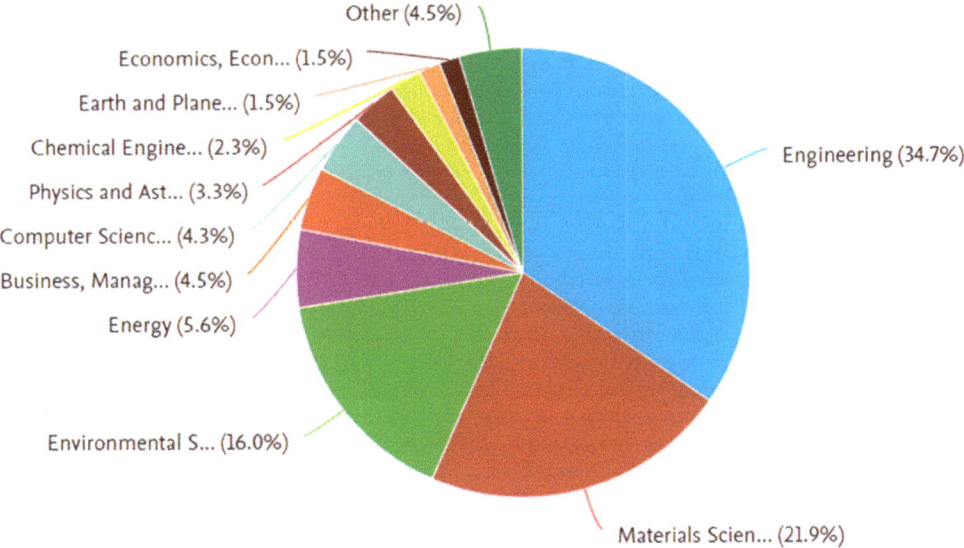

Figure 2. Subject area of articles.

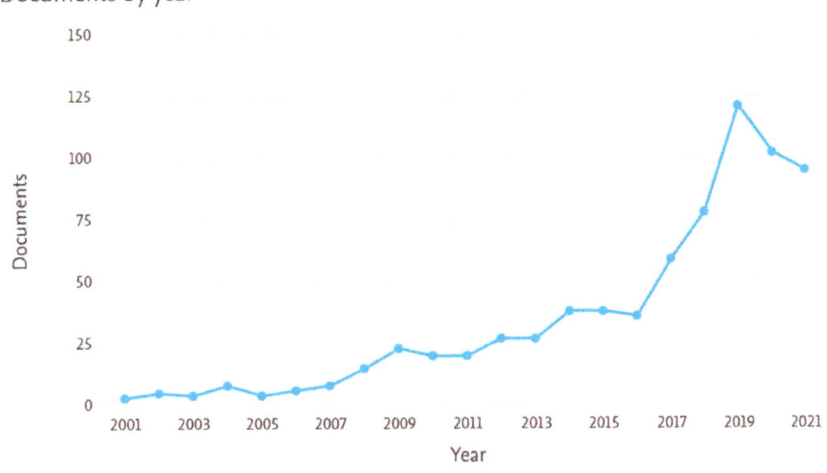

Figure 3. Annual publication pattern.

4.2. Sources Mapping

Mapping sources enables the analysis of development and innovation to be visualized. These sources make data available within the confines of predefined, unique constraints. By initializing the mapping of research origins, it is possible to apply the research pattern sequentially in the analysis area. This analysis was conducted in VOS viewer using Scopus bibliometric data. The "type of analysis" was selected to be "bibliographic coupling," and the "unit of analysis" was selected to be "sources". A source's minimum document count was set at 10, and 10 of the 256 sources met this criterion. Table 1 lists the leading sources/journals that publish at least 10 documents containing data on WG in concrete for sustainable development, along with their citations and total link strength. Based on the number of documents, the top 3 journals are construction and building materials, the journal of cleaner production, and the international journal of civil engineering and technology, containing 112, 56, and 25 documents, respectively. The highest citations are of construction and building materials (3989), followed by cement and concrete research (1935) and journal of cleaner production (1878). Figure 4 illustrates the annual publication trend of sources and their scientific mapping. This data was gathered as part of the process of establishing a network of research sources. It is worth noting here that this type of research would lay the groundwork for upcoming scientometric reviews in the current study field. Additionally, previous manual reviews lacked sufficient detail regarding science mapping. Figure 4a depicts the yearly publication trend of the top journals. The contribution of resources, recycling and conservation is from 2002. While the contribution of construction and building materials, journal of cleaner production, and materials is from 2007, 2009, and 2016, respectively, in the current study field. It can be seen that up to 2016, the number of publications was insignificant, while an abrupt hike in the last 5 years was observed especially for construction and building materials. The network visualization of journals containing at least 10 documents has been displayed in Figure 4b. The size of the frame in the figure corresponds to the journal's contribution in terms of citation and documents count; a larger frame size indicates a higher contribution. For example, construction and building materials has a bigger frame size, indicating that this journal has the greatest influence in the current study area relative to the other. Additionally, frames (sources) with identical colors display clusters of related frames developed through VOSviewer analysis. For instance, the red color indicates a cluster containing construction and building materials, journal of cleaner production, materials, journal of building engineering, and applied sciences (Switzerland). Clusters are formed based on the scope of research outlets or their co-citations. [72]. The number of co-citations in the articles in the current study area is indicated by the connection links between the research sources. Additionally, the link strength indicates the number of mutually cited references between two journals. For instance, the journal of cleaner production (total link strength: 9941) contained the greatest number of references to other research sources. Closely spaced frames (sources) in a cluster have stronger connections than those that are further apart. For example, the construction and building materials is more connected with cement and concrete composites than it is with other sources.

Table 1. Documents sources.

S/N	Source	Documents	Citations	Total Link Strength
1	Construction and building materials	112	3989	9941
2	Journal of cleaner production	56	1878	5526
3	International journal of civil engineering and technology	25	70	437
4	Materials	23	163	2785
5	Resources, conservation and recycling	21	1770	1915
6	Journal of building engineering	20	260	3931
7	Waste management	18	1792	1290
8	Cement and concrete composites	13	1261	2580
9	Applied sciences (switzerland)	10	35	1474
10	Cement and concrete research	10	1935	513

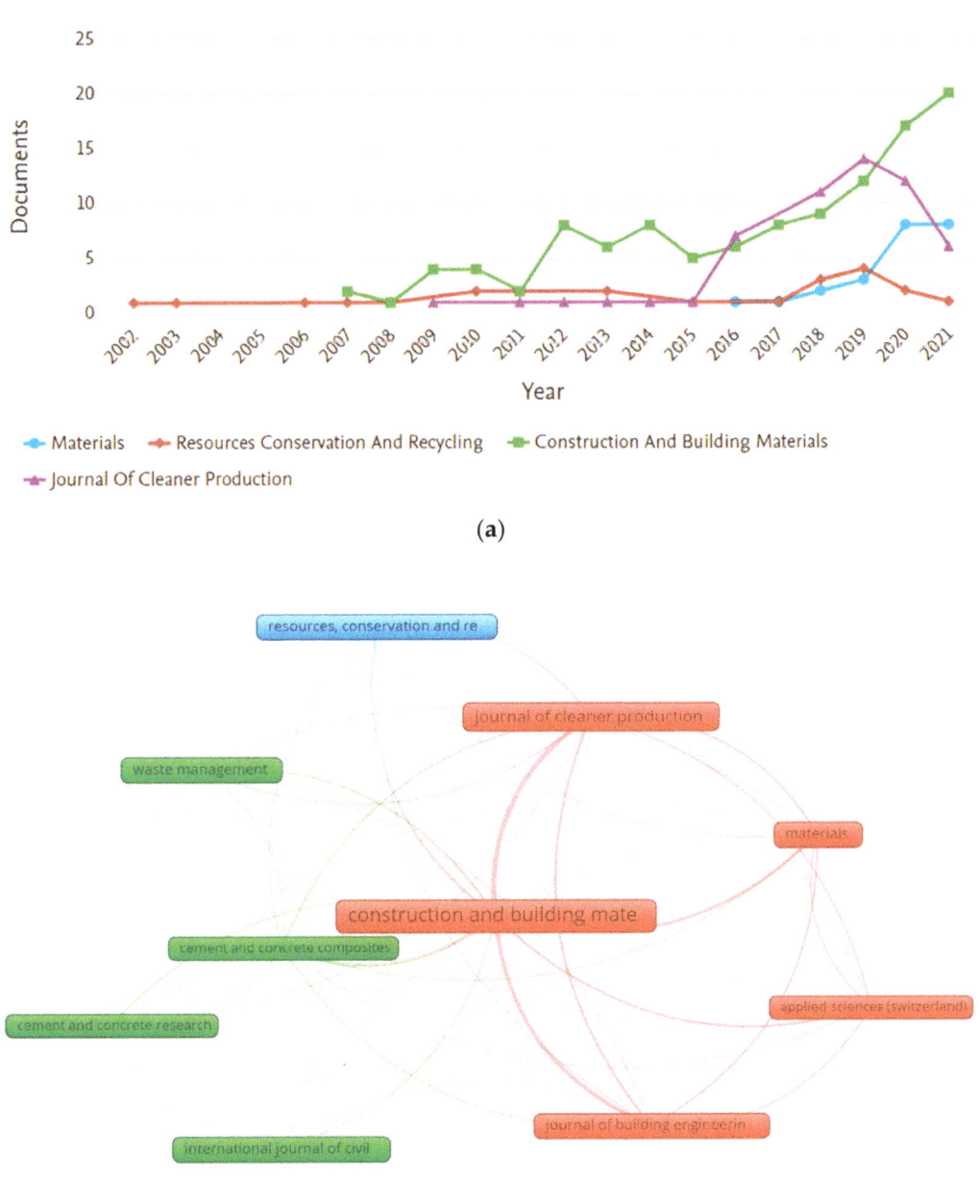

Figure 4. Sources of research articles: (**a**) yearly publication trend; (**b**) network visualization.

4.3. Keywords Mapping

Keywords are essential research materials because they identify and represent the research domain's fundamental field [73]. For that analysis, the "type of analysis" was chosen as "co-occurrence" and "unit of analysis" as "all keywords". The minimum number of occurrence of a keyword was kept to 20. These constraints indicated that only

100 keywords from 5576 satisfied the criteria. Table 2 lists the top 20 keywords having the most occurrence in the research articles used in the present study field. According to the researcher's study, the most commonly occurring keywords include glass, compressive strength, concretes, recycling, and concrete, making the top 5 mostly occurred keywords. Figure 5 illustrates the co-occurrence of keywords networks, their visualization, their connections to one another, and the density associated with their correlation frequency. The size of the keyword node in Figure 5a indicates its frequency, whereas its location indicates its co-occurrence in publications. Additionally, the visualization demonstrates that the aforementioned keywords have bigger nodes than the others, indicating that these are the most important keywords in the study of WG utilization in concrete for sustainable construction. Clusters of keywords have been colored differently in the network to indicate their co-occurrence in various publications. A total of five clusters were observed, represented by green, red, blue, yellow, and purple. For example, a cluster represented by green color contains glass, silica, cements, durability, silica fume, cement replacement, concrete mixtures, etc. As illustrated in Figure 5b, the density concentration of keywords is denoted by distinct colors. Red, yellow, green, and blue are the colors in ascending order of density. For example, glass, compressive strength, and concretes have red marks in the density visualization, indicating a higher density. This finding will aid writers in the future when selecting keywords to make it easier to locate published data in a particular domain. Figure 6 illustrates the connections of glass (Figure 6a), waste management (Figure 6b), and recycling (Figure 6c) with other keywords. The connection network demonstrates that glass and recycling have a sizeable impact on waste management. Thus, recycling WG would reduce the burden from waste management authorities.

Table 2. Top 20 most occurred keywords.

S/N	Keyword	Occurrences	Total Link Strength
1	Glass	407	3435
2	Compressive strength	257	2102
3	Concretes	232	2028
4	Recycling	225	1938
5	Concrete	172	1374
6	Aggregates	171	1576
7	Concrete aggregates	145	1298
8	Waste glass	111	925
9	Cements	110	984
10	Fly ash	99	869
11	Durability	96	867
12	Mechanical properties	93	758
13	Silica	91	913
14	Waste management	87	726
15	Mortar	75	707
16	Glass powder	73	538
17	Portland cement	73	700
18	Water absorption	72	667
19	Slags	68	666
20	Tensile strength	67	638

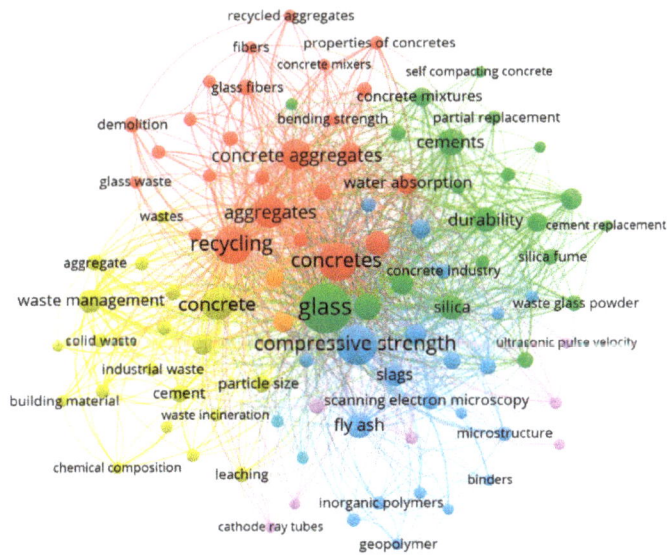

(a)

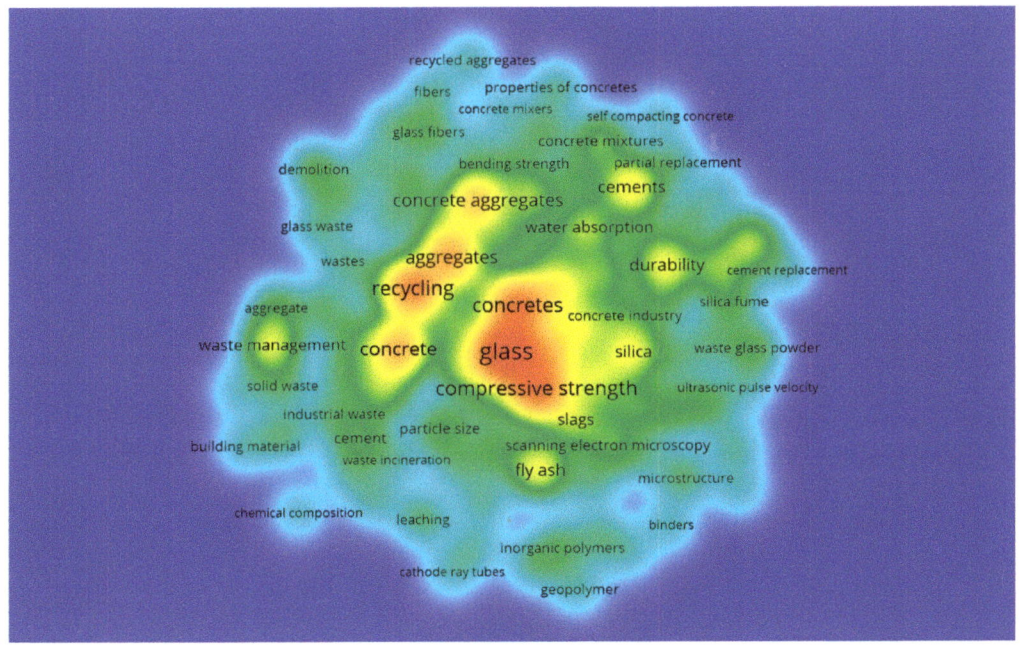

(b)

Figure 5. Mapping of keywords: (**a**) co-occurrence visualization; (**b**) density visualization.

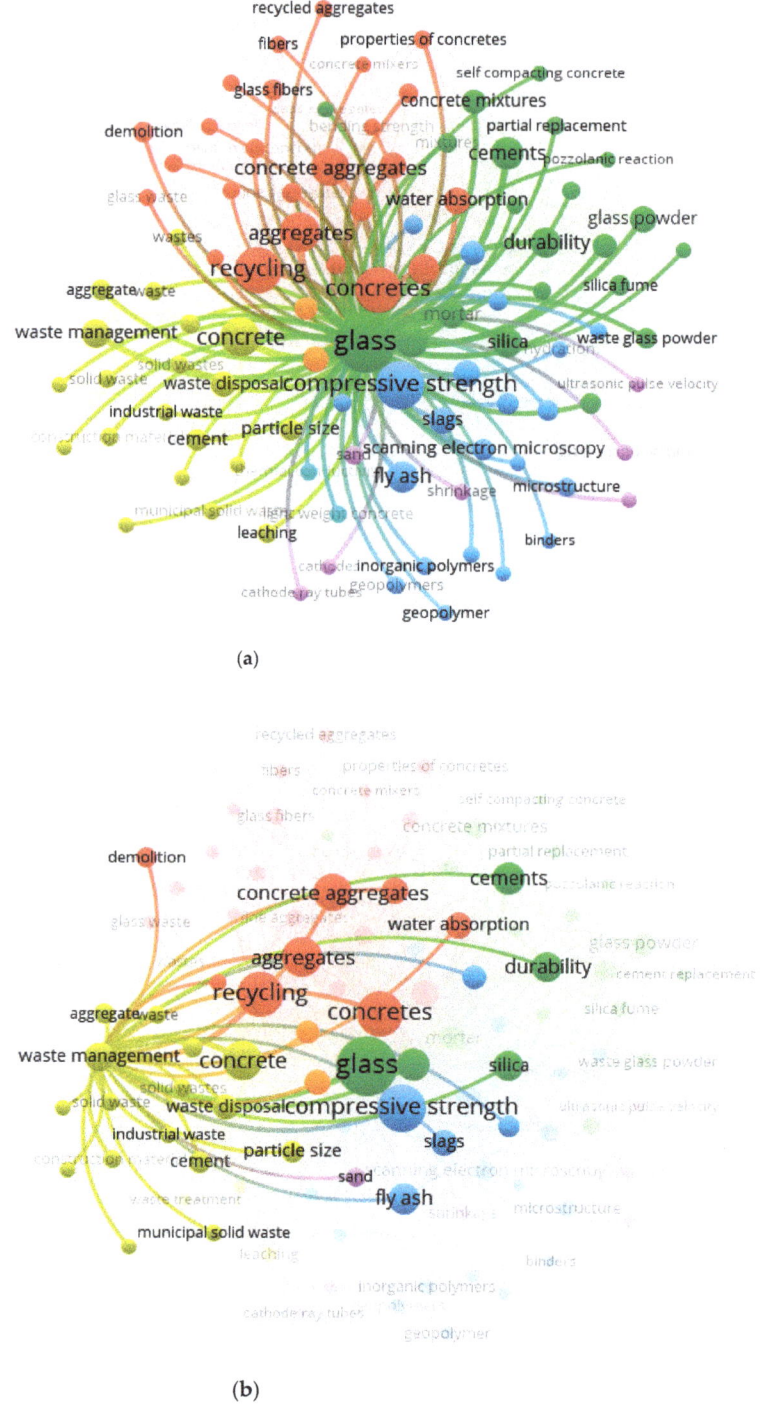

Figure 6. *Cont.*

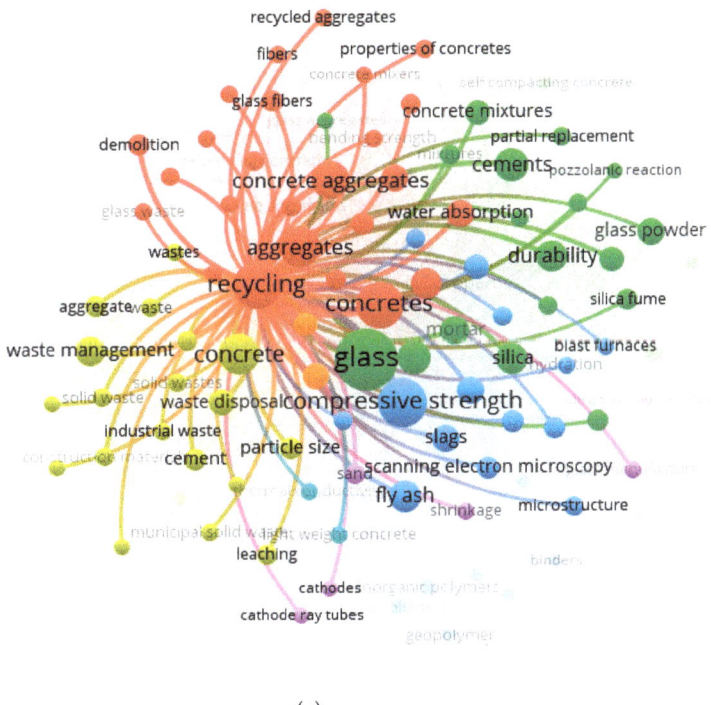

(c)

Figure 6. Connection of a keyword with others: (**a**) glass; (**b**) waste management; (**c**) recycling.

4.4. Co-Authorship Mapping

Citation counts indicate a researcher's influence in a particular field [74]. In the VOSviewer, "co-authorship" was selected as the "type of analysis", while "authors" were chosen as the "unit of analysis". The minimum number of documents required for an author was kept at 5, which resulted in 20 of the 2077 authors meeting the constraints. The top 20 authors in the field of WG in concrete for sustainable growth with the most documents and citations are listed in Table 3, according to the data retrieved from the Scopus database. The average citation count was calculated by dividing the total number of citations by the number of publications by each author. Poon C.S. was the author of the most publications (16), while Arulrajah A. was the author of the most citations (656). It will be difficult to independently assess a researcher's effectiveness. However, the author's rating will be determined by comparing all variables individually or in conjunction with one another. For instance, if the total number of documents is compared, the top three authors are Poon C.S with 16, Arulrajah A. with 13, and Tagnit-Hamou A. with 12 publications. Alternatively, if the number of citations is compared, the author's ranking would be Arulrajah A. with 656, Shi C. with 624, and Poon C.S. with 567 citations. Additionally, when comparing average citations, the authors are ranked as follows: Shi C. with 125, Poon C.-S. with 75, and Ling T.-C. with 54 average citations. Figure 7 illustrates the visualization of authors with a minimum of 5 documents and the linkage of the most prominent author. Of the 20 authors, only 4 have been linked. It was observed that authors from different regions are not connected to each other based on citations in the field of WG utilization in concrete.

Table 3. Top researchers.

S/N	Author	Documents	Citations	Average Citations	Total Link Strength
1	Poon C.S.	16	567	35	11
2	Arulrajah A.	13	656	50	16
3	Wang H.-Y.	12	321	27	4
4	Tagnit-Hamou A.	12	302	25	0
5	Horpibulsuk S.	11	454	41	15
6	Poon C.-S.	7	525	75	4
7	Ling T.-C.	7	378	54	5
8	Lu J.-X.	6	83	14	7
9	Shi C.	5	624	125	1
10	Brouwers H.J.H.	5	177	35	0
11	Dinis M.L.	5	150	30	15
12	Fiúza A.	5	150	30	15
13	Meixedo J.P.	5	150	30	15
14	Ribeiro M.C.S.	5	150	30	15
15	Mohammadinia A.	5	122	24	9
16	Lin K.-L.	5	108	22	0
17	Olofinnade O.M.	5	78	16	4
18	Ede A.N.	5	75	15	4
19	Xuan D.	5	69	14	6
20	Wang C.-C.	5	49	10	4

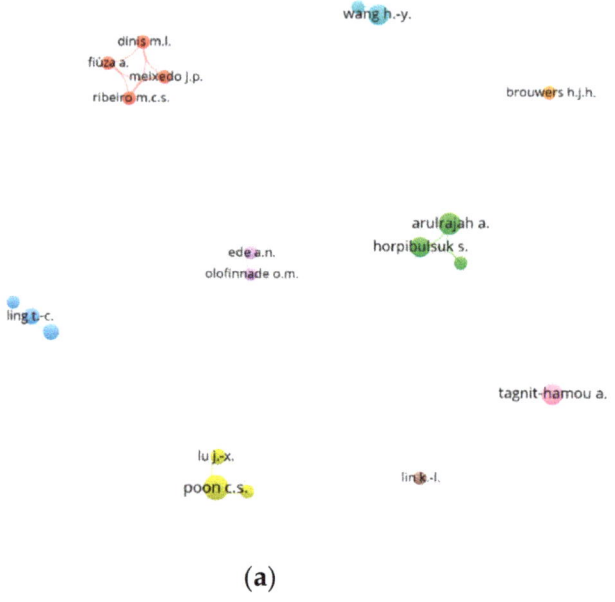

(a)

Figure 7. Cont.

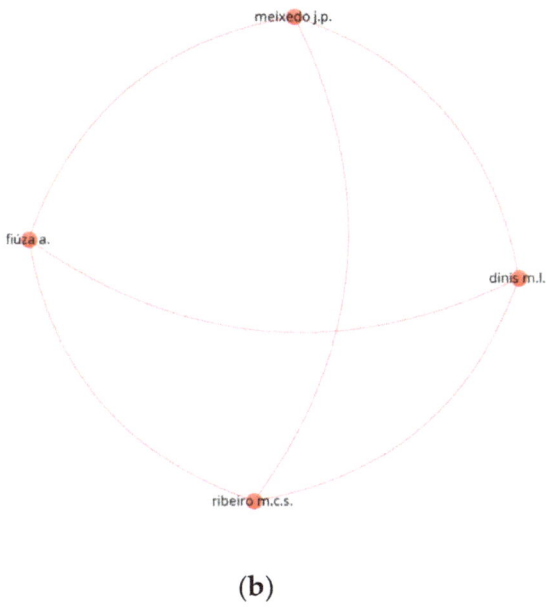

(b)

Figure 7. Visualization of co-authorship; (**a**) authors with minimum five documents; (**b**) connected authors.

4.5. Articles Mapping

The number of citations a research article receives indicates its impact on a particular field of study. Articles with a high citation count will be considered a landmark in the history of science. To analyze document citations, the "type of analysis" was set to "bibliographic coupling" and the "unit of analysis" to "document" in the VOSviewer. A document's minimum citation count was set to 50, and 102 of the 737 records adhered to these boundary requirements. The top 20 highly cited research articles, their authors, and the year of publication are listed in Table 4. Meyer C. [75] had the maximum citations of 691 on their article titled "The greening of the concrete industry". However, Shayan A. [76] and Topcu I.B [39] had 363 and 342 citations on their respective articles and have been ranked in the top three. Figure 8 shows the author's visualization having the most article citations on their respective publications (Figure 8a), the top connected articles (Figure 8b), and density visualization of connected articles (Figure 8c) in the present study field. It was observed that the connected documents were 88 out of 102 based on citations. The network of co-citations between the writers participating in the study of WG utilization in concrete for sustainable construction is depicted in this visualization. The closeness of the articles depicts how interconnected they are with each other in terms of citations.

Table 4. Top 20 publications based on citation count.

S/N	Author	Title	Citations	Total Link Strength
1	Meyer C. (2009)	The greening of the concrete industry	691	12
2	Shayan A. (2004)	Value-added utilisation of waste glass in concrete	363	27
3	Topçu I.B. (2004)	Properties of concrete containing waste glass	342	68
4	Batayneh M. (2007)	Use of selected waste materials in concrete mixes	341	2
5	Park S.B. (2004)	Studies on mechanical properties of concrete containing waste glass aggregate	264	0

Table 4. Cont.

S/N	Author	Title	Citations	Total Link Strength
6	Shi C. (2007)	A review on the use of waste glasses in the production of cement and concrete	260	189
7	Shayan A. (2006)	Performance of glass powder as a pozzolanic material in concrete: A field trial on concrete slabs	248	62
8	Tam V.W.Y. (2006)	A review on the viable technology for construction waste recycling	239	5
9	Ismail Z.Z. (2009)	Recycling of waste glass as a partial replacement for fine aggregate in concrete	211	108
10	Paris J.M. (2016)	A review of waste products utilized as supplements to Portland cement in concrete	189	98
11	Nassar R.-U.-D. (2012)	Strength and durability of recycled aggregate concrete containing milled glass as partial replacement for cement	181	100
12	Taha B. (2008)	Properties of concrete contains mixed colour waste recycled glass as sand and cement replacement	162	36
13	Puertas F. (2014)	Use of glass waste as an activator in the preparation of alkali-activated slag. Mechanical strength and paste characterisation	161	35
14	Matos A.M. (2012)	Durability of mortar using waste glass powder as cement replacement	159	149
15	Federico L.M. (2009)	Waste glass as a supplementary cementitious material in concrete: Critical review of treatment methods	149	165
16	Torres-Carrasco M. (2015)	Waste glass in the geopolymer preparation. Mechanical and microstructural characterisation	147	30
17	Aly M. (2012)	Effect of colloidal nano-silica on the mechanical and physical behaviour of waste-glass cement mortar	147	136
18	Jani Y. (2014)	Waste glass in the production of cement and concrete: A review	146	241
19	Pereira-De-Oliveira L.A. (2012)	The potential pozzolanic activity of glass and red-clay ceramic waste as cement mortars components	146	68
20	Idir R. (2010)	Use of fine glass as ASR inhibitor in glass aggregate mortars	145	6

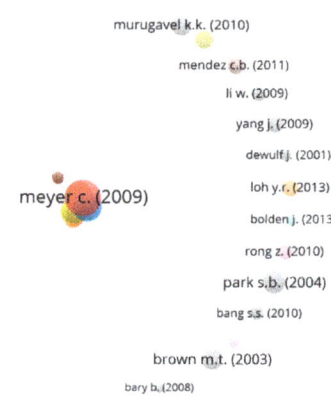

(a)

Figure 8. Cont.

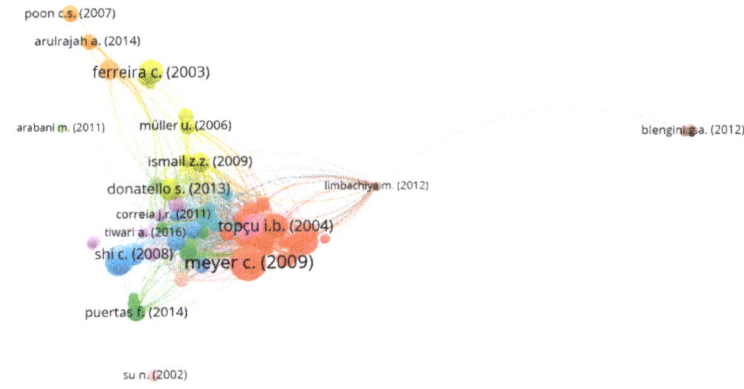

(b)

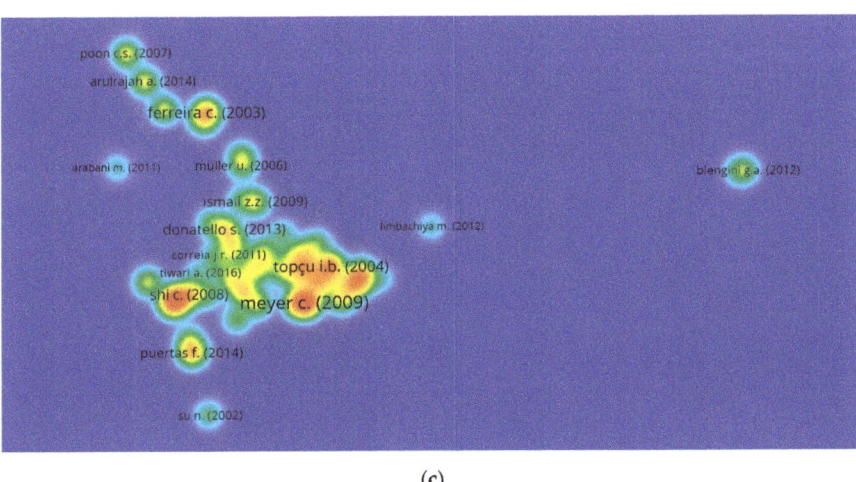

(c)

Figure 8. Mapping of documents; (**a**) documents with minimum 100 citations; (**b**) connected documents based on citations; (**c**) density of connected documents.

4.6. Countries Mapping

Certain nations have contributed more than others in the current research domain in the past and continue to do so. The visualization network was developed to help readers visualize regions that are deeply committed to sustainable construction. The "type of analysis" was "bibliographic coupling", and the "unit of analysis" was "countries". The criterion for a country's minimum number of documents was set at 5, and 37 of 87 countries met the criterion. The top 20 active nations are listed in Table 5 based on the number of documents and citations related to the present study area. India, United States, and China contributed the most documents overall, with 126, 63, and 52 documents, respectively. While the United States, Australia, and United Kingdom were the top three participating countries in terms of citation count, with 2778, 2098, and 1857 citations, respectively. The number of documents, citations, and total link strength indicates a nation's influence on the evolution of the current research domain. The total link strength indicates the extent to which a country's documents have influenced the other countries participating in these studies. The United States had the strongest total link strength in comparison to other countries, followed by Australia and China. As a result, the aforementioned countries were determined to have the greatest influence on the utilization of WG in concrete for sustainable construction. Figure 9a,b illustrates the countries' connectivity and density visualization of countries that are connected through citations. The frame's size indicates the country's contribution to the field of study. Additionally, the density visualization demonstrates that the countries with the highest participation had a higher density. Future researchers will be aided in establishing scientific collaborations, producing joint venture reports, and sharing innovative techniques and ideas by the graphical representation of participating countries.

Table 5. Top 20 contributing countries.

S/N	Country	Documents	Citations	Total Link Strength
1	India	126	1581	15,912
2	United States	63	2778	19,741
3	China	52	1622	16,421
4	Australia	49	2098	17,897
5	South Korea	34	691	9220
6	United Kingdom	34	1857	9041
7	Iran	33	730	13,483
8	Taiwan	32	825	9171
9	Canada	28	1095	11,043
10	Hong Kong	26	1425	10,994
11	Portugal	22	1167	6173
12	Spain	22	893	4421
13	Turkey	22	721	4795
14	Iraq	21	454	4291
15	Malaysia	21	424	8844
16	Egypt	20	816	7705
17	Italy	20	498	5085
18	Thailand	20	809	6157
19	Poland	17	99	5883
20	Nigeria	16	162	10,632

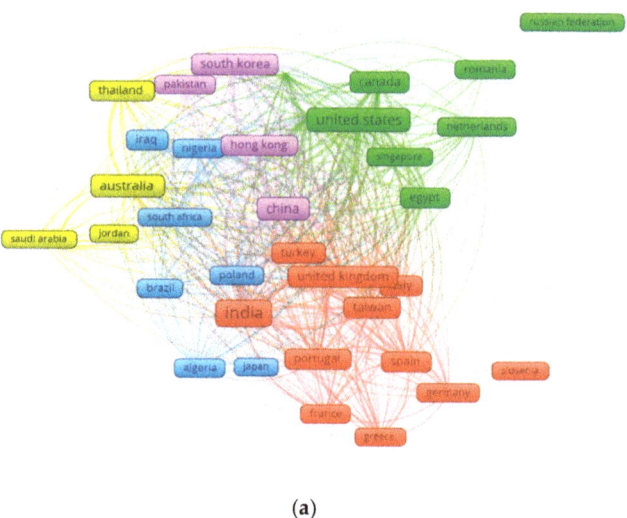

(a)

Figure 9. *Cont.*

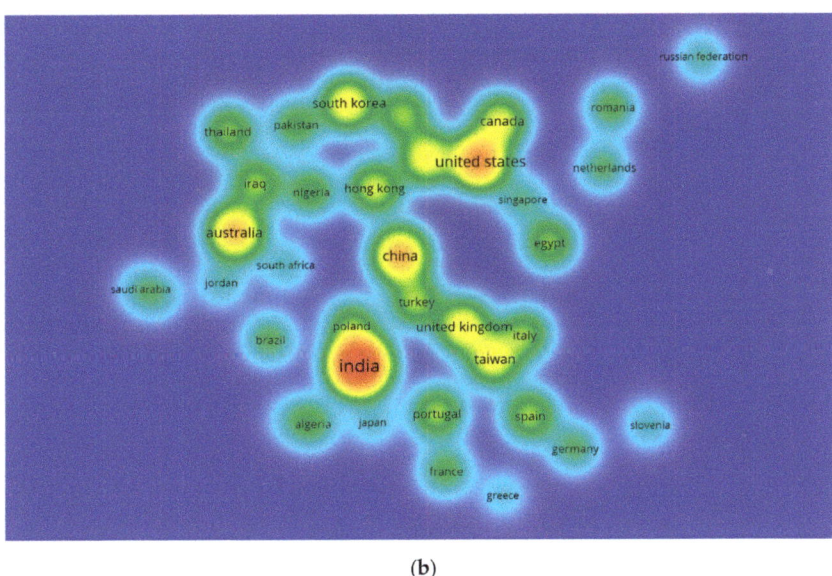

(b)

Figure 9. Countries mapping: (a) network visualization: (b) density visualization.

5. Types and Properties of Waste Glass

WG is classified chemically as lead, soda-lime, electric, and borosilicate glass. The most widely utilized type of glass is soda-lime glass. The main chemical composition of glass includes SiO_2, Na_2CO_3, and $CaCO_3$ [77]. Table 6 lists the percentage of various chemical compounds in different kinds of glass. Glass is classified according to its color into three categories: clear/flint, green, and brown/amber glass [78]. Glasses come in a variety of colors due to their chemical compositions, which relate to distinct levels of color impurity. Color impurity is limited to 4–6%, 5–30%, and 5–15% in clear, green, amber glass, respectively [77]. Glass is classified according to its application into six categories, namely, plate glass, container glass, continuous filament glass, mineral wool insulation, specialty glass, and domestic glass or tableware [77]. Between these types, plate and container glass are typically made of soda-lime glass; domestic glass is typically made of lead or soda-lime glass; continuous filament glass is typically made of electric glass; borosilicate glass is used to insulate mineral wool, and specialty glass is frequently made of borosilicate or soda-lime glass.

Table 6. Chemical composition of various kinds of waste glass [77].

Glass Type	Chemical Compound (%)					
	SiO_2	$Na_2O + K_2O$	CaO	Al_2O_3	B_2O_3	PbO
Soda-lime	71–75	12–16	10–15	-	-	-
Lead	54–65	13–15	-	-	-	25–30
Borosilicate glass	70–80	4–8	-	7	7–15	-
Electric	52–56	0–2	16–25	12–16	0–10	-

6. Waste Glass Utilization in Cement-Based Materials

WG can be utilized in CBMs as aggregate replacement and cement replacement. Thus, it conserves natural resources, solves waste management problems, reduces CO_2 emission by decreasing cement demand, protects the environment from toxic chemicals, and produces cost-effective composites. Therefore, the utilization of WG in construction materials is a better approach for sustainability in construction. In this section, the effect of utilizing WG on fresh and hardened properties of CBMs, including workability, compressive strength (CS), split-tensile strength (STS), and flexural strength (FS), are reviewed. Furthermore, the microstructure and durability of composites containing WG are investigated. Different limitations associated with WG use in CBMs and their possible solution are reported.

6.1. Mechanical Properties of Cement-Based Materials Containing Waste Glass as Natural Aggregate Replacement

6.1.1. Workability

WG's effect on the workability of fresh concrete has been reported to be inconsistent in previous studies. Partial substitution of sand with WG improved workability. For instance, Elaqra et al. [79] used a soda-lime glass (size: <20 µm) and found that increasing the replacement percentage from 0% to 30% increased the slump from 130 mm to 190 mm. However, the utilization of WG decreased the workability of concrete. For example, an electric glass (size: <150 µm) reduced the slump from 200 mm to 45 mm as the amount of WG increased from 0% to 40% [80]. The above-mentioned inconsistencies in WG's effect on the workability of composites are the result of two competing effects. Firstly, glass improves workability by reducing water adsorption and friction due to its dense microstructure and smooth surface. Indeed, WG has been used to create SCC that is vibration-free during construction [81–85]. Secondly, glass can reduce workability due to extremely fine glass particles; the surface-to-volume ratio increases, increasing water adsorption. Thus, the size of the glass particles is a critical parameter in determining their effect on workability. However, if coarser glass particles are utilized, the irregular shape of the glass particles improves mechanical interlocking among adjacent particles and ultimately reducing the workability [86]. This is consistent with Topcu and Canbaz's [39] findings, i.e., the addition of coarse glass particles (size: 16 mm) decreased the workability of the fresh mix. With increasing glass particle size, the surface-to-volume ratio decreases, resulting in a small quantity of paste or mortar adsorbed on the surface of glass particles for lubricating nearby glass particles. This effect of under-lubrication may have been facilitated by the glass particle's smooth and dense surface.

6.1.2. Compressive Strength

According to the literature, there is a detrimental effect on CS of composites containing WG as a partial or complete substitute of NA, as shown in Figure 10. Mostly, a decreasing trend is observed with the increasing replacement ratio of WG. However, using a smaller size WG at a lower replacement ratio can enhance the CS by filling the voids in the matrix. Liu et al. [87] found a decrease in CS by 2.9%, 10%, and 15.7% when coarser WG (5–10 mm) was used at 10%, 20%, and 30% content, respectively. Though, using finer WG (size: <4.75 mm), the reduction in CS was minimal. They reported two primary explanations for this decline. Firstly, the WG aggregate has a lower strength than the NA aggregate, and secondly, the smooth surface of the WG aggregate affects its binding with cement paste, reducing the strength of the concrete. A similar pattern of reduction in CS with the increasing amount of WG in composites as an aggregate replacement was also noted by other researchers [88–90]. Conversely, Ismail and Hashmi [91] and Abdallah and Fan [92] found a 4.3% and 4.9% increase in CS, respectively, when fine aggregate was replaced at 20% by WG. Similarly, Malik et al. [93] investigated the CS of composites containing finer WG particles (size: <1.18 mm) at various replacement ratios. The results revealed that composites containing 10%, 20%, and 30% WG enhanced the CS by 20.0%, 25.1%, and 9.8%, respectively, compared to the reference mix. However, at 40% replacement ratio, the CS decreased by 8.5% than the reference mix, as shown in the figure. Thus, the use of finer WG at a lower replacement ratio could enhance the CS by filling voids in the matrix, while coarser WG reduces the CS because of weak ITZ between WG aggregate and cement matrix.

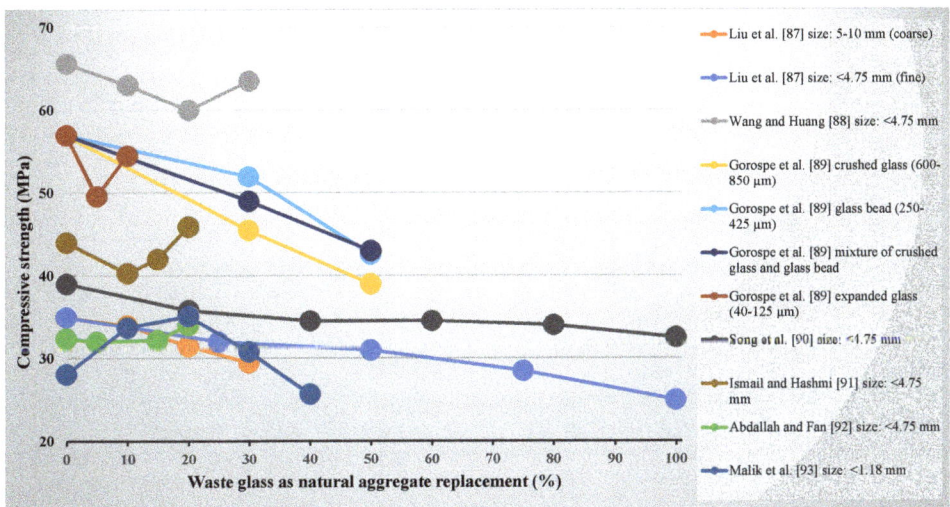

Figure 10. Effect of waste glass as a natural aggregate replacement on 28-days compressive strength.

6.1.3. Split-Tensile Strength

The WG usage as an NA replacement also has an unfavorable effect on the STS of composites, as depicted in Figure 11. The STS decreases with the increasing replacement ratio of WG, as reported by most researchers [87,90,93,94]. Liu et al. [87] reported a decrease in STS than that of reference mix by 8.5%, 14.1%, 21.1%, and 25.4% when WG replaced natural fine aggregate at 25%, 50%, 75%, and 100% ratios, respectively. Song et al. [90] found a decrease in STS compared to control mix by 2.3%, 2.3%, 5.2%, 6.8%, and 10% when NA was replaced by 20%, 40%, 60%, 80%, and 100% WG, respectively. However, Abdallah and Fan [92] and Malek et al. [95] found an increase in STS of WG composites than NA composites using lower content (up to 20%) of WG. It can be concluded that WG with smaller particle size and used in lower proportions positively influences the STS of composites.

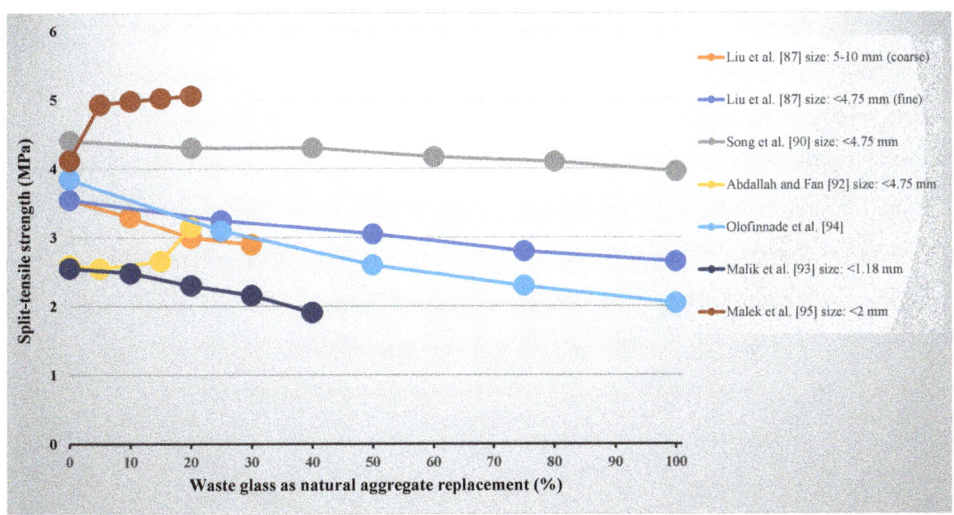

Figure 11. Effect of waste glass as a natural aggregate replacement on 28-days split-tensile strength.

6.1.4. Flexural Strength

Figure 12 depicts the influence on FS of composites with increasing WG content as the NA replacement. It shows that at lower WG content, the FS of composites can be improved, while higher WG content results in decreasing FS. However, the reduction in FS is minimal compared to CS and STS. Wang and Huang [88] found improvement in FS of 14.7% at a 10% replacement ratio, while a further increase in WG decreased FS compared to the reference sample without WG. The study of Ismail and Hashmi [91] and Abdallah and Fan [92] reported improvement in FS of WG aggregate composites in comparison with the NA composites. The improvement in FS with WG addition may be attributed to the pozzolanic properties of glass, which helped to improve the microstructure of the matrix [91]. Kim et al. [96] and Sikora et al. [97] observed a drop in FS of WG composites in comparison to the NA composites. The reduction in FS was more at higher WG content as aggregate replacement. This could be because the smooth surface of WG relative to the NA having lower adhesion to the surrounding matrix than that of NA. However, Malek et al. [95] reported increasing FS with increasing WG percentage up to 20% with an increment of 5%. At a 20% replacement ratio, the FS increased by 14.3%. The reason is the use of fine WG (size: <2 mm), which helped improve the microstructure by filling pores in the matrix. Thus, the use of WG can improve the properties of composites if used in smaller sizes and lower replacement ratios. This is because the finer WG can improve the microstructure by filling voids in the matrix [93]. In addition, a higher replacement ratio and a larger size of WG used as NA replacement reduced the properties of composites. This may be attributed to the more glass-matrix interfaces, and the smooth surface of WG may reduce the interfacial bond strength, resulting in decreased strength of composites [90].

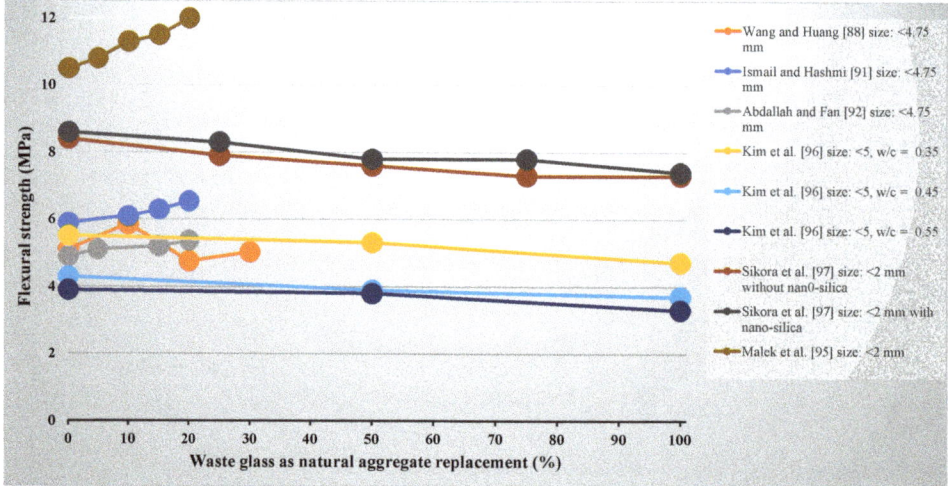

Figure 12. Effect of waste glass as a natural aggregate replacement on 28-days flexural strength.

6.2. Mechanical Properties of Cement-Based Materials Containing Waste Glass as Cement Replacement

6.2.1. Workability

The finer/powder WG is usually used as cement replacement. Islam et al. [98] performed a flow test on mixes of WG powder mortar. Water/binder ratio (w/b) was kept 0.5 for mix preparation. The findings indicated that as the percentage of WG as cement substitute increased, the flow diameter increased. The flow diameter of the reference mix was 132.5 mm, while the flow diameter of mortar samples containing 25% WG as cement replacement was 135 mm. As a result, a slight increase in flow was observed.

Aliabdo et al. [51] evaluated the workability of WG powder-modified concrete using a slump test. It was observed that the slump of mix containing WG powder as a substitute for cement improved as the WG powder content increased. The smooth surface and minimal water absorption capacity of WG powder may contribute to the slump increase. Additionally, WG powder contains coarser particles than cement, which might have caused the improvement in a slump. Soliman and Tagnit-Hamou [99] also demonstrated that incorporating WG powder in place of cement increased the workability of concrete, which may be due to the low water absorption and smooth texture of WG powder than the cement particles. Another factor contributing to increase the workability is the dilution of cement. The reasons outlined above account for the reduction in the formation of hydration products of cement during the initial time. As a result, there is an insufficient number of products available for combining disparate particles. As WG powder has a smaller specific surface area than cement, the total surface area of the cement and WG powder mixture is reduced. Therefore, it decreased the water requirement for particle surface lubrication and resulted in an increased slump.

6.2.2. Compressive Strength

Islam et al. [98] performed a CS test on mortar specimens containing recycled WG as cement replacement. Compared to controlled mortar specimens, recycled WG mortar had a lower CS at the age of 7, 14, 28, and 56 days. At 90 days, an increase in CS was observed; the highest CS was obtained with a 10% cement replacement. Similarly, 15% cement replacement at 180 days and 20% cement replacement at 365 days exhibited maximum CS. The reason could be the pozzolanic behavior of glass, which reacted slowly and improved the microstructure of the matrix at later ages and resulted in improved CS. Figure 13 is generated based on the past studies depicting the variation in 28-days CS with increasing WG content as cement replacement. A slight increase in CS can be observed at lower replacement ratios. Rehman et al. [82] used WG powder as cement replacement (20%, 30%, and 40%) and steel slag as a fine aggregate replacement (40%, 60%, and 80%) in SCC and investigated their influence on MPs. They observed an increase in CS when 20% cement is replaced by WG powder, but it decreased as the WG powder content is increased further. When the proportions of all other ingredients were constant, increasing the steel slag content increased the CS of SCC. At constant WG powder content, the CS of concrete improved as the steel slag content increased. The maximum increase in CS was observed by 11% in comparison with the control specimen when 20% WG powder was used in place of cement, and 80% steel slag was used in place of fine aggregate. On the other hand, there was a slight decrease in CS of SCC as the WG powder content increased while the steel slag content was kept constant. The minimum CS was 5.7% lower than the control specimen when WG powder and steel slag were used in place of 40% cement and 40% fine aggregate, respectively. The increase in CS with the addition of steel slag could be attributed to the pozzolanic action of steel slag or the difference in hardness between steel slag and the replaced aggregates. Al-Zubaid et al. [100] studied the effect of brown, green, and neon glass on MPs of concrete used as cement replacement by 11%, 13%, and 15%. The best results of CS were observed with neon glass at 13% content due to the high concentration of SiO_2 (68%) in neon glass, combined with the high CaO content (66.11%) in cement, and their combination with water formed a significant amount of $CaCO_3$ during the hydration process. Anwar [101] also observed improvement in CS at lower content of WG powder. At 10% WG powder content, the CS improved by 16.6% than the reference sample. The increase in CS occurred due to the pozzolanic reaction of glass powder. Because the glass powder acts as a pozzolanic material, it reduces the effect of carbonation and increases the strength of concrete. Thus, the smaller particle size of the glass powder interacts more readily with the lime in the cement, resulting in increased CS in the concrete. Aliabdo et al. [51] reported a 5.1% increase in CS for 33 MPa concrete containing 5% WG powder in place of cement when compared to the reference mix. Whereas CS decreased with further addition of WG powder, as shown in the figure.

Additionally, the CS of concrete mix grade 45 MPa increased by 2.5% at 5% replacement and 4.8% at 10% replacement compared to the reference mix. When more than 10% of the cement was replaced by WG powder, a decrease in CS was observed; this decrease could be attributed to the increased percentage of cement replacement, which resulted in cement dilution. A similar trend was also observed from various studies with the use of WG as cement replacement [102]. Hence, WG powder as cement replacement is preferable only at lower replacement ratios.

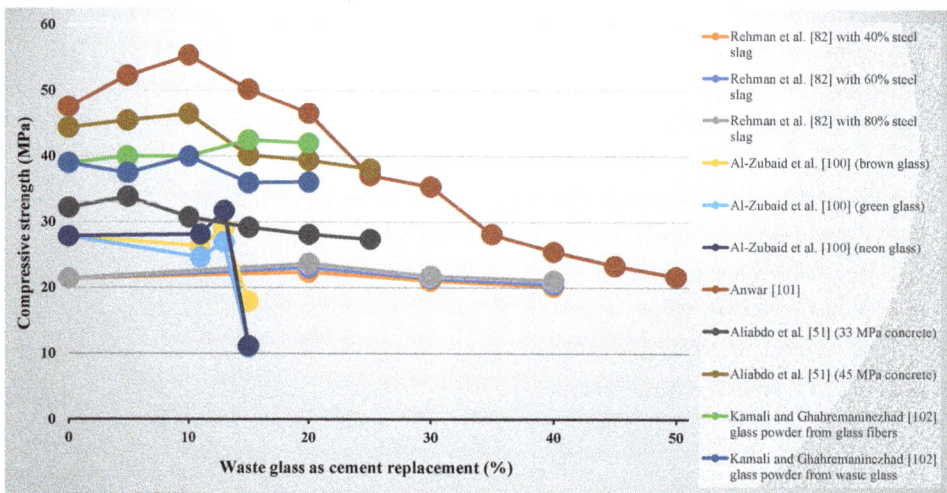

Figure 13. Effect of waste glass as cement replacement on 28-days compressive strength.

6.2.3. Split-Tensile Strength

The effect of using WG as cement replacement on STS has been shown in Figure 14. It also indicates that at lower content of WG, the STS can be increased while higher content of WG decreases the STS compared to the reference samples without WG. Similar to the CS, Rehman et al. [82] noted that the maximum improvement in STS was 13.2% when 20% of cement was replaced with WG powder, and 80% of cement was replaced with steel slag. The minimum STS was 5.6% less than that of the control mix when 40% steel slag and 40% glass powder were used as cement and aggregate replacements, respectively. Whereas, Al-Zubaid et al. [100] mostly found a decrease in STS with the addition of different types of WG in concrete. However, using green glass at 13% replacement of cement showed improvement in STS by 16.2% than the control mix. Aliabdo et al. [51] described enhancement in STS by 16.6%, 19.4%, and 5.9% for 33 MPa concrete containing 5%, 10%, and 15% WG powder, respectively, when compared to the control mix. Whereas STS decreased by 10% and 13.8% when 20% and 25% WG powder were substituted for cement in a 33 MPa concrete mix, respectively. Additionally, for 5%, 10%, and 15% replacement in 45 MPa grade concrete, the STS increased by 11.7%, 13.0%, and 18.1%, respectively. Whereas, at 20% and 25% replacement, a slight decrease in STS of 1.0% and 2.3%, respectively, was observed. STS decreases when more than 20% of cement is replaced with WG powder. The reasons for the improvement in STS at lower WG contents and reduction in STS at higher WG contents are the same as described earlier for CS.

6.2.4. Flexural Strength

The influence of WG powder as cement replacement on the FS of composites has been displayed in Figure 15. It also shows an almost similar trend as CS and STS. For instance, the results of Rehman et al. [82] showed enhanced FS at 20% and 30% contents of WG powder while at 40% content of WG powder, the FS reduced compared to the reference sample. The highest value of FS was observed at 20% content of WG powder as cement replacement and 80% steel slag as fine aggregate replacement. Similar to the CS, the maximum FS was observed with neon glass at a 13% replacement ratio [100]. Also, the addition of WG with up to 20% content exhibited improvement in FS while a further increment in WG content reduced the FS than the reference sample [101]. Hama [103] also reported maximum FS at 20% replacement of cement by WG powder. However, several studies reported a reduction in FS with the utilization of WG powder as cement replacement [100,104], as shown in the figure. The reasons for the CS behavior of composites with WG addition also apply to the flexural behavior.

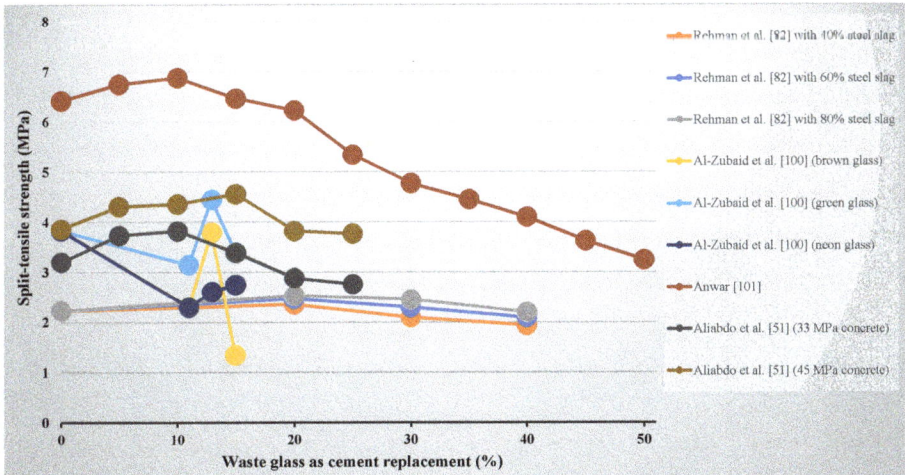

Figure 14. Effect of waste glass as cement replacement on 28-days split-tensile strength.

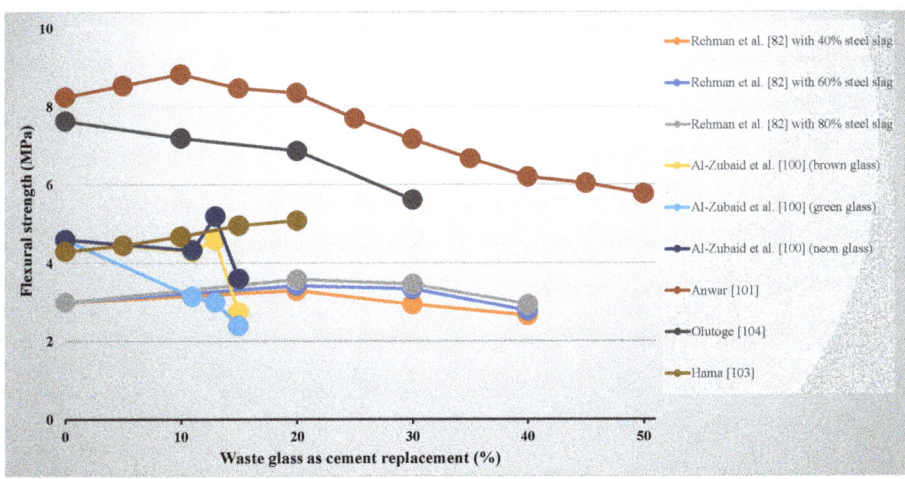

Figure 15. Effect of waste glass as cement replacement on 28-days flexural strength.

6.3. Microstructure of Cement-Based Materials Containing Waste Glass

The microstructure study of various past studies revealed that the size of WG used as aggregate replacement greatly influences the ITZ and porosity of the matrix. Afshinnia and

Rangaraju [105] found weak glass-matrix ITZ and high porosity when coarser WG was used as an aggregate replacement, compared to the NA concrete, as depicted in Figure 16. As illustrated in the figure, WG aggregates have a smooth particle surface that interfaces with the cement paste, resulting in minimal mechanical interlocking between the two phases. The weak bond between WG and cement matrix is also reflected in the mechanical characteristics of composites. Conversely, the use of finer WG as an aggregate replacement can improve the microstructure of composites [105]. Soliman and Hamou [106] performed SEM analysis to study the microstructure of composite containing 50% quartz sand and 50% recycled WG with a mean particle size of 275 μm. They stated that the bond among WG and cement matrix is comparable with that among quartz and cement matrix, as shown in Figure 17. Thus, the microstructure study also supports the use of finer WG as aggregate replacement. Kong et al. [107] examined the effect of WG powder on the microstructure under various curing conditions. The study demonstrates that WG powder exhibits strong pozzolanic reactions when cured in microwave or steam rather than under standard curing conditions. Matos and Sousa-Coutinho [108] conducted SEM analysis to study the microstructure of mortar containing 10% WG powder as cement replacement and compared it with the control mix as depicted in Figure 18. The glass particles appear to have been completely compressed and scattered within the hydration products of a compact, dense, and mature gel containing needle-shaped ettringite crystals (Figure 18c,d). It was seen that the C–S–H gel in samples containing WG powder has more calcium as well as more alkalis compared to the reference sample due to the pozzolanic properties of WG powder. This is the reason for having improved MPs of composites with WG powder.

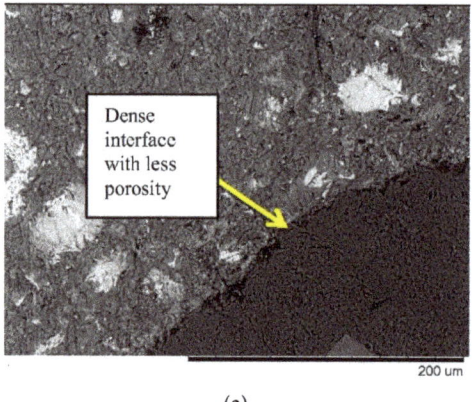
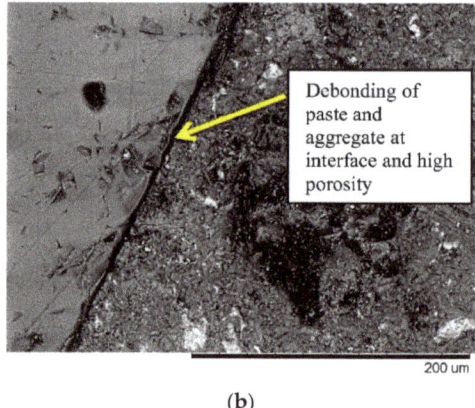

(a) (b)

Figure 16. Microstructure of composites with: (a) natural aggregate; (b) waste glass aggregate (size: 4.75–9.5 mm) [105].

6.4. Durability of Cement-Based Materials Containing Waste Glass

The durability of concrete is inextricably linked to its permeability [109]. Water permeability was increased by partially replacing cement with WG. A concrete sample having a 25 mm penetration depth of water when 30% of the cement was substituted with soda-lime WG (size: <120 μm), decreased the penetration depth of water to 9 mm; however, when 60% of the cement was substituted with WG, the penetration depth of water decreased it to 5 mm [110]. By substituting 60% natural sand for flint WG (size: <4 mm) in concrete with a 20 mm penetration depth of water, the penetration depth of water was decreased to 16 mm [111]. The use of finer WG reduced chloride penetration [79,102,112,113], while coarser WG showed less resistance to chloride penetration [114]. The percentage of water absorption reduced as the amount of WG in CBMs increased. When 20% WG (size: 100 μm) was used in place of cement, the percentage of water absorption ratio decreased from 4.6% to 3.2% [115]. Likewise, by substituting 25% cement for WG (size: <80 μm), the percentage water absorption was decreased from 6.2% to 3.8% [116]. When CRT WG (size: <5 mm) was used as sand substitute at 0%, 50%, and 100% the water absorption ratios were 7.3%, 7.0%, and 6.4%, respectively [96]. To replace coarse aggregate, WG (size: 3–16 mm) was used, and the percentage of water absorption decreased from 6.0% to 2.5% as the proportion of WG increased from 0 to 75%. Hence, consistent results revealed that proper usage of WG contributes to the reduction of CBMs' permeability and the impediment of detrimental elements transport in CBM. This enhancement is a result of the synergy of several effects, including the pozzolanic and filler effects that enhance the hydration process and improve the microstructure, thereby decreasing permeability. Simultaneously, appropriate glass gradation can increase the packing density of particles, thereby lowering the permeability even further.

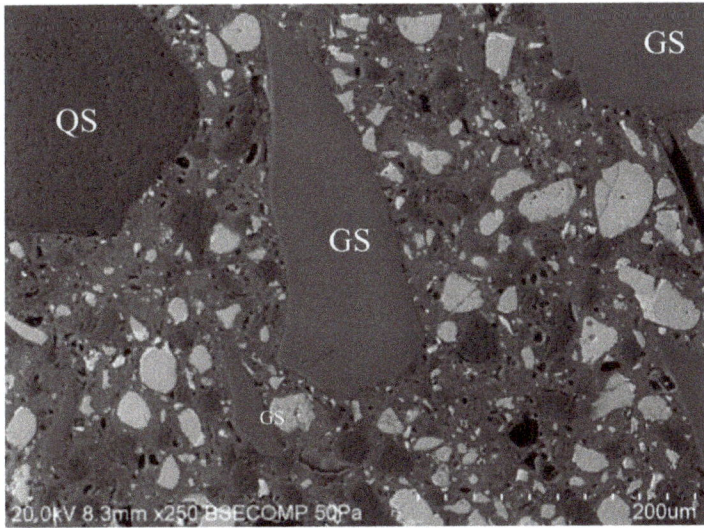

Figure 17. Microstructure of composite containing quartz and glass sand (glass mean particle size = 275 μm) [106].

Figure 18. *Cont.*

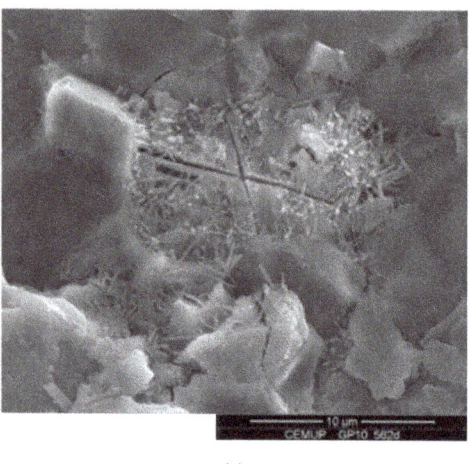 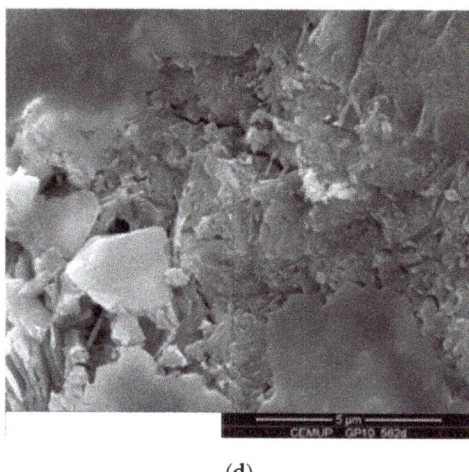

(c)   (d)

Figure 18. Microstructure of composites: (**a**) control mix, 10,000 times enlarged; (**b**) control mix, 20,000 times enlarged; (**c**) containing 10% waste glass powder, 10,000 times enlarged; (**d**) containing 10% waste glass powder, 20,000 times enlarged [108].

The sulphate immersion test was used to observe the resistance of CBM to sulphate attack. After five wet/dry cycles, the mass loss of concrete was determined. The mass loss was measured to be 0.8% for concrete with a w/b of 0.68; however, when 40% cement in concrete was replaced by WG (size: <150 μm), the loss in mass was decreased to 0.2% [81]. The anti-sulfate attack test revealed a consistent beneficial phenomenon. A WG (size: <4.75 mm) was used in place of natural sand in concrete, having a w/b ratio of 0.55. With increasing proportion of WG from 0% to 80%, the concrete's five-cycle mass loss decreased from 11% to 4% [117]. This increase in resistance to sulphate attack can be attributed to the concrete's refined microstructure, resulting from the pozzolanic reaction and filler effect of finer WG. The sulphate ion degrades CBMs by reacting with portlandite and forming gypsum, which sequentially forms expansive ettringite, which can crack the matrix of CBM [117]. A compact and improved microstructure prevents the passage of the sulphate ion, thereby increasing the resistance to sulphate attack.

Concrete with a w/b of 0.5 and cement was replaced by WG (size: 0.1–40 μm) at 0%, 10%, and 20%, proportions used to investigate the influence of WG on the depth of carbonation. At four months, the carbonation depth increased from 3 mm to 8 mm as the proportion of WG increased from 0% to 20% [108]. Similar observations were described in [118] when CRT WG (size: <5 mm) replaced the natural sand up to 100% in the production of heavyweight barite concrete with a w/b of 0.48. Depth of carbonation rose from 7.5 mm to 11.5 mm as the proportion of WG increased from 25% to 100%. The available data indicate that the use of WG degrades carbonation resistance. This is primarily due to the fact that glass reacts with $Ca(OH)_2$ contained within concrete. The $Ca(OH)_2$ aids in the delay of CO_2 diffusion, and glass consumption of $Ca(OH)_2$ speeds up carbonation. It is important to mention here that the studies cited above were performed on standard concrete. It is reasonable to consider that the detrimental impact of WG for resistance to carbonation can be alleviated or eliminated through microstructure refinement and low porosity, as concrete's diffusivity is decreased. There are numerous ways to decrease concrete's diffusivity, including increasing the packing density, decreasing the w/b, and using fillers.

Concrete's freezing-thawing resistance was increased by partially replacing cement with WG. Cement in concrete was replaced by electric WG (size: 13 μm), and loss in mass of the sample was reduced by 30% after 310 freezing-thawing cycles [119]. Similar findings were published in [120,121]. Cement was replaced by WG (size: 75 μm) at proportions of 5%, 10%, and 15%. It was found that as the proportion of WG increased from 0% to 10%, the loss in mass decreased monotonically, implying that the WG enhances the resistance to freezing-thawing, most possibly because of the pozzolanic reaction and filler effect. However, as the amount of WG increased from 10% to 15%, the loss in mass increased because of the dilution effect [120]. The dilution effect is related to the concrete's w/b. Due to the significant increase in the w/b caused by substituting an extreme quantity of cement with WG, the additional water raises the porosity and results in the dilution effect. Though it is envisioned that in high-performance concrete or ultra-high-performance concrete with a very low w/b, a greater amount of cement can be substituted by WG without a dilution effect.

7. Utilization of Waste Glass Powder in 3D Printing and Geopolimerization

7.1. 3D Printing

Three-dimensional (3D) printing of CBMs, also referred to as additive manufacturing in the construction sector, is the process of combining CBM extrusion in layers with robotic motion control [122]. Interest in 3D printing technology has increased significantly over the last few decades, both in academia and industry [123]. This is largely because it eliminates the need for formwork and human interference due to automation, which significantly decreases the time required to construct a structure. Additionally, because this technology is capable of creating complex structures, it enables structural optimization [124]. Mostly, fine aggregates were used in the 3D printing of CBMs [125]. The aggregate size is constrained by the material delivery system. The most frequently used material in 3D printing is natural sand [122]. By substituting recycled WG for natural sand in the material, a new market for WG can be created while also reducing demand for natural sand, a finite resource. With the rapid growth in the popularity of 3D printing technology, the widespread use of recycled WG as a raw material for such applications has the potential to be very beneficial. As a result, the WG recycling rate can be increased, thereby reducing the need for landfill space. Ting et al. [126] evaluated the properties of natural sand and recycled WG aggregate mortar 3D printing applications. The mixes were subjected to rheological and mechanical characterizations to compare the various aggregates' effects. The rheology results demonstrated the benefit of recycling WG in CBMs for 3D printing. The fact that the fresh material has a lower plastic viscosity and dynamic yield stress show that the recycled WG mix has better flow characteristics than the natural sand mix. This was most likely due to excess water in the recycled WG mix because of the WG particles' reduced capacity of water absorption than the natural sand, as observed in another study [114]. Other possible explanations include the smooth surface of recycled WG particles related to the natural sand particles [127]. However, the CS, STS, and flexural strength of recycled WG mixes were significantly less than those of natural sand mixes. Numerous publications have attributed these findings to the low adhesive strength amongst WG particles and the surrounding matrix at ITZ [48,114].

7.2. Geopolimerization

Cement is the primary constituent of concrete, acting as a binder for the aggregate particles contained in any concrete mixture. However, producing cement consumes a significant amount of energy and CO_2 emission. To address this issue, most studies and research have focused on developing a more environmentally friendly alternative binder. Geopolymer is a suitable alternative to ordinary concrete. Geopolymer is a novel binder that is currently being developed. This is a type of binder that was developed to replace cement in the manufacture of concrete. The objective is to develop sustainable and environmentally friendly concrete that does not contain cement as a binder. The geopolymer contains a binder that is rich in silica and alumina. WG is an amorphous material, and its chemical composition has been listed in Table 6. Recently, WG powder has been investigated as a potential source of alumina for geopolymer production [128].

Geopolymers derived from WG are a relatively new field of study. According to a study, WG-based geopolymer achieved comparable CS to fly ash-based geopolymer, and the mechanical properties were highly dependent on the WG particle size, curing conditions, and alkali solution concentration [128]. Novais et al. [129] conducted a study to partially replace metakaolin with WG powder in the production of geopolymer and studied its influence on the mechanical properties. The results indicated that adding 12.5% WG increased CS by nearly 46%, whereas a further increase in the amount of WG has the opposite effect when compared to the pure metakaolin-based geopolymer. Additionally, the results demonstrated the enormous influence of curing conditions on the CS of geopolymers containing WG. The CS loss was reduced considerably when geopolymers with a high WG content were cured at ambient conditions rather than in sealed bags. This allows for the addition of up to 37.5% WG without compromising CS.

8. Conclusions and Future Recommendations

8.1. Conclusions

In this study, a scientometric review was performed on the utilization of waste glass (WG) in concrete for sustainable construction, along with a comprehensive discussion. Scientometric analysis was carried out to evaluate the relevant study fields, publication trend of articles, most contributing sources, keywords co-occurrence, most cited articles and authors, and active countries contributing to the field of WG utilization in concrete for sustainable construction. Moreover, the sustainable aspects of WG utilization in construction materials were reviewed, and the influence of WG on the performance of cement-based materials (CBMs) was assessed. Particularly, the effect of WG on workability, compressive strength, split-tensile strength, flexural strength, microstructure, and durability of the resulting composite was evaluated. The following conclusions have been made:

- Scientometric analysis on the available data retrieved from the Scopus database exposed that the top four fields based on the number of documents were Engineering, Environmental Science, Materials Science, and Energy, containing 26.4%, 22.4%, 11.6%, and 9.9% of the total documents, respectively. A slight increase in the number of publications on the utilization of WG in construction materials was observed up to 2015. However, a remarkable increase was observed in the last 5 years. The top 3 journals based on the number of documents were found to be the journal of cleaner production, construction and building materials, and resources, conservation, and recycling, containing 241, 202, and 79 documents, respectively, of the 2066 total documents. Sustainable development, recycling, construction industry, compressive strength, and sustainability were observed to be the top 5 most occurred keywords. Additionally, the connection network demonstrated that waste management and recycling have a strong link with sustainable development and may have a sizable impact on sustainability in construction. Furthermore, India, China, and United Kingdom contributed the most documents in the current study field.
- WG can be used in CBMs as aggregate replacement and cement replacement. Thus, it conserves natural resources, solves waste management problems, reduces CO_2 emission by decreasing cement demand, protects the environment from toxic chemicals, and produces cost-effective composites. Therefore, the utilization of WG in construction materials is a better approach for sustainability in construction.
- The influence of WG as aggregate/cement replacement in CBMs on the mechanical properties (MPs) was found to be inconsistent. Usually, finer glass particles improved, while coarser glass particles decreased the MPs of composites. In additions, its use in higher proportions has a negative influence on the MPs of composites.
- Due to the pozzolanic reaction and filler effect of WG powder, it is possible that substituting WG for up to 25% of the cement or 20% of natural aggregate enhanced the MPs. However, using an excessive amount of WG may decrease the mechanical strength because of the dilution effect. The optimum content of WG in concrete is related to the water/binder ratio (w/b), as the w/b affects the amount of $Ca(OH)_2$

reacting with glass. By substituting coarser WG for natural aggregates, the mechanical strength may be reduced. The influence of WG depends on both particle size and the proportion of the replacement.
- The addition of WG can help improve the microstructure and reduce the permeability of CBMs, thereby increasing their resistance to sulphate attack and freeze-thaw and eventually enhancing the durability of composites. However, glass may impair the resistance to carbonation because of $Ca(OH)_2$ consumption by glass.

Appropriate WG selection is critical to the success of applications. It is suggested that the amount, size, and type of WG used in CBMs are appropriate for achieving adequate MPs and durability, reliant on the anticipated applications.

8.2. Future Recommendations

This review demonstrated the importance of developing a complete knowledge of the impacts of WG on the MPs of CBMs in order to ensure the long-term viability and durability of structures. The subsequent research needs can be found as a result of the foregoing discussions:
- Further research is essential to clarify the impact of WG on the rheological properties of CBMs in terms of amount, type, particle size, and morphology of WG particles. This is an assumption, but it is possible that the rheology of the CBMs prior to incorporating WG particles influences the glass particle's role.
- There is a necessity to explore potential coupling effects between the w/b of concrete, the size of the WG particles, and the percentage replacement. This is an assumption, but it is possible that the w/b of CBMs and the size of the WG particles affect the optimal content of WG, resulting in a loss of mechanical strength.
- There is a dearth of comprehensive knowledge regarding the effects of WG content and particle size on the durability of CBMs at various w/b. This is an assumption that the w/b has a considerable influence on the durability-related roles of WG content and particle size.
- Computer software tools s the machine learning approach and numerical modeling need to be applied to predict the various properties of CBMs containing WG.

Author Contributions: D.Q.: conceptualization, methodology, investigation, formal analysis, visualization, and writing-original draft preparation. Y.H.: conceptualization, methodology, investigation formal analysis, writing-reviewing and editing, and supervision. X.L.: supervision, resources, funding acquisition, and writing-reviewing and editing. All authors have read and agreed to the published version of the manuscript.

Funding: Jilin Provincial Department of Science and Technology Project (20180201027SF).

Institutional Review Board Statement: Not applicable.

Informed Consent Statement: Not applicable.

Data Availability Statement: Not applicable.

Acknowledgments: This work was sponsored in part by Jilin Provincial Department of Science and Technology Project (20180201027SF).

Conflicts of Interest: The authors declare no conflict of interest.

References

1. Alwan, Z.; Jones, P.; Holgate, P. Strategic sustainable development in the UK construction industry, through the framework for strategic sustainable development, using Building Information Modelling. *J. Clean. Prod.* **2017**, *140*, 349–358. [CrossRef]
2. Doan, D.T.; Ghaffarianhoseini, A.; Naismith, N.; Zhang, T.; Ghaffarianhoseini, A.; Tookey, J. A critical comparison of green building rating systems. *Build. Environ.* **2017**, *123*, 243–260. [CrossRef]
3. Hwang, B.-G.; Zhu, L.; Tan, J.S.H. Green business park project management: Barriers and solutions for sustainable development. *J. Clean. Prod.* **2017**, *153*, 209–219. [CrossRef]

4. Escamilla, E.Z.; Habert, G.; Wohlmuth, E. When CO_2 counts: Sustainability assessment of industrialized bamboo as an alternative for social housing programs in the Philippines. *Build. Environ.* **2016**, *103*, 44–53. [CrossRef]
5. Alencar, M.H.; Priori, L., Jr.; Alencar, L.H. Structuring objectives based on value-focused thinking methodology: Creating alternatives for sustainability in the built environment. *J. Clean. Prod.* **2017**, *156*, 62–73. [CrossRef]
6. Kang, S.; Zhao, Y.; Wang, W.; Zhang, T.; Chen, T.; Yi, H.; Rao, F.; Song, S. Removal of methylene blue from water with montmorillonite nanosheets/chitosan hydrogels as adsorbent. *Appl. Surf. Sci.* **2018**, *448*, 203–211. [CrossRef]
7. Wang, W.; Zhao, Y.; Bai, H.; Zhang, T.; Ibarra-Galvan, V.; Song, S. Methylene blue removal from water using the hydrogel beads of poly (vinyl alcohol)-sodium alginate-chitosan-montmorillonite. *Carbohydr. Polym.* **2018**, *198*, 518–528. [CrossRef]
8. Ren, B.; Zhao, Y.; Bai, H.; Kang, S.; Zhang, T.; Song, S. Eco-friendly geopolymer prepared from solid wastes: A critical review. *Chemosphere* **2020**, *267*, 128900. [CrossRef]
9. Smarzewski, P. Mechanical properties of ultra-high performance concrete with partial utilization of waste foundry sand. *Buildings* **2020**, *10*, 11. [CrossRef]
10. Smarzewski, P. Influence of basalt-polypropylene fibres on fracture properties of high performance concrete. *Compos. Struct.* **2019**, *209*, 23–33. [CrossRef]
11. Smarzewski, P. Influence of silica fume on mechanical and fracture properties of high performance concrete. *Procedia Struct. Integr.* **2019**, *17*, 5–12. [CrossRef]
12. Smarzewski, P. Study of Bond Strength of Steel Bars in Basalt Fibre Reinforced High Performance Concrete. *Crystals* **2020**, *10*, 436. [CrossRef]
13. Smarzewski, P. Flexural toughness of high-performance concrete with basalt and polypropylene short fibres. *Adv. Civ. Eng.* **2018**, *2018*, 5024353. [CrossRef]
14. Smarzewski, P. Comparative Fracture Properties of Four Fibre Reinforced High Performance Cementitious Composites. *Materials* **2020**, *13*, 2612. [CrossRef]
15. Ahmad, W.; Farooq, S.H.; Usman, M.; Khan, M.; Ahmad, A.; Aslam, F.; Yousef, R.A.; Abduljabbar, H.A.; Sufian, M. Effect of coconut fiber length and content on properties of high strength concrete. *Materials* **2020**, *13*, 1075. [CrossRef]
16. Smarzewski, P. Flexural toughness evaluation of basalt fibre reinforced HPC beams with and without initial notch. *Compos. Struct.* **2020**, *235*, 111769. [CrossRef]
17. Colangelo, F.; Cioffi, R.; Liguori, B.; Iucolano, F. Recycled polyolefins waste as aggregates for lightweight concrete. *Compos. Part B Eng.* **2016**, *106*, 234–241. [CrossRef]
18. Asutkar, P.; Shinde, S.; Patel, R. Study on the behaviour of rubber aggregates concrete beams using analytical approach. *Eng. Sci. Technol. Int. J.* **2017**, *20*, 151–159. [CrossRef]
19. Ferreira, L.; de Brito, J.; Saikia, N. Influence of curing conditions on the mechanical performance of concrete containing recycled plastic aggregate. *Constr. Build. Mater.* **2012**, *36*, 196–204. [CrossRef]
20. Saikia, N.; Brito, J.D. Waste polyethylene terephthalate as an aggregate in concrete. *Mater. Res.* **2013**, *16*, 341–350. [CrossRef]
21. Rashad, A.M. Recycled waste glass as fine aggregate replacement in cementitious materials based on Portland cement. *Constr. Build. Mater.* **2014**, *72*, 340–357. [CrossRef]
22. Smarzewski, P.; Barnat-Hunek, D. Mechanical and durability related properties of high performance concrete made with coal cinder and waste foundry sand. *Constr. Build. Mater.* **2016**, *121*, 9–17. [CrossRef]
23. Arshad, S.; Sharif, M.B.; Irfan-ul-Hassan, M.; Khan, M.; Zhang, J.-L. Efficiency of Supplementary Cementitious Materials and Natural Fiber on Mechanical Performance of Concrete. *Arab. J. Sci. Eng.* **2020**, *45*, 8577–8589. [CrossRef]
24. Khan, M.; Ali, M. Effect of super plasticizer on the properties of medium strength concrete prepared with coconut fiber. *Constr. Build. Mater.* **2018**, *182*, 703–715. [CrossRef]
25. Khan, M.; Rehman, A.; Ali, M. Efficiency of silica-fume content in plain and natural fiber reinforced concrete for concrete road. *Constr. Build. Mater.* **2020**, *244*, 118382. [CrossRef]
26. Khan, U.A.; Jahanzaib, H.M.; Khan, M.; Ali, M. *Improving the Tensile Energy Absorption of High Strength Natural Fiber Reinforced Concrete with Fly-Ash for Bridge Girders, Key Engineering Materials*; Trans Tech Publications: Stafa-Zurich, Switzerland, 2018; Volume 765, pp. 335–342.
27. Pelisser, F.; Montedo, O.R.K.; Gleize, P.J.P.; Roman, H.R. Mechanical properties of recycled PET fibers in concrete. *Mater. Res.* **2012**, *15*, 679–686. [CrossRef]
28. Merli, R.; Preziosi, M.; Acampora, A.; Lucchetti, M.C.; Petrucci, E. Recycled fibers in reinforced concrete: A systematic literature review. *J. Clean. Prod.* **2020**, *248*, 119207. [CrossRef]
29. Liu, J.; Liu, J.; Huang, Z.; Zhu, J.; Liu, W.; Zhang, W. Effect of Fly Ash as Cement Replacement on Chloride Diffusion, Chloride Binding Capacity, and Micro-Properties of Concrete in a Water Soaking Environment. *Appl. Sci.* **2020**, *10*, 6271. [CrossRef]
30. Bueno, E.T.; Paris, J.M.; Clavier, K.A.; Spreadbury, C.; Ferraro, C.C.; Townsend, T.G. A review of ground waste glass as a supplementary cementitious material: A focus on alkali-silica reaction. *J. Clean. Prod.* **2020**, *257*, 120180. [CrossRef]
31. Hamada, H.M.; Thomas, B.S.; Yahaya, F.M.; Muthusamy, K.; Yang, J.; Abdalla, J.A.; Hawileh, R.A. Sustainable use of palm oil fuel ash as a supplementary cementitious material: A comprehensive review. *J. Build. Eng.* **2021**, *40*, 102286. [CrossRef]
32. Chore, H.S.; Joshi, M.P. Strength characterization of concrete using industrial waste as cement replacing materials for rigid pavement. *Innov. Infrastruct. Solut.* **2020**, *5*, 1–9. [CrossRef]

33. Lim, N.H.A.S.; Shaari, F.H.; Shaari, E.H.; Sam, A.R.M.; Khalid, N.H.A.; Ariffin, N.F.; Muthusamy, K. Properties of Concrete Containing Bamboo Waste as Cement Replacement. *J. Comput. Theor. Nanosci.* **2020**, *17*, 1306–1310.
34. Rattanachu, P.; Toolkasikorn, P.; Tangchirapat, W.; Chindaprasirt, P.; Jaturapitakkul, C. Performance of recycled aggregate concrete with rice husk ash as cement binder. *Cem. Concr. Compos.* **2020**, *108*, 103533. [CrossRef]
35. Khan, M.; Ali, M. Improvement in concrete behavior with fly ash, silica-fume and coconut fibres. *Constr. Build. Mater.* **2019**, *203*, 174–187. [CrossRef]
36. Saikia, N.; De Brito, J. Use of plastic waste as aggregate in cement mortar and concrete preparation: A review. *Constr. Build. Mater.* **2012**, *34*, 385–401. [CrossRef]
37. Jain, K.L.; Sancheti, G.; Gupta, L.K. Durability performance of waste granite and glass powder added concrete. *Constr. Build. Mater.* **2020**, *252*, 119075. [CrossRef]
38. Martínez-Barrera, G.; del Coz-Díaz, J.J.; Álvarez-Rabanal, F.P.; Gayarre, F.L.; Martínez-López, M.; Cruz-Olivares, J. Waste tire rubber particles modified by gamma radiation and their use as modifiers of concrete. *Case Stud. Constr. Mater.* **2020**, *12*, e00321. [CrossRef]
39. Topcu, I.B.; Canbaz, M. Properties of concrete containing waste glass. *Cem. Concr. Res.* **2004**, *34*, 267–274. [CrossRef]
40. Meena, A.; Singh, R. *Comparative Study of Waste Glass Powder as Pozzolanic Material in Concrete*; Department of Civil Engineering, National Institute of Technology: Rourkela, India, 2012.
41. Pant, D.; Singh, P. Pollution due to hazardous glass waste. *Environ. Sci. Pollut. Res.* **2014**, *21*, 2414–2436. [CrossRef] [PubMed]
42. Singh, N.; Li, J.; Zeng, X. Solutions and challenges in recycling waste cathode-ray tubes. *J. Clean. Prod.* **2016**, *133*, 188–200. [CrossRef]
43. Olofinnade, O.M.; Ndambuki, J.M.; Ede, A.N.; Booth, C. Application of waste glass powder as a partial cement substitute towards more sustainable concrete production. *Int. J. Eng. Res. Africa.* **2017**, *31*, 77–93. [CrossRef]
44. Shelby, J. *Introduction to Glass Science and Technology*; The Royal Society of Chemistry: Cambridge, UK, 2005.
45. Schmitz, A.; Kamiński, J.; Scalet, B.M.; Soria, A. Energy consumption and CO_2 emissions of the European glass industry. *Energy Policy* **2011**, *39*, 142–155. [CrossRef]
46. Jani, Y.; Hogland, W. Waste glass in the production of cement and concrete—A review. *J. Environ. Chem. Eng.* **2014**, *2*, 1767–1775. [CrossRef]
47. Federico, L.; Chidiac, S. Waste glass as a supplementary cementitious material in concrete–critical review of treatment methods. *Cem. Concr. Compos.* **2009**, *31*, 606–610. [CrossRef]
48. Mohajerani, A.; Vajna, J.; Cheung, T.H.H.; Kurmus, H.; Arulrajah, A.; Horpibulsuk, S. Practical recycling applications of crushed waste glass in construction materials: A review. *Constr. Build. Mater.* **2017**, *156*, 443–467. [CrossRef]
49. Rashad, A.M. Recycled cathode ray tube and liquid crystal display glass as fine aggregate replacement in cementitious materials. *Constr. Build. Mater.* **2015**, *93*, 1236–1248. [CrossRef]
50. Paul, S.C.; Šavija, B.; Babafemi, A.J. A comprehensive review on mechanical and durability properties of cement-based materials containing waste recycled glass. *J. Clean. Prod.* **2018**, *198*, 891–906. [CrossRef]
51. Aliabdo, A.A.; Elmoaty, M.A.; Aboshama, A.Y. Utilization of waste glass powder in the production of cement and concrete. *Constr. Build. Mater.* **2016**, *124*, 866–877. [CrossRef]
52. Xu, Y.; Zeng, J.; Chen, W.; Jin, R.; Li, B.; Pan, Z. A holistic review of cement composites reinforced with graphene oxide. *Constr. Build. Mater.* **2018**, *171*, 291–302. [CrossRef]
53. Xiao, X.; Skitmore, M.; Li, H.; Xia, B. Mapping knowledge in the economic areas of green building using scientometric analysis. *Energies* **2019**, *12*, 3011. [CrossRef]
54. Mryglod, O.; Holovatch, Y.; Kenna, R. Data mining in scientometrics: Usage analysis for academic publications. In Proceedings of the 2018 IEEE Second International Conference on Data Stream Mining & Processing (DSMP), Lviv, Ukraine, 21–25 August 2018; IEEE: Piscataway, NJ, USA, 2018; pp. 241–246.
55. Darko, A.; Chan, A.P.; Huo, X.; Owusu-Manu, D.-G. A scientometric analysis and visualization of global green building research. *Build. Environ.* **2019**, *149*, 501–511. [CrossRef]
56. Song, J.; Zhang, H.; Dong, W. A review of emerging trends in global PPP research: Analysis and visualization. *Scientometrics* **2016**, *107*, 1111–1147. [CrossRef]
57. Hosseini, M.R.; Martek, I.; Zavadskas, E.K.; Aibinu, A.A.; Arashpour, M.; Chileshe, N. Critical evaluation of off-site construction research: A Scientometric analysis. *Autom. Constr.* **2018**, *87*, 235–247. [CrossRef]
58. Liao, H.; Tang, M.; Luo, L.; Li, C.; Chiclana, F.; Zeng, X.-J. A bibliometric analysis and visualization of medical big data research. *Sustainability* **2018**, *10*, 166. [CrossRef]
59. Chadegani, A.A.; Salehi, H.; Yunus, M.; Farhadi, H.; Fooladi, M.; Farhadi, M.; Ebrahim, N.A. A comparison between two main academic literature collections: Web of Science and Scopus databases. *Asian Soc. Sci.* **2013**, *9*, 18–26. [CrossRef]
60. Bergman, E.M.L. Finding citations to social work literature: The relative benefits of using Web of Science, Scopus, or Google Scholar. *J. Acad. Librariansh.* **2012**, *38*, 370–379. [CrossRef]
61. Meho, L.I. Using Scopus's CiteScore for assessing the quality of computer science conferences. *J. Informetr.* **2019**, *13*, 419–433. [CrossRef]
62. Zuo, J.; Zhao, Z.-Y. Green building research–current status and future agenda: A review. *Renew. Sustain. Energy Rev.* **2014**, *30*, 271–281. [CrossRef]

63. Darko, A.; Zhang, C.; Chan, A.P. Drivers for green building: A review of empirical studies. *Habitat Int.* **2017**, *60*, 34–49. [CrossRef]
64. Ahmad, W.; Khan, M.; Smarzewski, P. Effect of Short Fiber Reinforcements on Fracture Performance of Cement-Based Materials: A Systematic Review Approach. *Materials* **2021**, *14*, 1745. [CrossRef] [PubMed]
65. Markoulli, M.P.; Lee, C.I.; Byington, E.; Felps, W.A. Mapping Human Resource Management: Reviewing the field and charting future directions. *Hum. Resour. Manag. Rev.* **2017**, *27*, 367–396. [CrossRef]
66. Saka, A.B.; Chan, D.W. A scientometric review and metasynthesis of building information modelling (BIM) research in Africa. *Buildings* **2019**, *9*, 85. [CrossRef]
67. Goulden, S.; Erell, E.; Garb, Y.; Pearlmutter, D. Green building standards as socio-technical actors in municipal environmental policy. *Build. Res. Inf.* **2017**, *45*, 414–425. [CrossRef]
68. Jin, R.; Gao, S.; Cheshmehzangi, A.; Aboagye-Nimo, E. A holistic review of off-site construction literature published between 2008 and 2018. *J. Clean. Prod.* **2018**, *202*, 1202–1219. [CrossRef]
69. Park, J.Y.; Nagy, Z. Comprehensive analysis of the relationship between thermal comfort and building control research-A data-driven literature review. *Renew. Sustain. Energy Rev.* **2018**, *82*, 2664–2679. [CrossRef]
70. Oraee, M.; Hosseini, M.R.; Papadonikolaki, E.; Palliyaguru, R.; Arashpour, M. Collaboration in BIM-based construction networks: A bibliometric-qualitative literature review. *Int. J. Proj. Manag.* **2017**, *35*, 1288–1301. [CrossRef]
71. Van Eck, N.J.; Waltman, L. Software survey: VOSviewer, a computer program for bibliometric mapping. *Scientometrics* **2010**, *84*, 523–538. [CrossRef]
72. Wuni, I.Y.; Shen, G.Q.; Osei-Kyei, R. Scientometric review of global research trends on green buildings in construction journals from 1992 to 2018. *Energy Build.* **2019**, *190*, 69–85. [CrossRef]
73. Su, H.-N.; Lee, P.-C. Mapping knowledge structure by keyword co-occurrence: A first look at journal papers in Technology Foresight. *Scientometrics* **2010**, *85*, 65–79. [CrossRef]
74. Yu, F.; Hayes, B.E. Applying data analytics and visualization to assessing the research impact of the Cancer Cell Biology (CCB) Program at the University of North Carolina at Chapel Hill. *J. Sci. Librariansh.* **2018**, *7*, 4. [CrossRef]
75. Meyer, C. The greening of the concrete industry. *Cem. Concr. Compos.* **2009**, *31*, 601–605. [CrossRef]
76. Shayan, A.; Xu, A. Value-added utilisation of waste glass in concrete. *Cem. Concr. Res.* **2004**, *34*, 81–89. [CrossRef]
77. Vieitez, E.R.; Eder, P.; Villanueva, A.; Saveyn, H. *End-of-Waste Criteria for Glass Cullet: Technical Proposals*; JRC Scientific and Technical Reports; Publications Office of the European Union: Mercier, Luxembourg, 2011.
78. Jin, W.; Meyer, C.; Baxter, S. Glascrete-Concrete with glass aggregate. *ACI Mater. J.* **2000**, *97*, 208–213.
79. Elaqra, H.A.; Haloub, M.A.A.; Rustom, R.N. Effect of new mixing method of glass powder as cement replacement on mechanical behavior of concrete. *Constr. Build. Mater.* **2019**, *203*, 75–82. [CrossRef]
80. Chen, C.; Huang, R.; Wu, J.; Yang, C.-C. Waste E-glass particles used in cementitious mixtures. *Cem. Concr. Res.* **2006**, *36*, 449–456. [CrossRef]
81. Kou, S.; Poon, C.S. Properties of self-compacting concrete prepared with recycled glass aggregate. *Cem. Concr. Compos.* **2009**, *31*, 107–113. [CrossRef]
82. Rehman, S.; Iqbal, S.; Ali, A. Combined influence of glass powder and granular steel slag on fresh and mechanical properties of self-compacting concrete. *Constr. Build. Mater.* **2018**, *178*, 153–160. [CrossRef]
83. Hendi, A.; Mostofinejad, D.; Sedaghatdoost, A.; Zohrabi, M.; Naeimi, N.; Tavakolinia, A. Mix design of the green self-consolidating concrete: Incorporating the waste glass powder. *Constr. Build. Mater.* **2019**, *199*, 369–384. [CrossRef]
84. Wang, H.-Y.; Huang, W.-L. A study on the properties of fresh self-consolidating glass concrete (SCGC). *Constr. Build. Mater.* **2010**, *24*, 619–624. [CrossRef]
85. Tuaum, A.; Shitote, S.; Oyawa, W. Experimental study of self-compacting mortar incorporating recycled glass aggregate. *Buildings* **2018**, *8*, 15. [CrossRef]
86. Taha, B.; Nounu, G. Utilizing waste recycled glass as sand/cement replacement in concrete. *J. Mater. Civ. Eng.* **2009**, *21*, 709–721. [CrossRef]
87. Liu, H.; Shi, J.; Qu, H.; Ding, D. An investigation on physical, mechanical, leaching and radiation shielding behaviors of barite concrete containing recycled cathode ray tube funnel glass aggregate. *Constr. Build. Mater.* **2019**, *201*, 818–827. [CrossRef]
88. Wang, H.-Y.; Huang, W.-L. Durability of self-consolidating concrete using waste LCD glass. *Constr. Build. Mater.* **2010**, *24*, 1008–1013. [CrossRef]
89. Gorospe, K.; Booya, E.; Ghaednia, H.; Das, S.J.C.; Materials, B. Effect of various glass aggregates on the shrinkage and expansion of cement mortar. *Constr. Build. Mater.* **2019**, *210*, 301–311. [CrossRef]
90. Song, W.; Zou, D.; Liu, T.; Teng, J.; Li, L. Effects of recycled CRT glass fine aggregate size and content on mechanical and damping properties of concrete. *Constr. Build. Mater.* **2019**, *202*, 332–340. [CrossRef]
91. Ismail, Z.Z.; Al-Hashmi, E.A. Recycling of waste glass as a partial replacement for fine aggregate in concrete. *Waste Manag.* **2009**, *29*, 655–659. [CrossRef]
92. Abdallah, S.; Fan, M. Characteristics of concrete with waste glass as fine aggregate replacement. *Int. J. Eng. Tech. Res.* **2014**, *2*, 11–17.
93. Malik, M.I.; Bashir, M.; Ahmad, S.; Tariq, T.; Chowdhary, U. Study of concrete involving use of waste glass as partial replacement of fine aggregates. *IOSR J. Eng.* **2013**, *3*, 8–13. [CrossRef]

94. Olofinnade, O.M.; Ndambuki, J.M.; Ede, A.N.; Olukanni, D.O. Effect of substitution of crushed waste glass as partial replacement for natural fine and coarse aggregate in concrete. *Mater. Sci. Forum* **2016**, *866*, 58–62. [CrossRef]
95. Małek, M.; Łasica, W.; Jackowski, M.; Kadela, M. Effect of waste glass addition as a replacement for fine aggregate on properties of mortar. *Materials* **2020**, *13*, 3189. [CrossRef]
96. Kim, I.S.; Choi, S.Y.; Yang, E.I. Evaluation of durability of concrete substituted heavyweight waste glass as fine aggregate. *Constr. Build. Mater.* **2018**, *184*, 269–277. [CrossRef]
97. Sikora, P.; Augustyniak, A.; Cendrowski, K.; Horszczaruk, E.; Rucinska, T.; Nawrotek, P.; Mijowska, E. Characterization of mechanical and bactericidal properties of cement mortars containing waste glass aggregate and nanomaterials. *Materials* **2016**, *9*, 701. [CrossRef] [PubMed]
98. Islam, G.S.; Rahman, M.; Kazi, N. Waste glass powder as partial replacement of cement for sustainable concrete practice. *Int. J. Sustain. Built Environ.* **2017**, *6*, 37–44. [CrossRef]
99. Soliman, N.; Tagnit-Hamou, A. Partial substitution of silica fume with fine glass powder in UHPC: Filling the micro gap. *Constr. Build. Mater.* **2017**, *139*, 374–383. [CrossRef]
100. L-Zubaid, A.B.A.; Shabeeb, K.M.; Ali, A.I. Study the effect of recycled glass on the mechanical properties of green concrete. *Energy Procedia* **2017**, *119*, 680–692. [CrossRef]
101. Anwar, A. The influence of waste glass powder as a pozzolanic material in concrete. *Int. J. Civ. Eng. Technol* **2016**, *7*, 131–148.
102. Kamali, M.; Ghahremaninezhad, A. Effect of glass powders on the mechanical and durability properties of cementitious materials. *Constr. Build. Mater.* **2015**, *98*, 407–416. [CrossRef]
103. Hama, S.M. Improving mechanical properties of lightweight Porcelanite aggregate concrete using different waste material. *Int. J. Sustain. Built Environ.* **2017**, *6*, 81–90. [CrossRef]
104. Olutoge, F. Effect of waste glass powder (WGP) on the mechanical properties of concrete. *Am. J. Eng. Res.* **2016**, *5*, 213–220.
105. Afshinnia, K.; Rangaraju, P.R. Impact of combined use of ground glass powder and crushed glass aggregate on selected properties of Portland cement concrete. *Constr. Build. Mater.* **2016**, *117*, 263–272. [CrossRef]
106. Soliman, N.A.; Tagnit-Hamou, A. Using glass sand as an alternative for quartz sand in UHPC. *Constr. Build. Mater.* **2017**, *145*, 243–252. [CrossRef]
107. Kong, Y.; Wang, P.; Liu, S.; Gao, Z.; Rao, M. Effect of microwave curing on the hydration properties of cement-based material containing glass powder. *Constr. Build. Mater.* **2018**, *158*, 563–573. [CrossRef]
108. Matos, A.M.; Sousa-Coutinho, J. Durability of mortar using waste glass powder as cement replacement. *Constr. Build. Mater.* **2012**, *36*, 205–215. [CrossRef]
109. Tony, S.; Ion, D.; Bob, B. Long Term Durability Properties of Field Concretes with Glass Sand and Glass Powder. In Proceedings of the Fourth International Conference on Sustainable Construction Materials and Technologies, Las Vegas, NV, USA, 7–11 August 2016.
110. Du, H.; Tan, K. Durability performance of concrete with glass powder as supplementary cementitious material. *ACI Mater. J.* **2015**, *112*, 429–438.
111. Mardani-Aghabaglou, A.; Tuyan, M.; Ramyar, K. Mechanical and durability performance of concrete incorporating fine recycled concrete and glass aggregates. *Mater. Struct.* **2015**, *48*, 2629–2640. [CrossRef]
112. Patel, D.; Shrivastava, R.; Tiwari, R.; Yadav, R. Properties of cement mortar in substitution with waste fine glass powder and environmental impact study. *J. Build. Eng.* **2020**, *27*, 100940. [CrossRef]
113. Lalitha, G.; Sasidhar, C.; Ramachandrudu, C. Durability performance of concrete (M-60) fine aggregate partially replaced with crushed waste glass. *Technology* **2020**, *11*, 1–9.
114. Tan, K.H.; Du, H. Use of waste glass as sand in mortar: Part I–Fresh, mechanical and durability properties. *Cem. Concr. Compos.* **2013**, *35*, 109–117. [CrossRef]
115. Patel, D.; Tiwari, R.; Shrivastava, R.; Yadav, R. Effective utilization of waste glass powder as the substitution of cement in making paste and mortar. *Constr. Build. Mater.* **2019**, *199*, 406–415. [CrossRef]
116. Disfani, M.; Arulrajah, A.; Bo, M.; Sivakugan, N. Environmental risks of using recycled crushed glass in road applications. *J. Clean. Prod.* **2012**, *20*, 170–179. [CrossRef]
117. Wang, H.-Y. A study of the effects of LCD glass sand on the properties of concrete. *Waste Manag.* **2009**, *29*, 335–341. [CrossRef]
118. Ling, T.-C.; Poon, C.-S. Feasible use of recycled CRT funnel glass as heavyweight fine aggregate in barite concrete. *J. Clean. Prod.* **2012**, *33*, 42–49. [CrossRef]
119. Soroushian, P. Strength and durability of recycled aggregate concrete containing milled glass as partial replacement for cement. *Constr. Build. Mater.* **2012**, *29*, 368–377.
120. Abendeh, R.; Baker, M.B.; Salem, Z.A.; Ahmad, H. The feasibility of using milled glass wastes in concrete to resist freezing-thawing action. *Int. J. Civ. Environ. Eng.* **2015**, *9*, 1026–1029.
121. Al-Akhras, N.M. Performance of glass concrete subjected to freeze-thaw cycling. *Constr. Build. Technol. J.* **2012**, *6*, 392–397. [CrossRef]
122. Tay, Y.W.D.; Ting, G.H.A.; Qian, Y.; Panda, B.; He, L.; Tan, M.J. Time gap effect on bond strength of 3D-printed concrete. *Virtual Phys. Prototyp.* **2019**, *14*, 104–113. [CrossRef]
123. Tay, Y.W.D.; Panda, B.; Paul, S.C.; Noor Mohamed, N.A.; Tan, M.J.; Leong, K.F. 3D printing trends in building and construction industry: A review. *Virtual Phys. Prototyp.* **2017**, *12*, 261–276. [CrossRef]

124. Bos, F.; Wolfs, R.; Ahmed, Z.; Salet, T. Additive manufacturing of concrete in construction: Potentials and challenges of 3D concrete printing. *Virtual Phys. Prototyp.* **2016**, *11*, 209–225. [CrossRef]
125. Gosselin, C.; Duballet, R.; Roux, P.; Gaudillière, N.; Dirrenberger, J.; Morel, P. Large-scale 3D printing of ultra-high performance concrete—A new processing route for architects and builders. *Mater. Des.* **2016**, *100*, 102–109. [CrossRef]
126. Ting, G.H.A.; Tay, Y.W.D.; Qian, Y.; Tan, M.J. Utilization of recycled glass for 3D concrete printing: Rheological and mechanical properties. *J. Mater. Cycles Waste Manag.* **2019**, *21*, 994–1003. [CrossRef]
127. Jiao, D.; Shi, C.; Yuan, Q.; An, X.; Liu, Y.; Li, H. Effect of constituents on rheological properties of fresh concrete—A review. *Cem. Concr. Compos.* **2017**, *83*, 146–159. [CrossRef]
128. Cyr, M.; Idir, R.; Poinot, T. Properties of inorganic polymer (geopolymer) mortars made of glass cullet. *J. Mater. Sci.* **2012**, *47*, 2782–2797. [CrossRef]
129. Novais, R.M.; Ascensão, G.; Seabra, M.; Labrincha, J. Waste glass from end-of-life fluorescent lamps as raw material in geopolymers. *Waste Manag.* **2016**, *52*, 245–255. [CrossRef]

Article

Studying the C–H Crystals and Mechanical Properties of Sustainable Concrete Containing Recycled Coarse Aggregate with Used Nano-Silica

Shahriar Shahbazpanahi [1], Moslem Khalili Tajara [1], Rabar H. Faraj [2] and Amir Mosavi [3,4,5,6,7,*]

1. Department of Civil Engineering, Sanandaj Branch, Islamic Azad University, Sanandaj Kurdistan 66169, Iran; sh.shahbazpanahi@iausdj.ac.ir (S.S.); moslemtj@gmail.com (M.K.T.)
2. Civil Engineering Department, University of Halabja, Halabja 46006, Iraq; rabar.faraj@uoh.edu.iq
3. Faculty of Civil Engineering, Technische Universität Dresden, 01069 Dresden, Germany
4. School of the Built Environment, Oxford Brookes University, Oxford OX3 0BP, UK
5. John von Neumann Faculty of Informatics, Obuda University, 1034 Budapest, Hungary
6. Faculty of Economics and Informatics, J. Selye University, 94501 Komarno, Slovakia
7. School of Economics and Business, Norwegian University of Life Sciences, 1430 Ås, Norway
* Correspondence: amir.mosavi@mailbox.tu-dresden.de

Abstract: The present study aims to replace 30%, 40%, and 50% of the natural coarse aggregate (NCA) of concrete with recycled coarse aggregate containing used nano-silica (RCA-UNS) to produce a new sustainable concrete. Three groups of concrete are made and their mechanical properties and microstructure are studied. In the first group, which was the control group, normal concrete was used. In the second group, 30%, 40%, and 50% of the NCA were replaced with coarse aggregate obtained from crushed concrete of the control samples and with 0.5% nano-silica as filler. In the third group, 30%, 40%, and 50% of the concrete samples' NCA were replaced with aggregates obtained from 90-day crushed samples of the second group. Water absorption, fresh concrete slump, and compressive strength of the three groups were investigated and compared through scanning electron microscopy (SEM), X-ray diffraction (XRD) and Fourier transform infrared spectroscopy (FT-IR) tests. The results show that the third group's compressive strengths increased by 12.8%, 10.9%, and 10% with replacing 30%, 40%, and 50% of NAC with RCA-NS at 28 days compared to the control samples, respectively. This could be due to the secondary production of calcium silicate hydrate due to the presence of new cement paste. The third group's microstructure was also improved due to the change in the C–H and the production of extra C–S–H. Therefore, the hydration of cement with water produces C–H crystals while reactions are induced by recycled aggregate containing used nano-silica.

Keywords: sustainable concrete; sustainable materials; recycled coarse aggregate; concrete; used nano-silica; compressive strength; sustainable construction materials; recycled concrete; materials design; composite materials

Citation: Shahbazpanahi, S.; Tajara, M.K.; Faraj, R.H.; Mosavi, A. Studying the C–H Crystals and Mechanical Properties of Sustainable Concrete Containing Recycled Coarse Aggregate with Used Nano-Silica. *Crystals* **2021**, *11*, 122. https://doi.org/10.3390/cryst11020122

Academic Editors: Adam Stolarski and Piotr Smarzewski
Received: 21 November 2020
Accepted: 24 January 2021
Published: 27 January 2021

Publisher's Note: MDPI stays neutral with regard to jurisdictional claims in published maps and institutional affiliations.

Copyright: © 2021 by the authors. Licensee MDPI, Basel, Switzerland. This article is an open access article distributed under the terms and conditions of the Creative Commons Attribution (CC BY) license (https://creativecommons.org/licenses/by/4.0/).

1. Introduction

An average of 40 billion tons of natural aggregates (NA) is used to make concrete in fine and coarse form worldwide [1,2] leading to scarcity of these resources. In addition, the production and processing of NA play an essential role in the generation of dust, noise and greenhouse gases which have a serious negative impact on the environment. Therefore, the need to create an alternative source of NA is of concern to the engineering community today. Furthermore, because of industrial waste in the current century, the need to use recycled materials has become of vital importance, particularly in the construction industry [3,4]. A significant amount of construction and demolition waste is generated annually from the construction industry in all countries of the world. The most influential producers of this waste are China, India and the United States, producing approximately three billion tons of waste [1,5]. Therefore, recycling and re-use of this degraded waste as recycled aggregates are considered sufficient to reduce natural resource scarcity and better manage waste and environmental resources. Some research has shown that the use of recycled aggregate in concrete not only has social benefits but also creates a sustainable concrete structure. Sustainability factors include conserving energy and natural resources, reducing adverse effects on the environment, saving construction costs, reducing waste storage space [6–8]. Recycled aggregates are produced by separating degraded concrete parts from other unwanted materials and crushing them with appropriate grading [9,10]. Despite the many environmental and economic benefits, the usage of recycled aggregates in construction engineering is generally not accepted and is limited to non-structural uses such as embankments, sub-base of roads and mortar. The main reason for the limited use of recycled aggregates is the low quality of these materials due to micro-cracks and voids compared to natural aggregates [11]. The voids cause low density and high water absorption [1]. It has been reported that various factors such as the amount and moisture of recycled coarse aggregates (RCA), water to cement ratio, source concrete strength, number of crushing steps, age of primary concrete, and the amount of adhesive mortar have an important effect on regulating the properties of these aggregates [12–15]. Furthermore, the use of small amounts of recycled fine aggregates, between 6% to 10%, was recommended [16]. Moreover, there are interphases between cement and natural aggregates called the interfacial transition zone (ITZ) in the normal concrete. Cracks may pass through the ITZ and cause the concrete to fail, if the concrete is subjected to load. Therefore, this zone is a weak bond in the concrete. In concrete containing recycled coarse aggregate, there are more types and numbers of these ITZs compared with normal concrete [17]. The percentage of RCA used in structural concrete varies in different studies [18]. Due to the high water absorption tendency of RCA, the presence of cracks in the RCA, the decrease of workability, reduction in interfacial transition zone bonding, and the increased porosity of concrete, the use of RCA was limited [12,16]. However, the recommended amount of RCA usage in structural concretes as replacement of natural coarse aggregates (NCA) is 30% to 50% [16].

On the other hand, hydration of cement is a chemical reaction between water, tricalcium silicate and dicalcium silicate. During the reaction of cement with water, calcium hydroxide (C–H) crystal and calcium silica hydrate (C–S–H) are produced. However, the C–H crystals are less important than C–S–H in terms of increasing compressive strength. The C–H crystals are beneficial in that case to keeps the concrete pH high, maintain a passive layer, and postpone the corrosion. The C–S–H is the main source for the strength of cement based materials. One way to decrease the C–H crystals in the cement matrix is by adding pozzolanic materials such nano-silica to react with C–H crystals present in the matrix. As a result of this pozzolanic reaction, additional C–S–H is produced and the amount of C–H crystals inside the matrix is decreased, this led to the improvement of the micro structure of cement based materials [6].

Today, the use of nanotechnology in concrete structures has increased due to its high impacts on stability [19,20], and the durability and strength of concrete materials. Furthermore, adding nano-materials such as nano-silica to the mix design of concrete provides sustainable concrete [21,22]. Nano-materials, which range between 1 to 100 nanometers in size, can have different positive chemical and physical effects on concrete [15,23,24]. For instance, the addition of nano-silica increases the possibility of reaction with C–H crystal and the production of C–S–H which enhances the strength of the cement structure and fills the pores in the concrete [25]. Suitable silica cementitious materials such as nano-silica are added to the concrete, react with the C–H crystal, and then produce secondary C–S–H, reducing pores and permeability and increasing the compressive strength of concrete [26]. Nano-silica is a material that accelerates the hydration reaction and reacts fast [26]. Nano-silica consists of very fine particles of approximately 10 to 100 nanometers, which are ball-shaped with a small diameter and are used as an additive in concrete water [27,28]. Researchers have shown that nano-silica materials react rapidly with calcium hydroxide and very small amounts of these materials have the same pozzolanic effect [29,30]. Nano-silica produces higher compressive strength, is more widely used and more environmentally friendly than other nano-materials such as nano-clay [31]. In cement paste, by adding nanosilica, the hydration acceleration of cement increases and in the first moments, calcium hydroxide is formed due to the increase in the contact surface of nano-silica with water [19].

A great deal of research has been carried out on the application of nano-silica in ordinary concrete [25,28,32]. Furthermore, several research studies have been conducted on the application of Nano-silica with RCA in concrete demonstrating a reduction in the slump of concrete [17,33,34]. Also, currently nano-silica gained a huge interest from the construction industry to improve the concrete properties and produce a high performance sustainable concrete [20]. In the near future, the amount of demolished concrete which includes nano-silica will also be increased. However, there is a lack of knowledge of the reuse of recycled concrete aggregates containing used nano-silica (RCA-UNS). This means that a concrete structure made from recycled coarse aggregates with nano-silica might be recycled again. No research has been carried out on these massive amounts of recycled coarse aggregates containing used nano-silica in concrete. Thus, no data on the properties of these recycled coarse aggregates containing used nano-silica exists. Accordingly, reuse of recycled coarse aggregates containing used nano-silica is important.

To do so, ordinary concrete samples were made as the control for the first group at the ages of 7, 28 and 90 days. In the second group, 30%, 40%, and 50% of the NCA of concrete samples were replaced with coarse aggregate obtained from crushing concrete of control samples and followed by the addition of nano-silica. In the third group, 30%, 40%, and 50% of the NCA of concrete samples were replaced with the aggregates obtained from 90-day crushing of the samples of the second group. Water absorption, fresh concrete slump, compressive strength, scanning electron microscopy (SEM), Fourier transform infrared (FTIR) spectroscopy, and X-ray diffraction (XRD) of the three groups were compared.

2. Materials and Methods

2.1. Cement, Water, Natural Aggregates and Superplasticizer

Type-I of Portland cement (Kurdistan Cement Co., Bijar, Iran) with a density of 3.20 g/cm^3 based on ASTM C 150A [35] was used in this research. For mixing water, regular drinking water with pH = 7 was utilized. The natural aggregates of all three groups in the concrete mixing design were based on ASTM C128 [36]. The natural coarse aggregates with water absorption of 0.71% had irregular and crushed shape with a maximum size of 13 mm and density of 2.45 according to ASTM C 127 [37]. The natural fine aggregate (NFA) used was river sand with water absorption of 0.60%, 4.75 mm maximum sizes and the density of 2.6. The natural fine aggregates were the same for all groups. The grading curve for NA for the three groups is shown in Figure 1. The black dashed line in Figure 1 represents the control sample aggregates (natural aggregates). A liquid superplasticizer additive based on naphthalene sulfonate was used to increase workability. Its density was 1.33 g/cm^3, and 0.5% of the weight of cement was used in concrete according to the superplasticizer catalogue. A constant amount of superplasticizer was used in all concrete mixes.

2.2. Recycled Coarse Aggregates (RCA and RCA Containing Used Nano-Silica (RCA-NS))

The second group of concrete specimens, recycled coarse aggregates containing nano-silica, was named RCA-NS. This group of recycled coarse aggregates (RCA) was obtained by crushing 90-day control samples using impact crusher because at the age of 90 days the control concrete has almost reached its final strength. To increase workability, the fine aggregates were separated by sieving and the old cement with air pressure. The largest and the smallest sizes of RCA aggregate were approximately 22.5 mm and 10 mm, respectively. The water absorption and the specific gravity of RCA were, in the order mentioned, approximately 2.2% and 2.45. The grading curve of RCA, is illustrated with a red solid line in Figure 1.

The third group of concrete samples was named RCA-UNS. By crushing the 90-day samples of the second group (RCA-NS), RCA-NS were obtained for use in the third group. The specific gravity and water absorption of RCA-NS were 2.35 and almost 2.9%, respectively. The largest and smallest sizes of RCA-NS aggregates were approximately 23 mm and 8 mm, respectively. The cement coat from the aggregates was removed by using ultrasonic cleaning method to improve the bonding between the new cement paste and RCA [38]. Its grading curve is represented by black dots in Figure 1.

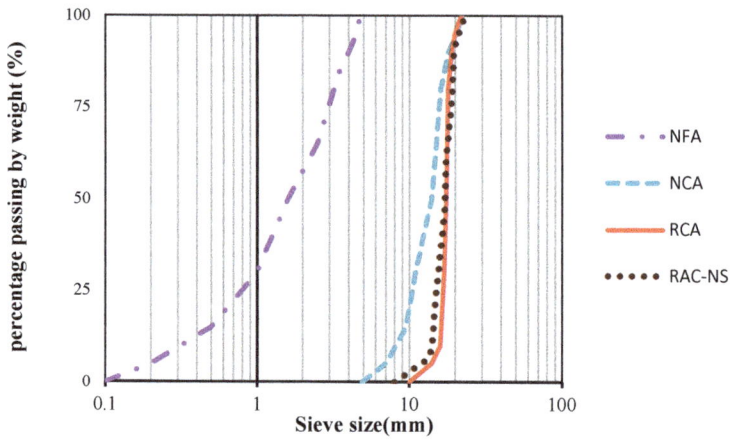

Figure 1. Grading curve of natural aggregates (NA) for the three groups, recycled coarse aggregates (RCA) of group two, recycled coarse aggregates containing nano-silica (RCA-NS) of group three.

2.3. Nano-Silica

Nano-silica (obtained from the EVONIK company, Germany) was available as powder with 99.9% Sio2 and 10–14 nm particle size. The color was white (Figure 2), pore size was 1.2 mL/g pore volume with 50 m²/g surface area, and density of 2.50. Figure 3 shows an image of the Nano-silica used in group two which was provided by the Iranian Nanomaterials Pioneers Company. The percentage of Nano silica used in concrete varies between 0.4 to 1.5 wt% of the cement as mentioned in previous research [15,21]. In the second group of this study, Nano-silica was used and added as 0.5% of the weight of cement [20]. Table 1 shows the chemical composition and physical properties of the cement and nano-silica.

Figure 2. The nano-silica powder.

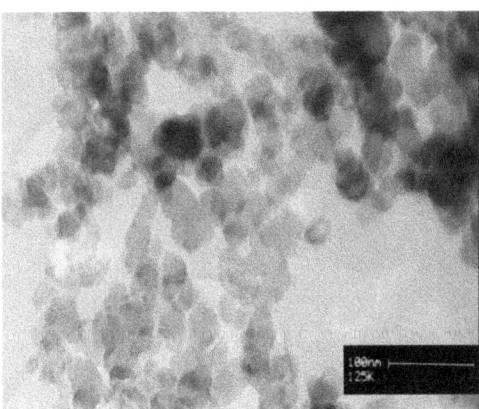

Figure 3. The scanning electron microscope (SEM) image of Nano-silica (provided by Iranian Nanomaterials Pioneers Company).

Table 1. Chemical and physical composition of cement and nano-silica.

Chemical Composition (%)	Cement	Nano-Silica
SiO_2	21.20	99.9
Al_2O_3	3.41	-
Fe_2O_3	2.78	-
CaO	62.32	-
MgO	1.91	-
K_2O	0.23	-
Other	8.15	0.1
Physical properties		
density	3.1	2.87
Average size	13.9 micron	10–14 Nm

2.4. Concrete Preparation and Mix Design

In order to achieve the objectives of the research, the amount of cement, the ratio of water to cement, and natural fine aggregates were considered constant in all mix designs. In this regard, an attempt was made to maintain constant conditions and only the amount of aggregate replacement differed. The water to cement ratio of concrete was 0.45 in all concrete mixtures. This ratio of water to cement was high because water might be absorbed in recycled concrete and concrete slump could be very low. In addition, a dark brown liquid superplasticizer (based on naphthalene sulfate) was added to increase workability. For the experiments, a concrete mix was made according to Table 2 to make a cubic sample for ages 7, 28 and 90 days (three samples for each age). RCA-NS standard cubic specimens (second group) were made using natural aggregates with 30%, 40% and 50% RCA and 0.5% Wt. of cement of added Nano-silica. These samples were tested at different ages and the 90-day coarse aggregates of these samples were used as recycled coarse aggregate containing used Nano-silica in the third group. In addition, RCA-UNS cubic samples (third group) were prepared using natural aggregates with 30%, 40% and 50% RCA-NS according to Table 2.

The steps for mixing the materials were as follows: before adding water, all aggregates were loaded and mixed in the mixer up to 2 min. Then, the amount of water needed for water absorption of aggregates was added and mixed with a high-speed mixer for up to 2 min. In the case of the second group, nano-silica was first added with half of the remaining water and added to the mixture. The cement was then added to the mixture and mixed for 3 min. Finally, the superplasticizer was mixed with the other half of the mixing water and added to the concrete within 2 min. Concrete mixing was stopped for one minute until the superplasticizer began to react. During this time, the mixer was covered so that the concrete water did not evaporate. The concrete was mixed at high speed for another 3 min until it reached the desired performance. This was followed by placing the concrete in standard cubic (10 × 10 × 10 cm) molds. To avoid water evaporation, a plastic sheet was used on cubic molds. The molds were opened after one day. After demolding, all samples were cured in a water tank at 20 °C according to ASTM C 511 [39]. For greater accuracy, 3 samples were made from each mold and the statistical average was calculated [39].

Table 2. Mix design of all samples.

Mix Design	w/c	Cement Kg/m^3	Nano-Silica (%Wt of Cement)	RCA and RCA-NS Kg/m^3	NCA Kg/m^3	Natural Fine Aggregates Kg/m^3
Control	0.45	400	0	0	1150	700
RCA-NS 30%	0.45	400	0.5	345	805	700
RCA-NS 40%	0.45	400	0.5	460	609	700
RCA-NS 50%	0.45	400	0.5	575	575	700
RCA-UNS 30%	0.45	400	0	345	805	700
RCA-UNS 40%	0.45	400	0	460	690	700
RCA-UNS 50%	0.45	400	0	575	575	700

3. Results and Discussion

3.1. Slump of Fresh Concrete

The slump of fresh concrete test of recycled concrete is one of the most important tests due to the adsorption of mixing water by recycled aggregates and old cement. One of the disadvantages of recycled concrete is slump loss [40]. The results of fresh concrete slump based on ASTM C143 [41] are presented in Figure 4. The slump of the control concrete was 79 mm. However, the slump of RCA-NS samples decreased with 30%, 40% and 50% of RCA by 30, 25 and 19 mm, respectively, this phenomenon was also observed in previous research [1,20,26]. The slump of RCA-UNS samples with 30%, 40% and 50% RCA-NS were reduced to 18.5, 18 and 18 mm, respectively. This was because the used nano-silica absorbed excess amounts of mixing water and reduced workability. Moreover, mortar attached to aggregates from the first and second groups was the cause of the low reduction of slump in the third group.

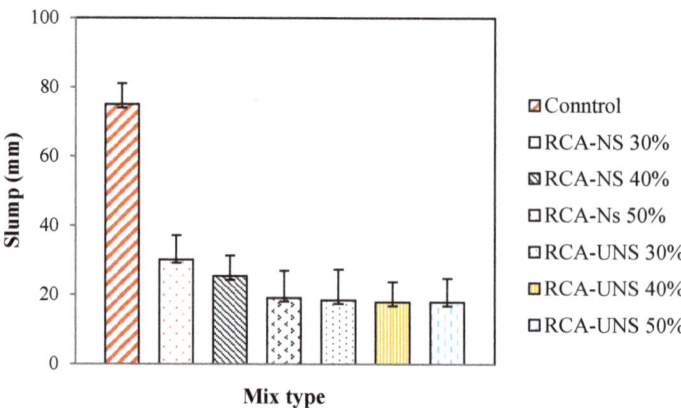

Figure 4. Slump of fresh concrete for all samples.

3.2. Water Absorption of Concrete Samples

Water absorption of all concrete samples in this study was performed following the method described in ASTM C642-13 standard test [42]. In the samples of this test, broken parts of the internal (core) of the concrete were used to measure water absorption. Figure 5 shows the water absorption of concrete samples at 28 days. The results show that the sample of 30% RCA-NS had greater water absorption than the control mixture. This could be as a result of recycled coarse aggregates which increase bleeding, segregation, and air voids [43]. The water absorption percentage of the control sample was 3.8%, while this amount for the RCA-NS sample with 30%, 40% and 50% of RCA were 4.3, 4.4 and 4.7, respectively. This increase in water absorption with rising percentage of RCA has also been reported by Younes et al. [19]. Furthermore, the high-water absorption of RCA-NS samples can be attributed to the inherent nature of the pores of recycled coarse particles and adsorption by the old mortar in them. The water penetrated into the cement matrix due to the ability of aggregate to absorb water, the presence of fine pores and capillary phenomena. The water absorption percentages of RCA-UNS samples with 30%, 40% and 50% recycled coarse aggregates containing used nano-silica were 3.7%, 3.6% and 3.6%, respectively. As can be observed in Figure 5, contrary to expectations, the water absorption of RCA-UNS samples decreased compared to control and RCA-NS samples. This phenomenon might be due to hydration of old cements and the formation of additional C–S–H which could play a positive role in filling pores and cracks of the samples.

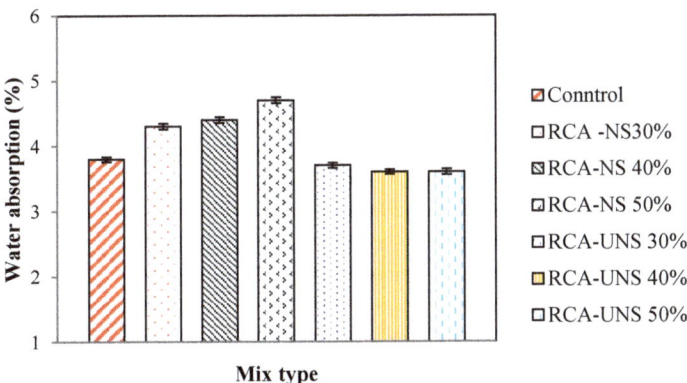

Figure 5. Water absorption of concrete samples at 28 days.

3.3. Compressive Strength of Samples

The compressive strength test using standard cubic specimens at 7, 28, and 90 days was in accordance with BS 1881 [44]. One way to make sustainable concrete is to increase its mechanical properties such as compressive strength. The results of the compressive strength tests for the ages mentioned are shown in Figure 6. By looking at the error bars, there are overlaps for different percentages of RCA. By considering a 95% confidence interval, it can be concluded that compressive strengths of all of the RCA samples are almost the same and a different percentage of RCA has no significant impact on the compressive strength regardless of being compared to the control sample. For all of the RCA samples, it is obvious that compressive strength has been dropped compared to the control sample. As illustrated in Figure 6, in the third group (RCA-UNS), there was a 5.7%, 4.6% and 4.4% increase in compressive strength at 7 days compared to the sample, respectively [45,46].

The compressive strength of the second group (RCA-NS) with the replacement of 30%, 40% and 50% of RCA aggregates were respectively 30.1, 29.3 and 28.7 MPa at 28 days of age, while the compressive strength of the control sample was 32 MPa. The decrease in compressive strength in the second group was due to the weak bond between recycled coarse aggregates and also the characteristics of these aggregates such as old mortar, the presence of impurities, cracks, and porosity (which in previous research decreased to a greater extent [17,22,40]); however, nano-silica partially compensated for this weakness. Furthermore, as shown in Figure 6, compressive strength was recorded as 36.1, 35.5, and 35.2 MPa at 28 days by replacing with 30%, 40%, and 50% of RCA-NS aggregates, respectively. This increase in compressive strength in the third group may be due to the presence of sufficient water to complete the hydration of the third amount of cement paste. This amount can act as filler and improve the ITZ of concrete. In addition, the replacement of 30% of natural coarse aggregate of concrete with recycled coarse aggregate containing used nano-silica (RCA-UNS 30%) increased the compressive strength by 15.7% compared to the control samples at 90 days. This increase in compressive strength at older ages (90 days) is also observed in this bar chart. This increase in compressive strength is consistent with the water absorption test of the samples. Therefore, the addition of 30% recycled coarse aggregate containing used nano-silica increased the compressive strength by approximately 12%.

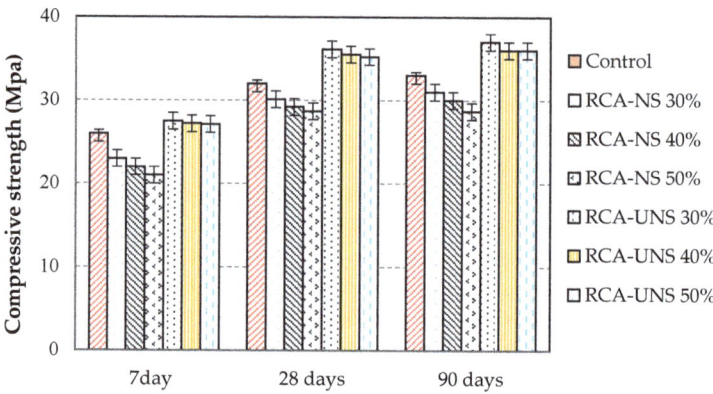

Figure 6. Compressive strength test of samples.

Figure 7 shows the failure modes of the samples. As can be observed in a, the desired failure of the control concrete sample occurred in such a way that with increasing pressure, the 4 sides evenly cracked and after failure, the remaining piece was stacked in two inverted pyramids or the two pyramids were separated. However, in the RCA-NS sample, fragmentation failure occurred only at the surfaces of the sample, which could be a plastic failure mode in the sample (Figure 7b). In the RCA-UNS example, the plastic failure mode is also shown in Figure 7c.

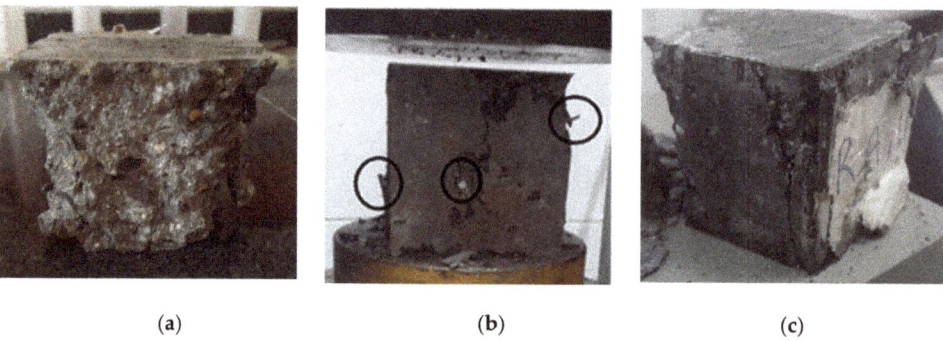

Figure 7. The failure modes of the (**a**) control, (**b**) RCA-NS, and (**c**) RCA-UNS samples.

3.4. Scanning Electron Microscopy (SEM) of Samples

Images of small-sized structures with SEM (TSCAN test machine made in Czech Republic) at 28 days are shown in Figures 8–10. Figures 8–10 show the microscopic structure images of the control sample, the RCA-NS 30%, and the RCA-UNS 30% sample, respectively. Figure 8 illustrates the significant amount of porosity, micro cracks, C–H, C–S–H, and most of the unreacted material. The unreacted material was due to the excessive amount of cement and some remaining cementitious materials [38]. Figure 9 demonstrates that RCA aggregates increased porosity and small cracks, and prevented secondary C–S–H formation. The C–S–H is a compound which is the main product of Portland cement hydration and a major factor in the strength of cement and all cement-based products. Approximately 50% to 60% of the volume of fully hydrated cement paste solids is C–S–H [47]. Figure 10 shows more C–S–H formation than Figures 8 and 9. The SEM results confirm previous research findings on water absorption and compressive strength and show that in the RCA-UNS 30% sample, C–S–H was produced in comparison to the control and RCA-NS 30% samples. The SEM images illustrate that used Nano-silica or new cement reduced porosity and micro cracks in the RCA-UNS 30% sample. Therefore, the size of the pores and C–H crystals are reduced creating a microstructure that is extra compressed when coarse aggregate containing used nano-silica was used. Also, Figure 10 showed that RCA-UNS 30% samples have larger ITZ compared to the RCA-NS 30% in Figure 9.

Figure 8. SEM of control sample at 28 days.

Figure 9. SEM of RCA-NS 30% sample at 28 days.

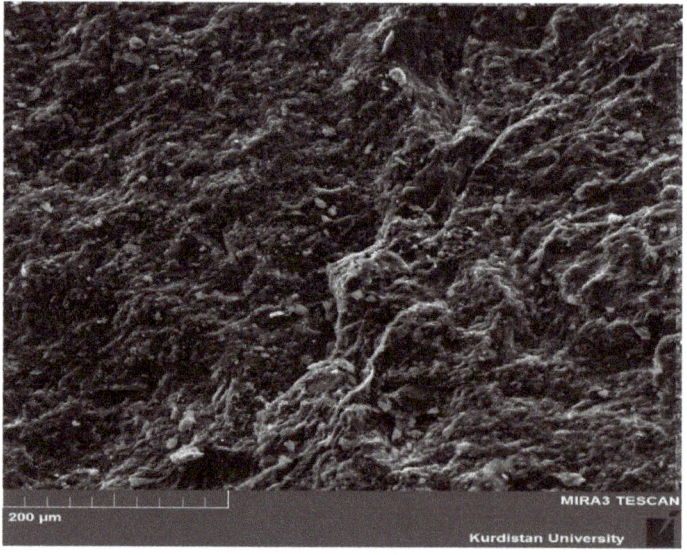

Figure 10. SEM of RCA-UNS 30% sample at 28 days.

3.5. X-ray Diffraction (XRD) Patterns of Samples

Figure 11 shows the XRD pattern (obtained from SAXess Co., Niederösterreich, Austria) for control samples, RCA-NS 30%, and RCA-UNS 30% at University of Kurdistan at 28 days. At 18° and 34° (2 Theta) from XRD patterns, intensity peaks of C–H crystal were traced [23,45]. To prepare the samples of this test, a powder form (300 microns) obtained from a combination of aggregate and mortar from broken parts of the compressive strength test was used. Intensity peaks of C–H crystal in the control sample were 67 and 59 at 18° and 34° (2Theta), respectively. In the RCA-NS 30%, intensity peaks of C–H crystal were 54 and 33 at 18° and 34° (2Theta), respectively. As shown in Figure 11, the intensity peaks of C–H crystal in the RCN-NS 30% samples were slightly reduced compared to the control sample due to low C–H crystal consumption and C–S–H formation. The reason may be that the specific gravity of recycled coarse aggregates (resulted from demolishing RCA-NS mixtures) is lower than that of natural coarse aggregates and the natural coarse aggregate was substituted based on weight and, considering the fact that the mixture design was not based on volumetric ratios, this would result in lowering the cement content by replacing natural coarse aggregate with RCA in the unit volume of concrete. Furthermore, it can be observed that the intensity peaks of C–H crystal in RCA-UNS 30% samples decreased by 65% and 67% at 18° and 34° (2Theta), respectively compared to the control sample. This decrease in intensity peaks of C–H crystal related to quartz (SiO2) was due to the production of secondary C–S–H, which has also been reported by other researchers [32,38,48]. Reaction and effect of coarse aggregate containing used nano-silica on C–H crystallization is more remarkable. Findings from the XRD test were consistent with the compressive strength, SEM test.

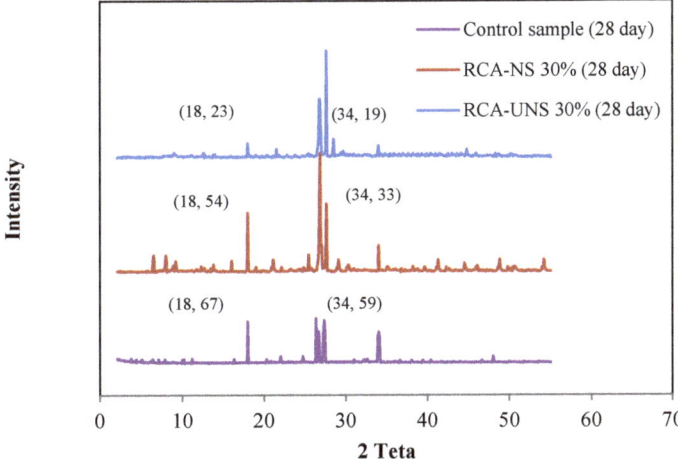

Figure 11. X-ray diffraction (XRD) of Control, RCA-NS 30%, and RCA-UNS 30% samples at 28 days.

3.6. Fourier Transform Infrared (FT-IR) Graphs of Samples

To discover the variations in the C–S–H chains related to Si–O–Si, Si–H, O–C–O and O–H bands, FT-IR graph tests were conducted. FT-IR test samples were in powder form. These powders were obtained by grinding the broken part of the compressive strength test, which includes cement and aggregates. After grinding, the powder was sieved with a 300 micron sieve. The FT-IR graphs (Bruker machine, made in the Netherlands, tested at University of Kurdistan) of the control, RCA-NS 30%, and RCA-UNS 30% samples at 28 days, between 600–2600 cm^{-1}, are shown in Figure 12. The stretching vibration of O–H at 1600–1640 cm^{-1} was due to the increased molecule of water in the samples [49].The 30% of natural coarse aggregates replacement with recycled coarse aggregate (RCA-NS 30%) increased the O–H stretching vibrations which were further augmented if replacement with recycled coarse aggregate contained used nano-silica (RCA-UNS 30%). Another strong band was the stretching vibration of O–C–O at 1300–1400 cm^{-1} which was due to atmospheric carbonation. The strongest band was Si–H at 1350–1450 cm^{-1} due to depolymerization. The Si–H bands increased in RCA-UNS 30% samples likely [49,50]. The changes in Si–O–Si transmittance band at 750–800 cm^{-1} indicate that a secondary C–S–H was formed, which has also been reported in other research [23,51]. The replacement of 30% natural coarse aggregates with recycled coarse aggregate containing used Nano-silica

(RCA-UNS 30%) increased the Si–O–Si transmittance bands. The results of FT-IR graphs of samples were in agreement with compressive strength, SEM, and XRD results.

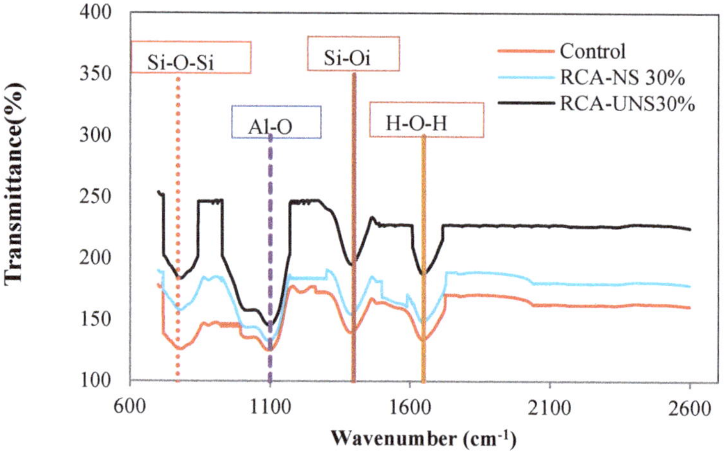

Figure 12. Fourier transform infrared (FT-IR) spectra of the control, RCA-NS 30%, and RCA-UNS 30% samples at 28 days.

4. Conclusions

In this study, for the first time, 30%, 40% and 50% of recycled coarse aggregates containing used Nano-silica (RCA-UNS) were reused in concrete. For this purpose, ordinary concrete samples were made and tested for the control group, also named the first group. In the second group (for comparison and use in third group samples), 30%, 40% and 50% of natural coarse aggregates were replaced by recycled coarse aggregates of the concrete obtained from the control samples. In the samples of the second group, nano-silica was added to the mixture at the rate of 0.5% by weight of cement (RCA-NS). Furthermore, in the samples of the third group, 30%, 40% and 50% of the natural coarse aggregates were replaced by recycled concrete from recycled coarse aggregates containing used nano-silica, obtained from the samples of the second group at 90 days. Water absorption, fresh concrete slump, compressive strength, SEM, FT-IR, and XRD tests of these three groups were compared.

- The findings demonstrate that the water absorption of RCA-UNS samples decreased compared to the control sample.
- Moreover, the findings of the compressive strength test illustrated that compressive strength in the third group increased 12.8%, 10.9%, and 10% by replacing 30%, 40%, and 50% of NAC with RCA-NS at 28 days compared to control samples.
- The SEM results confirm previous results such as water absorption and compressive strength and show that the RCA-UNS 30% sample produced extra C–S–H.
- In addition, the XRD and FT-IR graphs illustrate that in the RCA-UNS samples, more C–H crystal was consumed and converted to C–S–H.
- For the production of sustainable concrete, 30% of natural coarse aggregates can be replaced with recycled coarse aggregate containing used nano-silica. Therefore, the hydration of cement with water produces C–H crystals while reactions induced by coarse aggregate containing used nano-silica consume the C–H crystals.

Author Contributions: Conceptualization, M.K.T., S.S. and R.H.F.; methodology, M.K.T., S.S. and R.H.F.; software, M.K.T., S.S. and R.H.F.; validation, A.M.; formal analysis, M.K.T., S.S. and R.H.F.; investigation, M.K.T., S.S. and R.H.F.; resources, A.M.; data curation, M.K.T., S.S. and R.H.F.; writing—original draft preparation, M.K.T., S.S. and R.H.F.; writing—review and editing, M.K.T., S.S., R.H.F. and A.M.; visualization, M.K.T., S.S. and R.H.F.; supervision, S.S. and A.M.; project administration, A.M.; funding acquisition, A.M. All authors have read and agreed to the published version of the manuscript.

Funding: This research has received no funding.

Institutional Review Board Statement: Not applicable.

Informed Consent Statement: Not applicable.

Data Availability Statement: Not applicable.

Acknowledgments: Support of the Alexander von Humboldt Foundation is acknowledged.

Conflicts of Interest: The authors declare no conflict of interest.

References

1. Muduli, R.; Mukharjee, B.B. Performance assessment of concrete incorporating recycled coarse aggregates and metakaolin: A systematic approach. *Constr. Build. Mater.* **2020**, *233*, 117223. [CrossRef]
2. Li, Q.; Hu, J. Mechanical and Durability Properties of Cement-Stabilized Recycled Concrete Aggregate. *Sustainability* **2020**, *12*, 7380. [CrossRef]
3. Danraka, M.N.; Aziz, F.N.A.A.; Jaafar, M.S.; Nasir, N.M.; Abdulrashid, S. Application of Wood Waste Ash in Concrete Making: Revisited. *Lect. Notes Civ. Eng.* **2019**, 69–78. [CrossRef]
4. Bostanci, S.C. Use of waste marble dust and recycled glass for sustainable concrete production. *J. Clean. Prod.* **2020**, *251*, 119785. [CrossRef]
5. Nguyen, T.T.H.; Mai, H.H.; Phan, D.H.; Nguyen, D.L. Responses of Concrete Using Steel Slag as Coarse Aggregate Replacement under Splitting and Flexure. *Sustainability* **2020**, *12*, 4913. [CrossRef]
6. Thomas, B.S.; Hasan, S.K.; Arel, S. Sustainable concrete containing palm oil fuel ash as a supplementary cementitious material—A review. *Renew. Sustain. Energy Rev.* **2017**, *80*, 550–561. [CrossRef]
7. Memon, S.A.; Khan, M.K. Ash blended cement composites: Eco-friendly and sustainable option for utilization of corncob ash. *J. Clean. Prod.* **2018**, *175*, 442–455. [CrossRef]
8. Collivignarelli, M.C.; Cillari, G.; Ricciardi, P.; Miino, M.C.; Torretta, V.; Rada, E.C.; Abbà, A. The Production of Sustainable Concrete with the Use of Alternative Aggregates: A Review. *Sustainability* **2020**, *12*, 7903. [CrossRef]
9. Golewski, G.L. Generalized Fracture Toughness and Compressive Strength of Sustainable Concrete Including Low Calcium Fly Ash. *Materials* **2017**, *10*, 1393. [CrossRef]
10. Zhang, M.H.; Islam, J. Use of nano-silica to reduce setting time and increase early strength of concretes with high volumes of fly ash or slag. *Constr. Build. Mater.* **2012**, *29*, 573–580. [CrossRef]
11. Tam, V.W.; Soomro, Y.; Evangelista, A.C.J. A review of recycled aggregate in concrete applications (2000–2017). *Constr. Build. Mater.* **2018**, *172*, 272–292. [CrossRef]
12. Majhi, R.; Nayak, A.N.; Mukharjee, B.B. Development of sustainable concrete using recycled coarse aggregate and ground granulated blast furnace slag. *Constr. Build. Mater.* **2018**, *159*, 417–430. [CrossRef]
13. Katare, V.D.; Madurwar, M.V. Design and investigation of sustainable pozzolanic material. *J. Clean. Prod.* **2020**, *24*, 14–25. [CrossRef]
14. Nicoara, A.I.; Stoica, A.E.; Vrabec, M.; Rogan, N.S.; Sturm, S.; Ow-Yang, C.; Gulgun, M.A.; Bundur, Z.B.; Ciuca, I.; Vasile, B.S. End-of-Life Materials Used as Supplementary Cementitious Materials in the Concrete Industry. *Materials* **2020**, *13*, 1954. [CrossRef] [PubMed]
15. Golewski, G.L. Green concrete composite incorporating fly ash with high strength and fracture toughness. *J. Clean. Prod.* **2018**, *172*, 218–226. [CrossRef]
16. Zhang, L.W.; Sojobi, A.O.; Kodur, V.K.R.; Liew, K.M. Effective utilization and recycling of mixed recycled aggregates for a greener environment. *J. Clean. Prod.* **2019**, *236*, 117600. [CrossRef]
17. Mukharjee, B.B.; Barai, S.V. Influence of nano-silica on the properties of recycled aggregate concrete. *Constr. Build. Mater.* **2014**, *55*, 29–37. [CrossRef]
18. Omrane, M.; Kenai, S.; Kadri, E.; Aït-Mokhtar, A. Performance and durability of self compacting concrete using recycled concrete aggregates and natural pozzolan. *J. Clean. Prod.* **2017**, *165*, 415–430. [CrossRef]
19. Ling, T.C.; Poon, S.V. Use of recycled CRT funnel glass as fine aggregate in dry mix concrete paving blocks. *J. Cleaner Prod.* **2014**, *68*, 209–215. [CrossRef]

20. Wang, X.; Cheng, F.; Wang, Y.; Zhang, X.; Niu, H. Impact properties of recycled aggregate concrete with nanosilica modification. *Adv. Civ. Eng.* **2020**, 1–10. [CrossRef]
21. Prasada Rao, D.V.; Navaneethamma, V. Influence of nano-silica on strength properties of concrete containing rice husk ash. *Int. J. Adv. Res.* **2016**, *3*, 39–43.
22. Younis, K.H.; Mustafa, S. Feasibility of Using Nanoparticles of SiO$_2$ to Improve the Performance of Recycled Aggregate Concrete. *Adv. Mater. Sci. Eng.* **2018**, *2018*, 1–11. [CrossRef]
23. Farzadnia, N.; Noorvand, H.; Yasin, A.M.; Aziz, F.N.A. The effect of nano silica on short term drying shrinkage of POFA cement mortars. *Constr. Build. Mater.* **2015**, *95*, 636–646. [CrossRef]
24. Zhao, S.; Zhang, Q. Effect of Silica Fume in Concrete on Mechanical Properties and Dynamic Behaviors under Impact Loading. *Materials* **2019**, *12*, 3263. [CrossRef] [PubMed]
25. Sikora, P.; Horszczaruk, E.; Skoczylas, K.; Rucins, T. Thermal properties of cement mortars containing waste glass aggregate and nanosilica. *Procedia Eng.* **2017**, *196*, 159–166. [CrossRef]
26. Hosseini, P.; Booshehrian, A.; Madari, A. Developing Concrete Recycling Strategies by Utilizationof Nano-SiO$_2$ Particles. *Waste Biomass Valorization* **2011**, *2*, 347–355. [CrossRef]
27. Li, W.; Long, C.; Tam, V.W.Y.; Poon, C.S.; Duan, H. Effects of nano-particles on failure process and microstructural properties of recycled aggregate concrete. *Constr. Build. Mater.* **2017**, *142*, 42–50. [CrossRef]
28. Varghese, J.; Gopinath, A.; Bahurudeen, A.; Senthilkumar, R. Influence of nano-silica on characteristics of cement mortar and concrete. *Sustain. Constr. Build. Mater.* **2018**, *25*, 839–851.
29. Roychand, R.; Silva, S.D.; Setunge, S.; Law, D. A quantitative study on the effect of nano SiO$_2$, nano Al$_2$O$_3$ and nano CaCO$_3$ on the physicochemical properties of very high volume fly ash cement composite. *Eur. J. Environ. Civ. Eng.* **2017**, *2*, 1–16. [CrossRef]
30. Vishwakarma, V.; Ramachandran, D. Green Concrete mix using solid waste and nanoparticles as alternatives—A review. *Constr. Build. Mater.* **2018**, *162*, 96–103. [CrossRef]
31. Kawashima, S.; Hou, P.; Wang, K.; Corr, D.J.; Shah, S.P. Activation of fly ash through nanomodification. *Adv. Green Bind. Syst.* **2013**, *294*, 1–11.
32. Adak, D.; Sarkar, M.; Mandal, S. Structural performance of nano-silica modified fly-ash based geopolymer concrete. *Constr. Build. Mater.* **2017**, *135*, 430–439. [CrossRef]
33. Mukharjee, B.B.; Barai, S.V. Characteristics of sustainable concrete incorporating recycled coarse aggregates and colloidal nano-silica. *Adv. Concr. Constr.* **2015**, *3*, 187–202. [CrossRef]
34. Ying, J.; Zhou, B.; Xiao, J. Pore structure and chloride diffusivity of recycled aggregate concrete with nano-SiO$_2$ and nano-TiO$_2$. *Constr. Build. Mater.* **2017**, *150*, 49–55. [CrossRef]
35. ASTM C150A. *Standard Specification for Portland Cement*. In *Annual book of Standards*; ASTM International: West Conshohocken, PA, USA, 1999.
36. ASTM C128. *Standard Test Method for Density, Relative Density (Specific Gravity), and Absorption of Fine Aggregate*; ASTM International: West Conshohocken, PA, USA, 2004.
37. ASTM C127. *Standard Test Method for Specific Gravity and Absorption of Coarse Aggregate*; ASTM International: West Conshohocken, PA, USA, 1993.
38. Ibrahim, M.; Johari, M.A.M.; Rahman, M.K.; Maslehuddin, M. Effect of alkaline activators and binder content on the properties of natural pozzolan-based alkali activated concrete. *Constr. Build. Mater.* **2017**, *147*, 648–660. [CrossRef]
39. ASTM C511. *Standard Specification for Mixing Rooms, Moist Cabinets, Moist Rooms, and Water Storage Tanks Used in the Testing of Hydraulic Cements and Concretes*; ASTM International: Washington, DC, USA, 2013.
40. Mukharjee, B.B.; Barai, S.V. Influence of incorporation of nano-silica and recycled aggregates on compressive strength and microstructure of concrete. *Constr. Build. Mater.* **2014**, *71*, 570–578. [CrossRef]
41. ASTM C143. *Slump of Hydraulic Cement Concrete*; ASTM International: West Conshohocken, PA, USA, 1998.
42. ASTM C642-13. *Standard Test Method for Density, Absorption, and Voids in Hardened Concrete*; ASTM International: West Conshohocken, PA, USA, 2013.
43. Poon, C.S.; Kou, S.C.; Lam, L. Influence of recycled aggregate on slump and bleeding of fresh concrete. *Mater. Struct.* **2007**, *40*, 981–988. [CrossRef]
44. BS 1881. *Testing Concrete. Methods for Analysis of Hardened Concrete*; BSI: London, UK, 2014.
45. Farzadnia, N.; Ali, A.; Demirboga, R.; Anwar, M.P. Effect of halloysite nanoclay on mechanical properties, thermal behavior and microstructure of cement mortars. *Cem. Concr. Res.* **2013**, *48*, 97–104. [CrossRef]
46. Huang, Y.; He, X.; Wang, Q.; Sun, Y. Mechanical properties of sea sand recycled aggregate concrete under axial compression. *Constr. Build. Mater.* **2018**, *175*, 55–63. [CrossRef]
47. Tamanna, K.; Raman, S.N.; Jamil, M.; Hamid, R. Utilization of wood waste ash in construction technology: A review. *Constr. Build. Mater.* **2020**, *237*, 117654. [CrossRef]
48. Aghabaglou, A.M.; Tuyan, M.; Ramyar, K. Mechanical and durability performance of concrete incorporating fine recycled concrete and glass aggregates. *Mater. Struct.* **2015**, *48*, 2629–2640. [CrossRef]
49. Lei, B.; Li, W.; Tang, Z.; Tam, V.W.; Sun, Z. Durability of recycled aggregate concrete under coupling mechanical loading and freeze-thaw cycle in saltsolution. *Constr. Build. Mater.* **2018**, *163*, 840–849. [CrossRef]

50. Rudić, O.; Ducman, V.; Malešev, M.; Radonjanin, V.; Draganić, S.; Šupić, S.; Radeka, M. Aggregates Obtained by Alkali Activation of Fly Ash: The Effect of Granulation, Pelletization Methods and Curing Regimes. *Materials* **2019**, *12*, 776. [CrossRef] [PubMed]
51. Pan, X.; Shi, C.; Farzadnia, N.; Hu, X.; Zheng, J. Properties and microstructure of CO_2 surface treated cement mortars with subsequent lime-saturated water curing. *Cem. Concr. Compos.* **2019**, *99*, 89–99. [CrossRef]

Article

Incorporation of Recycled Tire Products in Pavement-Grade Concrete: An Experimental Study

Sayed Mohamad Soleimani *, Abdel Rahman Alaqqad, Adel Jumaah, Naser Mohammad and Alanoud Faheiman

Department of Civil Engineering, Australian College of Kuwait, P.O. Box 1411, Safat 13015, Kuwait; a.alaqqad@ack.edu.kw (A.R.A.); 1414700@go.ack.edu.kw (A.J.); n.mohammad@ack.edu.kw (N.M.); a.faheiman@ack.edu.kw (A.F.)
* Correspondence: s.soleimani@ack.edu.kw; Tel.: +965-1828225 (ext. 4046)

Citation: Soleimani, S.M.; Alaqqad, A.R.; Jumaah, A.; Mohammad, N.; Faheiman, A. Incorporation of Recycled Tire Products in Pavement-Grade Concrete: An Experimental Study. *Crystals* **2021**, *11*, 161. https://doi.org/10.3390/cryst11020161

Academic Editor: Piotr Smarzewski
Received: 19 January 2021
Accepted: 3 February 2021
Published: 6 February 2021

Publisher's Note: MDPI stays neutral with regard to jurisdictional claims in published maps and institutional affiliations.

Copyright: © 2021 by the authors. Licensee MDPI, Basel, Switzerland. This article is an open access article distributed under the terms and conditions of the Creative Commons Attribution (CC BY) license (https://creativecommons.org/licenses/by/4.0/).

Abstract: The phenomenon of dumping used tires in Kuwait has reached critical levels, with a landfill containing millions of tires being formed in a remote area, which is a major environmental hazard. Nowadays, recycled rubber is used as a suitable and useful material in civil engineering applications, particularly in the production of "green concrete". This study aims to see whether recycled tire by-products can be used to make "green concrete" for pavements. Each type of tire by-product was tested individually to examine its properties and effects on a benchmark mix before creating hybrid mixes that contain a combination of the materials. Eleven mixes containing different doses of shredded or crumbed rubber or steel fibers contained within the tires were made to evaluate their impact on the concrete's slump, compressive strength, split tensile strength, and modulus of rupture. Additionally, twelve hybrid concrete mixes containing different doses of various tire by-products were developed. Preliminary results show that the incorporation of rubber products has a reduction on the concrete's properties. The use of replacement materials sourced from recycled tires using the dosages investigated in this study does not detract from the usability of green pavement concrete suited for hot weather. The concrete produced in this study could be evaluated for specific properties relating to its road safety in further studies. Additionally, long-term effects of using the concrete can be studied using finite element analysis.

Keywords: green concrete; shredded rubber; crumbed rubber; steel fiber; recycled tire; pavement; waste materials; sustainability

1. Introduction

The management of used tires in Kuwait is considered as a significant environmental challenge [1]. Millions of used tires are dumped to open landfill in Kuwait, although the storage of this type of waste could be a major environmental hazard [2]. As the temperature in Kuwait commonly exceeds 50 °C during the summer, several massive fires previously destroyed the waste tires, causing serious environmental issues [3–6]. Additionally, the disposal of waste tires in general represents an important environmental concern as the natural degradation of rubber takes several years [7].

Due to an exponential growth in the use of cars in Kuwait, the accumulation of used tires poses a serious risk to the ecosystem; therefore, recycling waste products is vital [8–10]. Recently, great effort has been directed toward finding alternate ways to use waste materials emerging in the world [11–13]. Researchers have found that the recycling of waste rubber tires has several environmental and economic advantages [11,14–16]. Nowadays, the recycled rubber is considered as a suitable and useful material in civil engineering applications [7,17,18]. In general, the recycling of the waste tires goes through a process of shredding, separation of components, and granulation in order to convert the tires into ground tire rubber [19].

In most studies, the recycled rubber is classified into three main categories such as shredded rubber (also known as chipped rubber), crumb rubber, and ground rubber [18,20].

According to an analytical review done by Lavanga et al., the mechanical response of rubber concrete is affected by different factors such as the rubber content granulometry of both the substituted mineral aggregate (coarse or fine) and the added rubber, the introduction of additives, the quantities of all components within the formulation, the water-to-cement ratio, and the appropriate pre-treatments and surface coatings [13].

A study conducted by Aiello and Leuzzi [21] to investigate the properties of various concrete mixtures concluded that the waste rubber tire can be used toward making workable rubberized concrete. The same study found that the rubberized concrete mixtures showed lower unit weight compared to plain concrete [21]. The literature indicated that the compressive strength of rubberized concrete decreases by increasing the particle size and rubber content [11,17,18,22]. Moreover, several studies established an improvement in compressive and tensile strengths in the mixture containing coated rubber crumb and silica fume [23–25].

Meddah et al. [2] recognized that adding shredded rubber can improve the performance of concrete by modifying the roughness of rubber particle surfaces. In addition, the same study pointed that using rubber particles showed improvement in some characteristics such as porosity, ductility, and cracking resistance performance.

Khan and Singh [26] investigated the partial replacement of sand by tire crumb rubber with 5%, 10%, and 15% replacement. The compressive and tensile strengths of the rubberized concrete were reduced by 15% and 43% respectively.

Sofi [27] tested high-strength concrete specimens by replacing 5%, 7.5%, and 10% of aggregate with rubber and reported a reduction of 10–23% in the compressive strength. The same study showed that the modulus of elasticity of the rubberized concrete was reduced by 17–25% and recommended a maximum replacement of 12.5% of fine aggregate by rubber for high strength concrete.

Akinwonmi and Seckley [28] investigated the change in compressive strength of concrete when the aggregates were replaced with shredded and crumb rubber. The results showed that the replacement of 2.5% shredded rubber increased the compressive strength by 8.5%, while any replacement of more than 2.5% rubber decreased the compressive strength when compared to the control mix.

Issa and Salem [29] reported that partial replacement of the sand with crumb rubber up to 25% would result in an acceptable compressive strength for the concrete. He indicated that increasing the rubber content above this threshold would reduce the compressive strength substantially that the mix would not be acceptable for structural or even non-structural applications.

Recently, Irmawaty et al. [30] investigated the flexural behavior of the concrete made in part by using waste rubber tires. They tested specimens with 100 mm × 100 mm × 400 mm and with the replacement of 10%, 20%, and 30% of crumb rubber and tire chips. They concluded that rubberized concrete with 10% crumb rubber achieved the optimal energy absorption.

The literature made it clear that the greater the amount of steel fibers in the concrete, the greater the value of strength and flexural toughness [31–34]. Therefore, the addition of steel fiber appears to improve the tensile and compressive strengths [32,34,35].

Gul and Nasser [32] conducted a study to investigate the behavior of concrete by using the waste rubber tires as an alternative to steel fiber in fiber-reinforced concrete. They concluded that by increasing the percentage replacement of rubber, the compressive and split tensile strengths of concrete decrease compared to specimens containing steel fiber.

Manufactured steel fibers and steel fibers extracted from tire waste have been used in different concrete mixes to compare their effectiveness [33]. The increase in both reused and normal steel fibers showed less slump values for fresh concrete. The initial modulus of elasticity increased by 7–8% for the mixes with normal steel fibers and 2–3% for the mixes with reused steel fibers. The compressive strength of the mix with reused fibers increased by a maximum of 12%, while the mix with normal fibers increased by a maximum of 20%.

No significant increase was noticed in the splitting tensile strength test between the two mixes while increasing the fiber dosage.

Some researchers examined the hybrid concrete mixes: for example, Noaman et al. [34] investigated the mechanical properties of rubberized concrete combined with steel fiber. They combined rubberized concrete with different replacement ratios of crumb rubber in plain and steel fiber concrete mixes via the partial replacement of fine aggregate (17.5%, 20%, 22.5%, and 25%). The study indicated that a reduction in mechanical properties was observed by the increasing increment of crumb rubber in both mixes. The study suggested using a combination of steel fiber and crumb rubber due to the improvement of strain capacity under flexural loading.

As well, Eisa et al. [35] studied the effect of a combination of crumb rubber and steel fibers on the behavior of reinforced concrete beams under static loads. They concluded that an acceptable level of performance of reinforced concrete beams could be obtained by using crumb rubber as a partial replacement of fine aggregates by 5% and 10%. The study also recommended the use of steel fibers with rubberized concrete, with percentages of rubber over 10%, as this showed a significant improvement in the performance and toughness of these mixtures.

As per the above literature review, researchers directed great attention to investigating the utilization of recycled tire rubber in concrete in order to find a proper solution for minimizing tire waste and producing a green concrete. In spite of the reduction in mechanical properties due to increasing the rubber content in the mix [11,13,17,18,22,31], it is still recommended to use rubber particles in civil engineering applications, especially in pavement projects [2,11,15,17,20,35,36].

The objective of this study is to investigate whether recycled tire by-products can be used to make a suitable "green concrete" to be used for pavement construction in hot-weather climates. To achieve optimal results, each type of tire by-product was tested individually to observe its properties and effects on a benchmark mix before creating "hybrid" mixes that contain a combination of the materials; this is where the novelty of this study lies.

2. Materials and Methods

2.1. Benchmark Concrete Mix Design

In this study, a benchmark pavement-grade concrete mix with a 28-day compressive strength of about 35 MPa has been designed and cast. Concrete pavements are preferred in Kuwait due to their ability to resist high temperatures without causing permanent damage to the pavement itself. However, concrete used in hot climates is subjected to high rates of evaporation, loss of moisture, and quick setting times [37]. As a result, concrete mixes used in hot weather climates need to have a slump value at the higher end of the recommended range for pavements. This was achieved by introducing a superplasticizer in order to improve the workability of the mix and reach the targeted 28-day compressive strength while limiting the water/cement ratio to 0.55. The mix proportions of the benchmark concrete used for the study are shown in Table 1.

Table 1. Mix design propositions of benchmark pavement-grade concrete.

Material	Quantity
Cement (kg/m^3)	327
Sand (kg/m^3)	663
Coarse aggregates: 9.5 mm (kg/m^3)	740
Coarse aggregates: 12.5 mm (kg/m^3)	442
Water (kg/m^3)	180
SIKAment®-500 OM Superplasticizer (% of binder weight)	2.00
Water/Cement ratio	0.55

To achieve the desired workability, SIKAment®-500 OM superplasticizer is used. This is important for hot climate places.

2.2. Recycled Tire Products

Crumb rubber (CR), shredded rubber (SR), and steel fibers (SF) were obtained by recycling used tires (Figure 1). These products are easily available in the market, as tires are now recycled around the world. The tire rubber used in this study is varied in source but is believed to contain percentages of natural and synthetic hydrophobic rubber optimized for automobile use. For this study, the recycled tire products were provided by the Green Rubber Tire Recycling Plant in Kuwait.

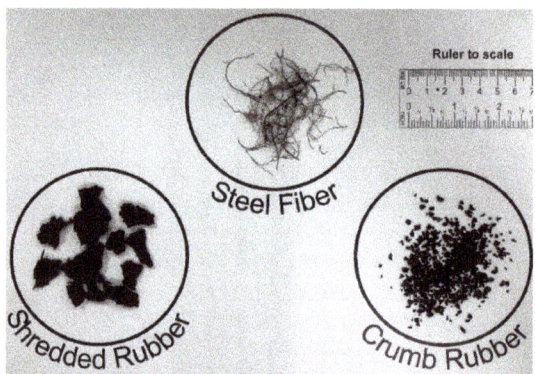

Figure 1. Crumb rubber, shredded rubber, and steel fibers obtained by recycling used tires.

CR and SR are used as a partial replacement for the sand and coarse aggregates respectively. SF is used as an extra ingredient in different concrete mixes.

2.2.1. Crumb Rubber (CR)

The density of CR used in this study is 552 kg/m^3. The size distribution, by sieve analysis, is shown in Figure 2 and is compared with the size distribution of the sand used in this project. The particle size distribution of the sand is close to being well-graded, whereas the CR is uniformly graded. Ideally, the CR would be introduced to replace a similarly sized portion of the sand; however, this is not feasible in real-life applications. Therefore, a direct replacement of sand with CR is used.

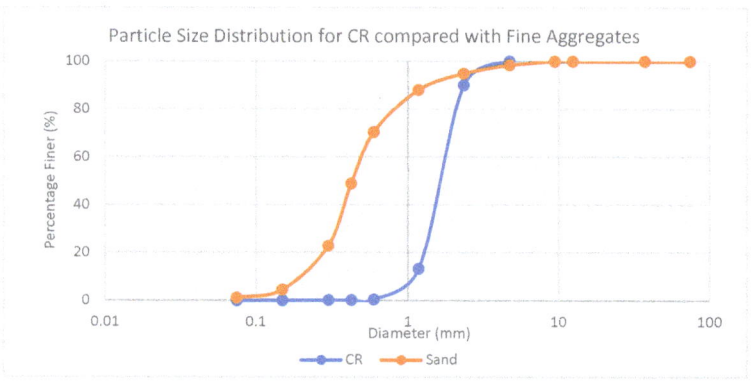

Figure 2. Size distribution of crumb rubber and sand used in this study.

2.2.2. Shredded Rubber (SR)

The density of SR used in this study is 566 kg/m^3. SR obtained from the recycling plant has been sieved, and particles smaller than 12.5 mm are used to make sure that similarly-sized SR particles are used in the partial replacement of the coarse aggregates. The size distribution of the SR, which is also compared with that of the coarse aggregates (9.5 mm and 12.5 mm), is shown in Figure 3. The particle size distribution for both sizes of coarse aggregate as well as the SR are more or less well-distributed, and interestingly, the particle size distribution of the SR fits in between the two sizes of coarse aggregates used in this study. Therefore, a direct replacement of both sizes of coarse aggregates with SR is used.

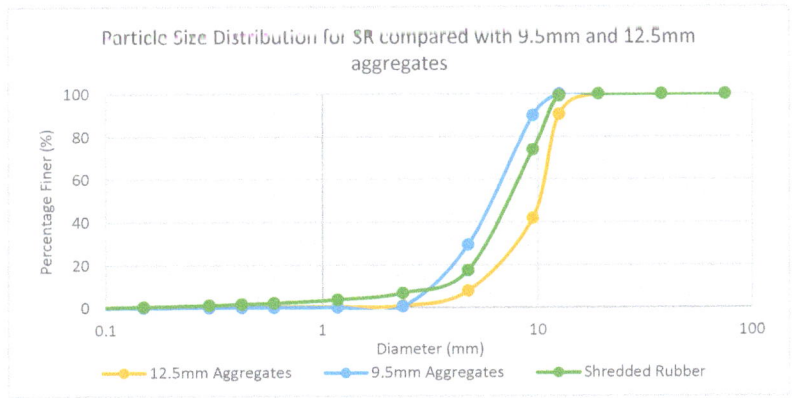

Figure 3. Size distribution of shredded rubber and coarse aggregates used in this study.

2.2.3. Steel Fibers (SF)

The density of SR used in this study is 7850 kg/m^3. The length and diameter of 100 randomly selected steel fibers have been measured, and the results are shown in Table 2.

Table 2. Steel fibers dimensions.

Dimension	Average	Standard Deviation
Length (mm)	24	8.6
Diameter (mm)	0.34	0.076

2.3. Concrete Mixes Containing Recycled Tire Products

A total of 24 concrete mixes are cast, including the benchmark. The mix codes and descriptions are shown in Table 3. SR, CR, and SF in the concrete mix codes show the presence of shredded rubber, crumb rubber, and steel fiber in the concrete mix respectively. When combinations of SF, CR, and SF are used in a concrete mix, the code starts with an "H."

For each concrete mix, a total of 14 cylinders (100 mm diameter by 200 mm height) and 3 beams (100 mm × 100 mm × 400 mm) are cast and sulfur-capped as per ASTM C617 [38]. A minimum of 3 cylinders are tested to calculate the tensile strength at 28 days as per ASTM C496 [39]. Similarly, a minimum of 3 cylinders are used to calculate the compressive strength at 7 and 28 days (Figure 4) as per ASTM C39 [40]. For the benchmark concrete only, 3 cylinders are tested at 3 days under compression.

Table 3. Concrete mixes and descriptions.

No.	Code	Description
1	BM	Benchmark concrete with no recycled tire products
2	SR-1	5% (by volume) of coarse aggregates (9.5 and 12.5 mm) is replaced by shredded rubber
3	SR-2	10% (by volume) of coarse aggregates (9.5 and 12.5 mm) is replaced by shredded rubber
4	SR-3	15% (by volume) of coarse aggregates (9.5 and 12.5 mm) is replaced by shredded rubber
5	SR-4	20% (by volume) of coarse aggregates (9.5 and 12.5 mm) is replaced by shredded rubber
6	CR-1	5% (by volume) of sand is replaced by crumb rubber
7	CR-2	10% (by volume) of sand is replaced by crumb rubber
8	CR-3	15% (by volume) of sand is replaced by crumb rubber
9	CR-4	20% (by volume) of sand is replaced by crumb rubber
10	SF-1	0.1% steel fibers are added to BM (by volume of concrete)
11	SF-2	0.2% steel fibers are added to BM (by volume of concrete)
12	SF-3	0.3% steel fibers are added to BM (by volume of concrete)
13	H-1	5% (by volume) of coarse aggregates (9.5 and 12.5 mm) is replaced by shredded rubber and 0.1% steel fibers (by volume of concrete) are added to the mix
14	H-2	10% (by volume) of coarse aggregates (9.5 and 12.5 mm) is replaced by shredded rubber and 0.1% steel fibers (by volume of concrete) are added to the mix
15	H-3	15% (by volume) of coarse aggregates (9.5 and 12.5 mm) is replaced by shredded rubber and 0.1% steel fibers (by volume of concrete) are added to the mix
16	H-4	20% (by volume) of coarse aggregates (9.5 and 12.5 mm) is replaced by shredded rubber and 0.1% steel fibers (by volume of concrete) are added to the mix
17	H-5	5% (by volume) of sand is replaced by crumb rubber and 0.1% steel fibers (by volume of concrete) are added to the mix
18	H-6	10% (by volume) of sand is replaced by crumb rubber and 0.1% steel fibers (by volume of concrete) are added to the mix
19	H-7	15% (by volume) of sand is replaced by crumb rubber and 0.1% steel fibers (by volume of concrete) are added to the mix
20	H-8	20% (by volume) of sand is replaced by crumb rubber and 0.1% steel fibers (by volume of concrete) are added to the mix
21	H-9	5% (by volume) of coarse aggregates (9.5 and 12.5 mm) is replaced by shredded rubber, 5% (by volume) of sand is replaced by crumb rubber, and 0.1% steel fibers (by volume of concrete) are added to the mix
22	H-10	10% (by volume) of coarse aggregates (9.5 and 12.5 mm) is replaced by shredded rubber, 10% (by volume) of sand is replaced by crumb rubber, and 0.1% steel fibers (by volume of concrete) are added to the mix
23	H-11	5% (by volume) of coarse aggregates (9.5 and 12.5 mm) is replaced by shredded rubber and 5% (by volume) of sand is replaced by crumb rubber
24	H-12	10% (by volume) of coarse aggregates (9.5 and 12.5 mm) is replaced by shredded rubber and 10% (by volume) of sand is replaced by crumb rubber

 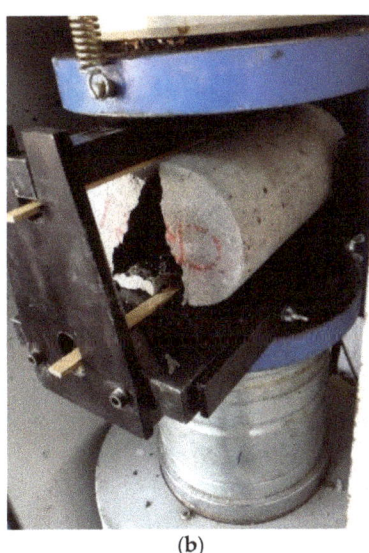

Figure 4. Test of a concrete cylindrical sample (**a**) Compression; and (**b**) Tension.

Third-point loading tests (Figure 5) are performed on the beams to obtain the modulus of rupture of the concrete mix as per ASTM C78 [41].

Figure 5. Third-point loading test to obtain the modulus of rupture.

3. Results

3.1. Slump Test Results

For a pavement-grade concrete mix in a hot climate, a target slump of 70 mm ± 30 mm is set. The slump of the benchmark concrete, BM, was 75 mm, and the slump of all other concrete mixes were within the targeted range. The slump test results are compared in Figure 6. It is worth mentioning that in all mixes, oven-dried sand and aggregates are used to ensure consistency in water-to-cement ratio. In addition, to improve the workability of concrete mixes, superplasticizer is used in all mixes (2.0% to 2.5% of the weight of cement). Since 2.0% is used for the benchmark, a minimum amount of 2.0% is used in all other mixes; if the workability is not desirable, an extra 0.5% is added to the mix.

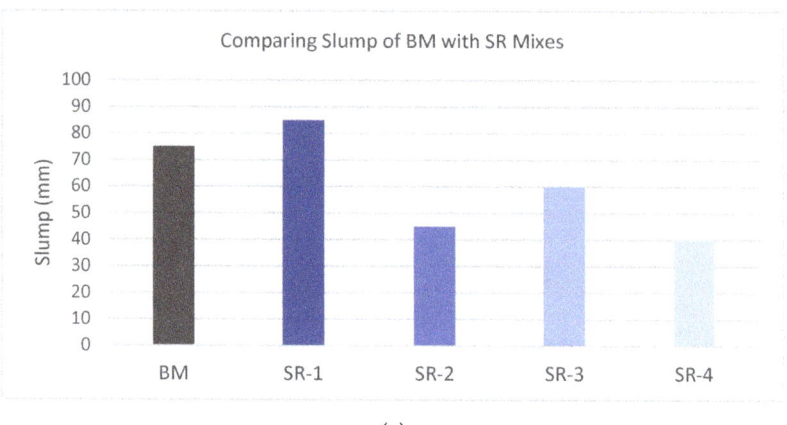

(a)

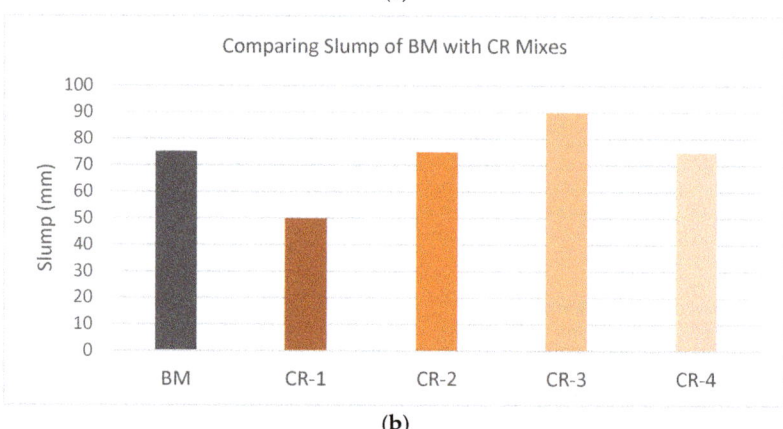

(b)

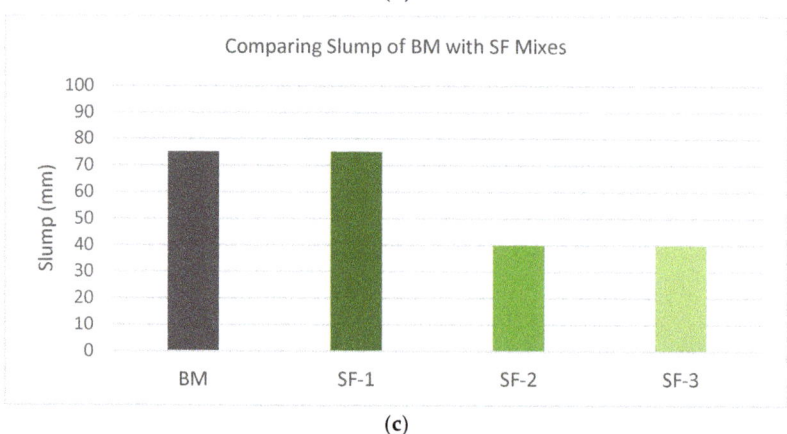

(c)

Figure 6. *Cont.*

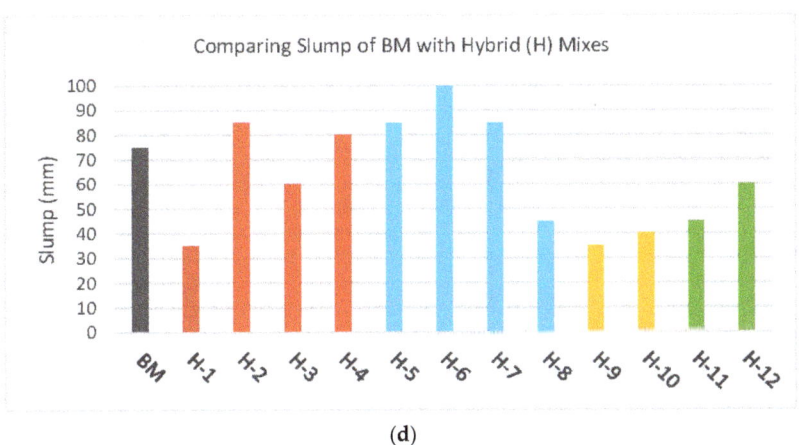

(**d**)

Figure 6. Comparing the slump test results of benchmark (75 mm) with (**a**) shredded rubber (SR); (**b**) crumb rubber (CR); (**c**) steel fibers (SF); and (**d**) Hybrid (i.e., H) mixes.

3.2. Density of Concrete

The BM's density is 2390 kg/m^3. Replacing the coarse aggregates and sand by SR and CR respectively reduces the density. For example, the density of SR-4 and CR-4 is 2317 kg/m^3 and 2362 kg/m^3, respectively. On the other hand, the addition of SF increases the density; for example, SF-3 has a density of 2425 kg/m^3. In the hybrid mixes (i.e., H-1 to H-12), the maximum density is seen in H-5 (2402 kg/m^3) and the minimum is seen in H-4 (2314 kg/m^3).

3.3. Compressive Strength

A minimum of 3 cylinders are tested in order to obtain the compressive strength. Figure 7 shows the compressive strength of BM mix at 3, 7, and 28 days. All other mixes are tested at 7 and 28 days. The compressive strength of SR, CR, and SF mixes at these two ages are compared with those of the BM mix in Figure 8. Similarly, the compressive strength of hybrid mixes (H-1 to H-12) are compared with BM in Figure 9.

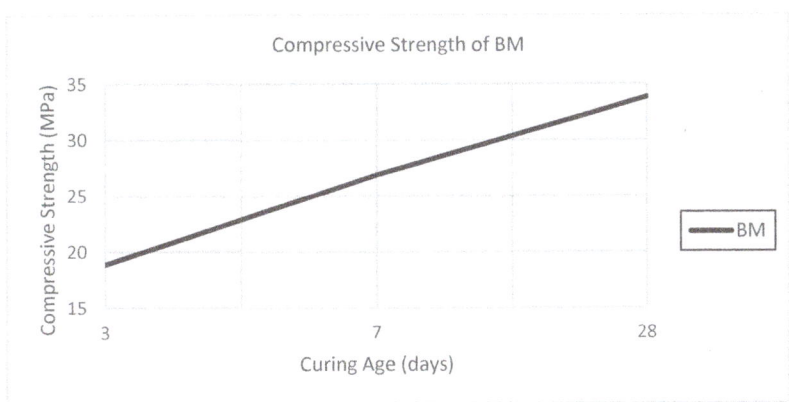

Figure 7. Compressive strength of the benchmark mix at 3, 7, and 28 days.

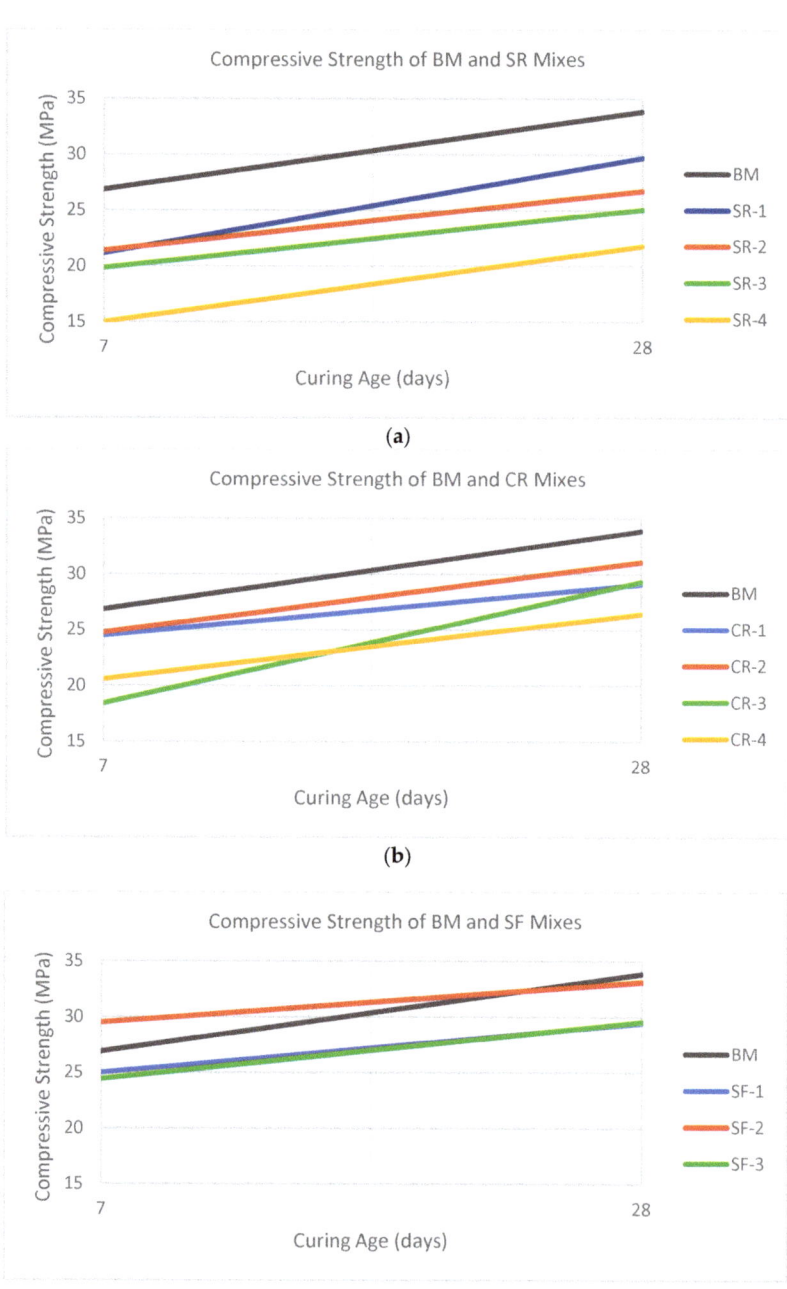

Figure 8. Comparing the compressive strength of the benchmark mix with (**a**) SR; (**b**) CR; and (**c**) SF mixes.

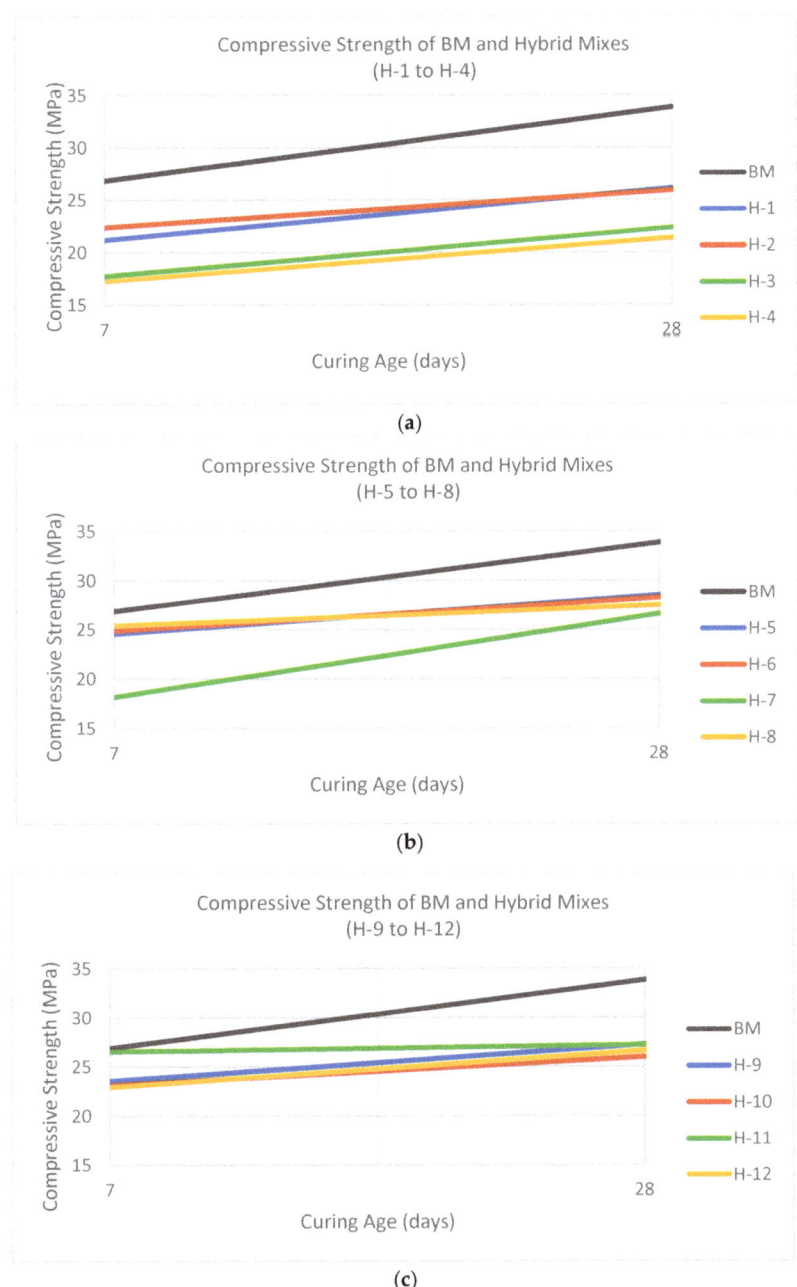

Figure 9. Comparing the compressive strength of the benchmark mix with hybrid mixes (**a**) H-1 to H-4; (**b**) H-5 to H-8; and (**c**) H-9 to H-12.

3.4. Tensile Strength

A splitting tensile test is performed on 3 cylinders for each mix, and the average tensile strength of the benchmark mix is compared with that of all other mixes in Figure 10.

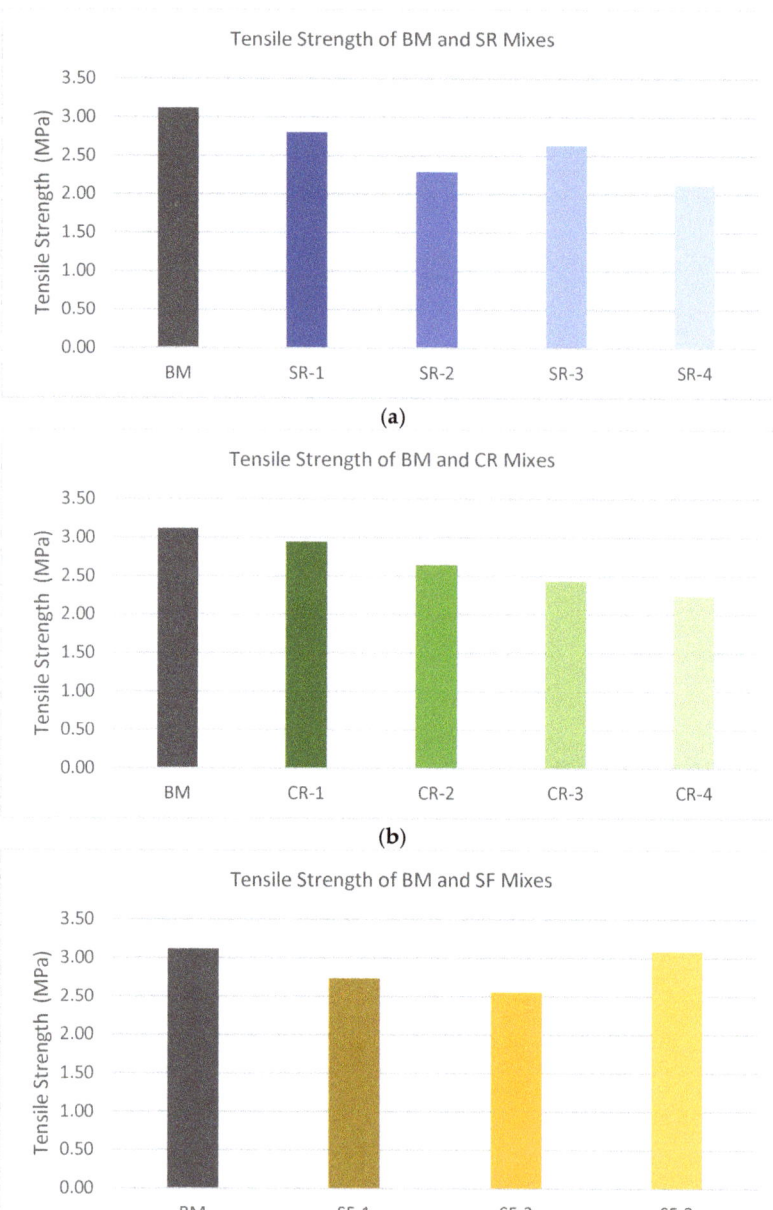

Figure 10. *Cont.*

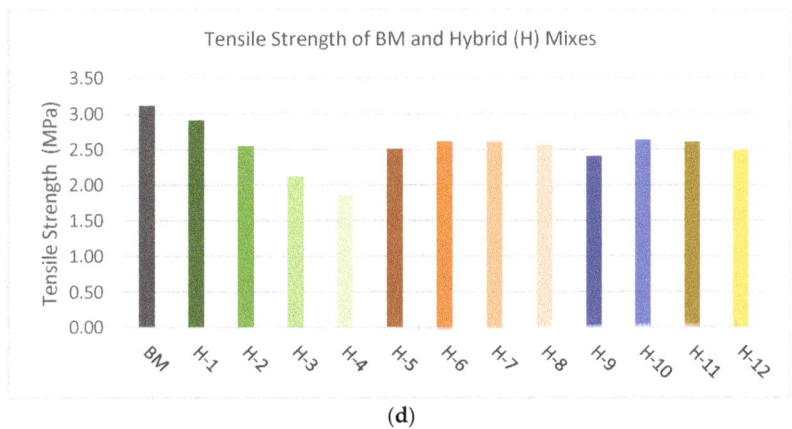

(d)

Figure 10. Comparing the tensile strength of the benchmark mix (3.11 MPa) with (**a**) SR; (**b**) CR; (**c**) SF; and (**d**) Hybrid (i.e., H) mixes.

3.5. Modulus of Rupture

A third-point loading test is performed on 3 beams for each mix, and the average value relating to the modulus of rupture of the benchmark mix is compared with that of all other mixes in Figure 11.

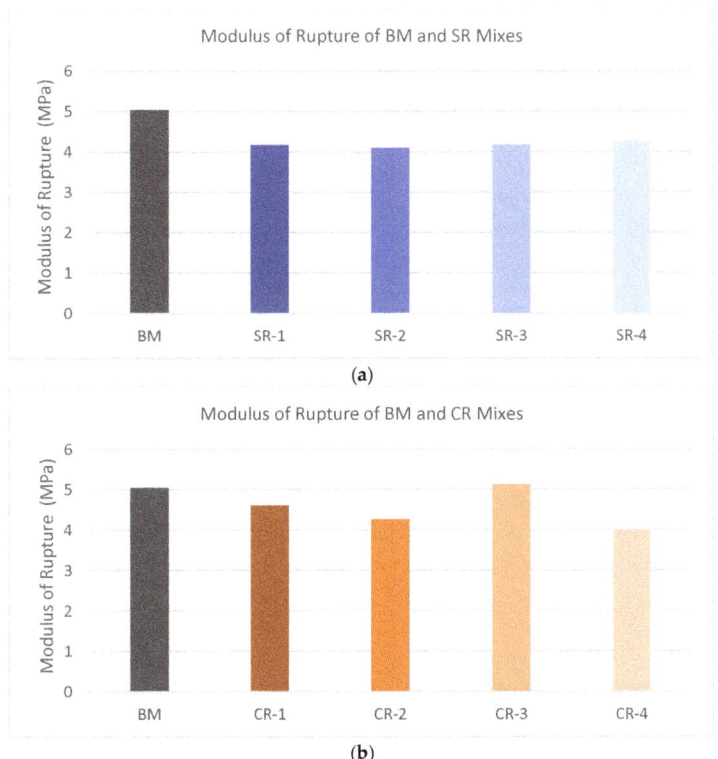

Figure 11. *Cont.*

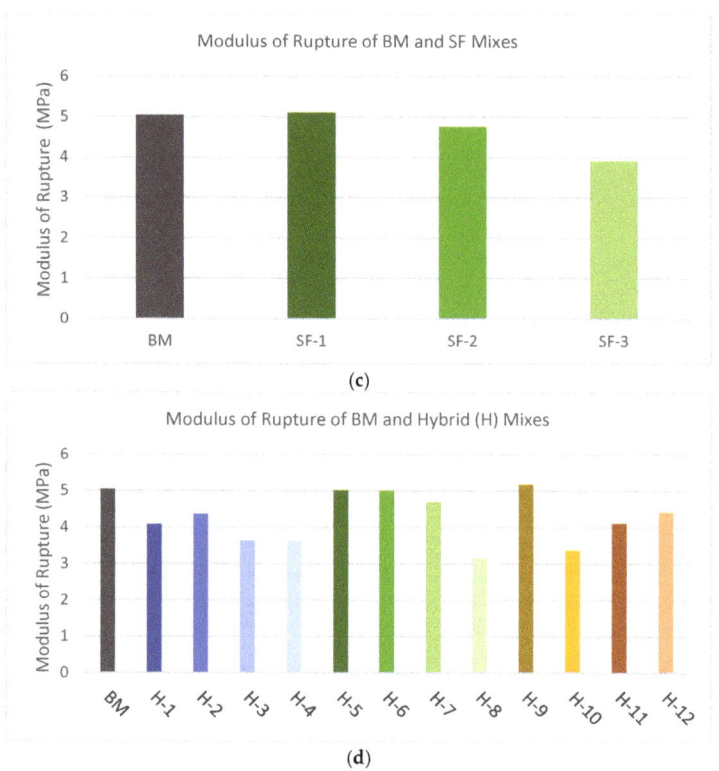

Figure 11. Comparing the modulus of rupture of the benchmark mix (5.04 MPa) with (**a**) SR; (**b**) CR; (**c**) SF; and (**d**) Hybrid (i.e., H) mixes.

4. Discussion

4.1. Slump Test

Figure 6a,c show the slump test results for mixes containing SR and SF, respectively. The results shown in the figures are inconclusive, as they do not clearly indicate a trend with the introduction of either of the materials. However, the reduction in slump due to the incorporation of SF into the concrete mix was corroborated by findings by other researchers [33]. Additionally, all SR and SF mixes presented acceptable slump values that are within the target slump range for pavements (70 mm ± 30 mm).

Figure 6b shows the slump test results for mixes containing CR. As shown in the figure, the introduction of CR initially decreased the slump but then increases with additional doses. This trend is the case for doses up to and including 15%. The introduction of 20% CR reverses the trend but still maintains a slump value similar to the benchmark. The conclusions of this family of concrete are similar to that of other researchers [34].

Figure 6d shows the slump test results for the hybrid mixes, containing combinations of SR, CR, and SF. The figure shows four "families" of hybrid mixes: combinations of SR and SF are shown in red, combinations of CR and SF are shown in blue, combinations of SR, CR, and SF are shown in yellow, and combinations of SR and CR are shown in green. The results in red show that with the exception of H2, there is a clear and increasing trend in the slump of concrete. Initially, the slump significantly decreased but then increases, with the last mix reaching the benchmark's slump. The results in blue show that the slump remained somewhat consistent, hovering in the range of 85–100 mm. This excludes the results of H8, which can be considered anomalous. The results in yellow show that there is a slight increase in the slump at the higher dosage of H10, but this remains significantly lower than the benchmark's slump. The results in green show that there is a slight increase in the slump at the higher dosage of H12, but this remains lower than the benchmark's slump and outperforms the results in yellow. All slump test results were within the target slump range for pavements.

4.2. Compressive Strength

Figure 8a shows the variation of compressive strength of samples containing SR compared with the benchmark at the 7- and 28-day marks. As shown in the figure, compressive strength is directly affected by the introduction of SR into the mix, where larger quantities of SR present in the mix lead to a decrease in the 7- and 28-day compressive strength in an almost proportional fashion. The samples also exhibit similar rates of compressive strength gain to the benchmark mix, as shown by the slope of the lines. Comparable studies stated that the introduction of SR would reduce the compressive strength by up to 23% [27,29], where the compressive strength reduction in the results of this study ranged between 13% and 36%. There is some correlation between the results of the study and existing literature, with variations in the upper/lower bounds of compressive strength reduction. However, the results stand in contrast to the findings of some other researchers [13] who found that compressive strength, among other properties, was improved by the introduction of SR into the mix.

Figure 8b shows the variation of compressive strength of samples containing CR compared with the benchmark at the 7- and 28-day marks. As shown in the figure, there is no clear correlation between the amount of CR present in the sample and the compressive strength at any age. However, the introduction of CR lowers both 7- and 28-day compressive strength. It is worth mentioning that all concrete mixes in this figure have acceptable 28-day compressive strengths, ranging from 26.4 MPa (for CR4) to 31.1 MPa (for CR2). Comparable studies stated that the introduction of CR would reduce the compressive strength by 15% [26], where the compressive strength reduction in the results of this study ranged between 8% and 30%. It is worth mentioning that the replacements considered in this study are within the bounds of acceptable replacement of conventional materials [29].

Figure 8c shows the variation of compressive strength of samples containing SF compared with the benchmark at the 7- and 28-day marks. As shown in the figure, an interesting observation occurs at SF2, where the 7-day compressive strength was higher than the benchmark and the 28-day compressive strength was very similar to the benchmark. Other than that, both SF1 and SF3 performed similarly to each other, both being slightly lower than the benchmark and having similar rates of compressive strength gain to each other. This comes at odds with the conclusions of other researchers who had success in increasing mechanical properties (such as compressive strength) with increasing dosages of SF [30].

Figure 9a shows the variation of compressive strength of samples containing a combination of SR and SF compared with the benchmark at the 7- and 28-day marks. As shown in the figure, higher doses of SR with the introduction of 0.1% SF into the mix decrease the 7- and 28-day compressive strength. However, there is no impact on the rate of compressive strength gain in any of the samples, exhibiting similar slopes to that of the benchmark.

Figure 9b shows the variation of compressive strength of samples containing a combination of CR and SF compared with the benchmark at the 7- and 28-day marks, and Figure 9c shows the variation of compressive strength of samples containing a combination of SR, CR, and SF compared with the benchmark at the 7- and 28-day marks. An interesting observation can be seen across the two figures above, where the compressive strengths of the samples converge to the same value at the 28-day mark. The rate of compressive strength gain is expected to decrease as the proportion of the replacement materials increase in the mixes.

All mixes used in this study exhibit acceptable levels of 28-day compressive strength, with a minimum of 21.3 MPa (H-4) and a maximum of 33.1 MPa (SF-2).

4.3. Tensile Strength

Figure 10a shows the variation of split tensile strength of samples containing SR compared with the benchmark. The results show a very clear decreasing trend as the percentage of SR increases in the mix. The results of SR3 can be considered anomalous.

Figure 10b shows the variation of split tensile strength of samples containing CR compared with the benchmark. The results show a very clear decreasing trend as the percentage of CR increases in the mix. Comparable studies stated that the introduction of CR would reduce the tensile strength by up to 43% [2,26], where the tensile strength reduction in the results of this study ranged around 22%. While this is a marked improvement over existing literature, it still represents a significant drop in the tensile strength of concrete.

Figure 10c shows the variation of split tensile strength of samples containing SF compared with the benchmark. The results show very consistent tensile strength results that are close to the benchmark's results, regardless of the dosage of SF present in the concrete. The results of this test indicate that there is no advantage in incorporating additional doses of steel fibers beyond 0.1%. This, coupled with the fact that higher dosages of SF led to a significant decrease in the slump, justifies the use of this dosage in the hybrid mixes.

Figure 10d shows the tensile strength test results for the hybrid mixes, containing combinations of SR, CR, and SF. The figure shows four "families" of hybrid mixes: combinations of SR and SF are shown in shades of green, combinations of CR and SF are shown in shades of orange, combinations of SR, CR, and SF are shown in shades of blue, and combinations of SR and CR are shown in shades of yellow. The results in green show that there is a clear decreasing trend in the tensile strength of the samples as the proportion of SR increases in the samples. The results in orange indicate a clear trendline showing that the values are almost consistent throughout the samples tested. However, the samples have tensile strength values that are lower than the benchmark's. The results in blue show that there is a slight increase in the tensile strength at the higher dosage of H10. The results in yellow show that there is a slight decrease in the tensile strength at the higher dosage of H12.

4.4. Modulus of Rupture

Figure 11a shows the variation of modulus of rupture of samples containing SR compared with the benchmark. The results show a very clear trend where the modulus of rupture remains consistent for all four dosages of SR tested.

Figure 11b shows the variation of modulus of rupture of samples containing CR compared with the benchmark. The results show a very clear trend where the modulus of rupture decreases with increasing dosages of crumb rubber. The results of CR3 can be considered anomalous. The findings of the CR trials match closely with the findings of other researchers, who have stated that the incorporation of rubber products has a negative effect on mechanical properties [2].

Figure 11c shows the variation of modulus of rupture of samples containing SF compared with the benchmark. Initially, this does not have an impact on the modulus of rupture at a low dosage. However, with increasing dosages of SF, there is a clear trend of decreasing modulus of rupture. As was the case with the tensile test results, the results of this test also indicate that there is no advantage in incorporating additional doses of steel fibers beyond 0.1%. Again, this comes at odds with the conclusions of other researchers who had success in increasing mechanical properties (such as flexural strength) with increasing dosages of SF [30].

Figure 11d shows the test results for the modulus of rupture for the hybrid mixes, containing combinations of SR, CR, and SF. The figure shows four "families" of hybrid mixes: combinations of SR and SF are shown in shades of blue, combinations of CR and SF are shown in shades of green, combinations of SR, CR, and SF are shown in shades of yellow, and combinations of SR and CR are shown in shades of orange. The results in blue show that there is a clear decreasing trend in the modulus of rupture of the samples as the proportion of SR increases in the samples. The results in green show that there is a clear decreasing trend in the modulus of rupture of the samples as the proportion of CR increases in the samples. However, results up to and including H7 are very close to the modulus of rupture of the benchmark, which is similar to the findings of other researchers [2,35]. The results in yellow show that there is a sharp drop in the modulus of rupture values as the dosage of replacement materials increases, while noting the improvement in the modulus of rupture for H-9 compared with the benchmark mix, which proves the benefits of including all replacement materials in the concrete mix. The results in orange show that there is a sharp increase in the modulus of rupture values as the dosage of replacement materials increases.

5. Conclusions

After analyzing the results presented above, and despite the reduction in fresh properties due to the introduction of recycled tire products in the concrete mix, it has been shown that multiple tire by-products (shredded/crumb rubber and recovered steel fibers) can be successfully hybridized to produce a novel pavement-grade green concrete that is suited for use in hot-weather climates. This is important, as it achieves two goals in one: to be able to produce a sustainable construction material that is tailored for use in Kuwait's climate and to also reduce the number of tires going to landfill.

- The incorporation of CR has a detrimental effect on all concrete properties considered in this study. However, the effects on tensile strength and modulus of rupture were the most evident.
- The introduction of SR into a benchmark concrete mix has a noticeable, almost proportional effect on the 7 and 28-day compressive strength and tensile strength.
- The introduction of 0.1% and 0.2% SF managed to increase the modulus of rupture and the 7-day compressive strength of the benchmark, respectively. On the other hand, excessive use of SF leads to the concrete having significantly weaker properties compared to a benchmark mix.
- The preliminary results from the individual tire products were used to determine optimal dosages of each material for use in a hybrid mix, which has not been attempted by many researchers. It is interesting to note how the hybrid mixes exhibit very similar trends to their constituent replacement materials. The incorporation of each material at a low dosage (5% for rubber products and 0.1% for steel fibers) was able to maintain the properties of the benchmark mix, if not improve on them.

All mixes, whether individual replacement or hybrid mixes, exhibited acceptable properties for pavement-grade concrete for use in hot climates. The conclusions of this study show that it is possible to hybridize all materials sourced from recycled tires in the production of a feasible pavement-grade concrete suited for hot weather. Possible extensions of the project could include the investigation of other properties to evaluate the usefulness of the hybrid concrete, such as measuring skid resistance and modulus of elasticity. Further, long-term effects of using this concrete could be modeled using finite-element analysis software packages, which could be used to predict the behavior of the concrete while it is being used at a selected intersection in Kuwait.

Author Contributions: Conceptualization, S.M.S., A.R.A., A.J., N.M. and A.F.; methodology, S.M.S.; formal analysis, S.M.S., A.R.A., A.J. and N.M.; investigation, A.R.A., A.J. and N.M.; resources, A.R.A.; data curation, A.J.; writing—original draft preparation, S.M.S., A.R.A., A.J., N.M. and A.F.; writing—review and editing, S.M.S., A.R.A., A.J., N.M. and A.F.; supervision, S.M.S.; project administration, A.F.; funding acquisition, S.M.S. All authors have read and agreed to the published version of the manuscript.

Funding: This project was funded "partially" by Kuwait Foundation for the Advancement of Sciences under project code: PR19-15EV-02.

Institutional Review Board Statement: Not applicable.

Informed Consent Statement: Not applicable.

Data Availability Statement: The data presented in this study are available on request from the corresponding author.

Acknowledgments: The authors would like to acknowledge Tahir Afrasiab for his contributions, hard work and dedication while preparing the concrete mixes for this study. The authors would like to thank Sameer Hamoush of North Carolina A&T State University for his input and advice throughout the course of the project. The authors would also like to express their gratitude to Green Rubber Recycling Co. for providing samples of their recycled tire products to be used in this study. The authors are grateful to Sika for providing the team with the superplasticizer used in the study.

Conflicts of Interest: The authors declare no conflict of interest.

Abbreviations

CR	crumb rubber
Kg	kilogram
Mm	millimeter
MPa	megapascal
m^3	cubic meter
SF	steel fiber
SR	shredded rubber

References

1. Arab Times. Available online: https://www.arabtimesonline.com/news/red-tape-proper-planning-dearth-delays-shifting-of-huge-tire-dump/ (accessed on 1 January 2021).
2. Meddah, A.; Miloud, B.; Bali, A. Use of shredded rubber tire aggregates for roller compacted concrete pavement. *J. Clean. Prod.* **2014**, *72*, 187–192. [CrossRef]
3. Kuwait News Agency. Available online: https://www.kuna.net.kw/ArticleDetails.aspx?id=2931763&language=en (accessed on 1 January 2021).
4. Arab Times. Available online: https://www.arabtimesonline.com/news/millions-of-tires-destroyed-in-fire/ (accessed on 1 January 2021).
5. Kuwait News Agency. Available online: https://www.kuna.net.kw/ArticleDetails.aspx?id=2367062&Language=en (accessed on 1 January 2021).
6. Kuwait News Agency. Available online: https://www.kuna.net.kw/ArticleDetails.aspx?id=2234353&Language=en (accessed on 1 January 2021).
7. Fazli, A.; Rodrigue, D. Waste Rubber Recycling: A Review on the Evolution and Properties of Thermoplastic Elastomers. *Materials* **2020**, *13*, 782. [CrossRef] [PubMed]
8. Debrah, J.K.; Vidal, D.G.; Dinis, M.A.P. Raising Awareness on Solid Waste Management through Formal Education for Sustainability: A Developing Countries Evidence Review. *Recycling* **2021**, *6*, 6. [CrossRef]
9. Kassim, S.M. The Importance of Recycling in Solid Waste Management. *Macro. Symp.* **2012**, *320*, 34–50. [CrossRef]
10. Assadi, M.H.N.; Sahajwalla, V. Recycling End-of-Life Polycarbonate in Steelmaking: Ab Initio Study of Carbon Dissolution in Molten Iron. *Ind. Eng. Chem. Res.* **2014**, *53*, 3861–3864. [CrossRef]
11. Yazdi, M.A.; Yang, J.; Yihui, L.; Su, H. A Review on Application of Waste Tire in Concrete. *Int. Sch. Sci. Res. Innov.* **2015**, *9*, 1656–1661.
12. Toghroli, A.; Shariati, M.; Sajedi, F.; Ibrahim, Z.; Koting, S.; Mohamad, E.T.; Khorami, M. A review on pavement porous concrete using recycled waste materials. *Smart Struct. Syst.* **2018**, *22*, 433–440.
13. Lavagna, L.; Nisticò, R.; Sarasso, M.; Pavese, M. An Analytical Mini-Review on the Compression Strength of Rubberized Concrete as a Function of the Amount of Recycled Tires Crumb Rubber. *Materials* **2020**, *13*, 1234. [CrossRef] [PubMed]
14. Forrest, M.J. *Recycling and Re-Use of Waste Rubber*, 2nd ed.; De Gruyter: Berlin, Germany, 2019; pp. 1–20.
15. Bulei, C.; Todor, M.P.; Heput, T.; Kiss, I. Directions for material recovery of used tires and their use in the production of new products intended for the industry of civil construction and pavements. *IOP Conf. Ser. Mater. Sci. Eng.* **2018**, *294*, 012064. [CrossRef]
16. Oliveira Neto, G.C.d.; Chaves, L.E.C.; Pinto, L.F.R.; Santana, J.C.C.; Amorim, M.P.C.; Rodrigues, M.J.F. Economic, Environmental and Social Benefits of Adoption of Pyrolysis Process of Tires: A Feasible and Ecofriendly Mode to Reduce the Impacts of Scrap Tires in Brazil. *Sustainability* **2019**, *11*, 2076. [CrossRef]
17. Al-Fakih, A.; Mohammed, B.S.; Liew, M.S. Tires Rubber as a Useable Material in Civil Engineering Applications: An Overview. *Int. J. Adv. Res. Eng. Technol.* **2020**, *11*, 315–325.

18. Senin, M.S.; Shahidan, S.; Abdullah, S.R.; Guntor, N.A.; Leman, A.S. A review on the suitability of rubberized concrete for concrete bridge decks. *IOP Conf. Ser. Mater. Sci. Eng.* **2017**, *271*, 012074. [CrossRef]
19. Hejna, A.; Zedler, Ł.; Przybysz-Romatowska, M.; Cañavate, J.; Colom, X.; Formela, K. Reclaimed Rubber/Poly(ε-caprolactone) Blends: Structure, Mechanical, and Thermal Properties. *Polymers* **2020**, *12*, 1204. [CrossRef]
20. Alfayez, S.; Suleiman, A.R.; Nehdi, M.L. Recycling Tire Rubber in Asphalt Pavements: State of the Art. *Sustainability* **2020**, *12*, 9076. [CrossRef]
21. Aiello, M.A.; Leuzzi, F. Waste tyre rubberized concrete: Properties at fresh and hardened state. *Waste Manag.* **2010**, *30*, 1696–1704. [CrossRef] [PubMed]
22. Gupta, T.; Chaudhary, S.; Sharma, R.K. Mechanical and durability properties of waste rubber fiber concrete with and without silica fume. *J. Clean. Prod.* **2016**, *112*, 702–711. [CrossRef]
23. Obinna, O.; Panesar, D.K. Hardened properties of concrete mixtures containing pre-coated crumb rubber and silica fume. *J. Clean. Prod.* **2014**, *82*, 125–131.
24. Guneyisi, E.; Gesoglu, M.; Ozturan, T. Properties of rubberized concretes containing silica fume. *Cem. Concr. Res.* **2004**, *34*, 2309–2317. [CrossRef]
25. Kanmalai Williams, C.; Partheeban, P. Experimental investigation on recycled rubber filled concrete. *Int. J. Earth Sci. Eng.* **2016**, *9*, 170–175.
26. Khan, S.; Singh, A. Behavior of Crumb Rubber Concrete. *Int. J. Res. Eng.* **2018**, *8*, 86–92.
27. Sofi, A. Effect of Waste Tyre Rubber on Mechanical and Durability Properties of Concrete—A review. *Ain Shams Eng. J.* **2018**, *9*, 2691–2700. [CrossRef]
28. Akinwonmi, A.S.; Seckley, E. Mechanical Strength of Concrete with Crumb and Shredded Tyre As Aggregate Replacement. *Int. J. Eng. Res. Appl.* **2013**, *3*, 1098–1101.
29. Issa, C.A.; Salem, G. Utilization of Recycled Crumb Rubber as Fine Aggregates in Concrete Mix Design. *Constr. Build. Mater.* **2013**, *42*, 48–52. [CrossRef]
30. Irmawaty, R.; Parung, H.; Abdurrahman, M.A.; Nur Qalbi, I. Flexural toughness of concrete with aggregate substitution (steel fiber, crumb rubber and tire chips). *IOP Conf. Ser. Earth Environ. Sci.* **2020**, *419*, 012038. [CrossRef]
31. Chen, Y.; Liang, X.; Liang, L. Application research of steel fiber rubber regenerated concrete engineering. *IOP Conf. Ser. Earth Environ. Sci.* **2020**, *526*, 012087. [CrossRef]
32. Gul, S.; Nasser, S. Concrete Containing Recycled Rubber Steel Fiber. *Procedia Struct. Integr.* **2019**, *18*, 101–107. [CrossRef]
33. Samarakoon, S.M.S.M.K.; Ruben, P.; Pedersen, J.W.; Evangelista, L. Mechanical Performance of Concrete Made of Steel Fibers from Tire Waste. *Case Stud. Constr. Mater.* **2019**, *11*, e00259. [CrossRef]
34. Noaman, A.; Abu Bakar, B.H.; Akil, H.M. Investigation on the mechanical properties of rubberized steel fiber concrete. *Eng. Struct. Technol.* **2017**, *9*, 79–92. [CrossRef]
35. Eisa, A.S.; Elshazli, M.T.; Nawar, M.T. Experimental investigation on the effect of using crumb rubber and steel fibers on the structural behavior of reinforced concrete beams. *Constr. Build. Mater.* **2020**, *252*, 119078. [CrossRef]
36. Mohammad, B.M.; Adamu, M.; Shafiq, N. A review on the effect of crumb rubber on the properties of rubbercrete. *Int. J. Civ. Eng. Technol.* **2017**, *8*, 599–617.
37. Hot Weather Concrete—American Concrete Institute. Available online: https://www.concrete.org/topicsinconcrete/topicdetail/hot%20weather%20concrete?search=hot%20weather%20concrete (accessed on 29 January 2021).
38. ASTM C617/C617M-15. *Standard Practice for Capping Cylindrical Concrete Specimens*; ASTM International: West Conshohocken, PA, USA, 2015. Available online: www.astm.org (accessed on 3 February 2021).
39. ASTM C496/C496M-17. *Standard Test Method for Splitting Tensile Strength of Cylindrical Concrete Specimens*; ASTM International: West Conshohocken, PA, USA, 2017. Available online: www.astm.org (accessed on 3 February 2021).
40. ASTM C39/C39M-20. *Standard Test Method for Compressive Strength of Cylindrical Concrete Specimens*; ASTM International: West Conshohocken, PA, USA, 2020. Available online: www.astm.org (accessed on 3 February 2021).
41. ASTM C78/C78M-18. *Standard Test Method for Flexural Strength of Concrete (Using Simple Beam with Third-Point Loading)*; ASTM International: West Conshohocken, PA, USA, 2018. Available online: www.astm.org (accessed on 3 February 2021).

Article

High-Durability Concrete with Supplementary Cementitious Admixtures Used in Corrosive Environments

Shiming Liu [1], Miaomiao Zhu [1,*], Xinxin Ding [1,2], Zhiguo Ren [3], Shunbo Zhao [1,2,*], Mingshuang Zhao [1] and Juntao Dang [1]

[1] International Joint Research Lab for Eco-Building Materials and Engineering of Henan, School of Civil Engineering and Communications, North China University of Water Resources and Electric Power, Zhengzhou 450045, China; liushm@ncwu.edu.cn (S.L.); dingxinxin@ncwu.edu.cn (X.D.); zhaomingshuang@stu.ncwu.edu.cn (M.Z.); dangjuntao@ncwu.edu.cn (J.D.)

[2] Collaborative Innovation Center of New Urban Building Technology of Henan, North China University of Water Resources and Electric Power, Zhengzhou 450045, China

[3] Zhuzhou China Railway Electrical Materials Co. LTD., Zhuzhou 412001, China; zhaoms@stu.ncwu.edu.cn

* Correspondence: zhumm@stu.ncwu.edu.cn (M.Z.); sbzhao@ncwu.edu.cn (S.Z.); Tel.: +86-371-65665160 (S.Z.)

Citation: Liu, S.; Zhu, M.; Ding, X.; Ren, Z.; Zhao, S.; Zhao, M.; Dang, J. High-Durability Concrete with Supplementary Cementitious Admixtures Used in Corrosive Environments. *Crystals* **2021**, *11*, 196. https://doi.org/10.3390/cryst11020196

Academic editors: Sławomir J. Grabowski and Piotr Smarzewski
Received: 21 January 2021
Accepted: 15 February 2021
Published: 17 February 2021

Publisher's Note: MDPI stays neutral with regard to jurisdictional claims in published maps and institutional affiliations.

Copyright: © 2021 by the authors. Licensee MDPI, Basel, Switzerland. This article is an open access article distributed under the terms and conditions of the Creative Commons Attribution (CC BY) license (https://creativecommons.org/licenses/by/4.0/).

Abstract: Durability of concrete is of great significance to prolong the service life of concrete structures in corrosive environments. Aiming at the economical and environment-friendly production of concrete by comprehensive utilization of the supplementary cementitious materials made of industrial byproducts, the resistances to chloride penetration, sulfate attack, and frost of high-performance concrete were studied in this paper. Fifteen concretes were designed at different water–binder ratio with the changes of contents of fly ash (FA), silica fume (SF), ground granulated blast-furnace slag (GGBS), and admixture of sulfate corrosion-resistance (AS). The compressive strength, the total electric flux of chloride penetrability, the sulfate resistance coefficient, and the indices of freezing and thawing were measured. Results indicate that, depending on the chemical composition, fineness, and pozzolanic activity, the supplementary cementitious admixtures had different effects on the compressive strength and the durability of concrete; despite having a higher fineness and pozzolanic activity, the GGBS gave out a negative effect on concrete due to a similar chemical composition with cement; the SF and FA presented beneficial effects on concrete whether they were used singly with GGBS or jointly with GGBS; the AS improved the compressive strength and the sulfate corrosion resistance of concrete. In general, the grade of durability was positively related to the compressive strength of concrete. Except for the concretes admixed only with GGBS or with GGBS and FA, others had super durability with the compressive strength varying from 70 MPa to 113 MPa. The concretes with water to binder ratio of 0.29 and total binders of 500 kg/m^3 admixed with 7% FA + 8% SF + 8% GGBS or 7% FA + 8% SF + 8% GGBS + (10~12)% AS presented the highest grades of resistances specified in China codes to chloride penetration, sulfate corrosion, and frost, while the compressive strength was about 100 MPa.

Keywords: high-durability concrete; fly ash; silica fume; ground granulated blast-furnace slag; sulfate corrosion inhibitor; compressive strength; chloride penetration; sulfate corrosion; freezing and thawing

1. Introduction

Concrete is the most widely used building material in the world. The degradation of material properties caused by environmental effects will affect the durability of concrete in the process of using concrete structures. The lack of durability of concrete will not only affect the normal use of the project and prematurely terminate the service life of the structure, but also increase the maintenance cost during use and cause serious waste of resources [1]. Therefore, the preparation of high-performance concrete with excellent durability is of great significance to improve the service life of building structures. The additional benefit is to promote the transformation and development of the construction industry.

Normally, the issue of durability of concrete structures is caused by the chloride penetration, sulfate corrosion, or frost damage. The chloride that penetrates into concrete through pores and microcracks results in the corrosion of reinforcement, further damage, and the cracking and spalling of concrete cover, and will take place due to the volume expansion of corroded products. Finally, structural damage happens with the loss of bond between reinforcement and concrete [2,3]. Sulfate corrosion is a complicated process of physical, chemical, and mechanical changes. The sulfate ions enter firstly into the pores of concrete, then the chemical reaction comes up between sulfate ions and cement hydration products to cause the crystallization precipitation of corrosive substances. The expansion of corrosion products leads to the damage of cracking, spalling, and strength loss of concrete [4–6]. For concrete subjected to the action of freezing and thawing, the degradation of concrete happens due to the continuous extending of the pores with repeated freeze–thaw of the free moisture [7].

Different measures have been used to improve the durability of concrete, in which an effective method is the use of supplementary cementitious admixtures in concrete. Fly ash (FA), silica fume (SF), ground granulated blast-furnace slag (GGBS), and other mineral admixtures of industrial byproducts are always applied to concrete by replacing parts of Portland cement. This not only aims at the reduction of energy consumption of cement production, but also comprehensively realizes the industrial sustainability with lower carbon emission. The functions of supplementary cementitious admixtures are the pozzolanic effect and the filling effect. The benefits are the refining of the pore structure, the improving of the interface transition zone, and the increase in density of concrete [8,9]. However, due to the different mineral compositions, chemical activity, and physical properties, as well as forming processes of different cementitious admixtures, the concrete with single mineral admixture has some shortcomings [10]. It is well known that concrete with FA has such advantages as reduction of hydration heat, restraint of alkali–silicate reaction, long-term development of strength, and improvement of durability [6,11]. However, the problems are difficult to avoid, including the easier carbonation, the necessity of long curing duration, the slower development of strength, and the heterogeneity [12,13]. The strength and long-term durability of concrete can be effectively improved by admixing SF, while the risk exists in the increasing of water demand and shrinkage of concrete [3,14–17]. GGBS, with a main component of vitreous, is a kind of auxiliary cementitious material with potential activity. The potential activity can be motivated by $Ca(OH)_2$ produced by cement hydration, and well developed in the later hydration stage. Meanwhile, the supplementary cementitious admixtures have a significant impact on the resistance of concrete to chloride penetration, due to the chloride-binding capacity of these materials. That is, chloride ions can react with C_3A and C_4AF to the stable forms $3CaO·Al_2O_3·CaCl_2·10H_2O$ and $3CaO·Fe_2O_3·Al_2O_3·CaCl_2·10H_2O$ to decrease the free chlorides [18].

In the design of high-performance concrete, it is necessary to take into account the optimization of a variety of performance indicators. This provides opportunities for the hybrid uses of different kinds of supplementary cementitious admixtures. The synergistic action of multi-component cement is significant, if the blender could have a wider particle size distribution, a highly reactive pozzolan that would consume the $Ca(OH)_2$ released by the early hydration of ordinary Portland cement, and a latent pozzolan that would consume the $Ca(OH)_2$ released at a later stage [10]. As reported for a ternary concrete containing 20% FA + 7.5% SF with the water to binder ratio (w/b) of 0.3, the resistance of chloride penetration increased significantly, with the index of total electric flux reduced by 78.0%; however, the compressive strength reduced by 5.7% compared with the reference concrete [14]. As studied by Elahi et al. [11], the total electric flux of chloride penetration test was reduced by 88.7% without affecting the compressive strength for the ternary concrete containing 40% FA + 10% SF with w/b = 0.32, and the total electric flux was decreased by 92.7% with the increase of compressive strength by 5.8% for the ternary concrete containing

7.5% SF + 50% GGBS. The study of Wu et al. [19] indicated that the total electric flux of chloride penetration test for ternary concrete with 10% FA + 20% GGBS, 15% FA + 15% GGBS, and 20% FA + 10% GGBS decreased by 44.8%, 60.5% and 55.8%, respectively, and the relative dynamic elastic modulus after 300 freezing and thawing cycles of frost test increased to 87.3%, 90.1%, and 84.9% from 58.4% of reference concrete. Additionally, the quaternary concrete with three mineral admixtures was studied by Yan et al. [20]. In condition of the w/b = 0.33, the compressive strength and the sulfate resistance of concrete with 10% FA + 10% SF + 5% GGBS was the same as those of reference group; the compressive strength and the corrosion resistance of concrete with 15% FA + 15% SF + 10% GGBS increased by 15.2% and 5.0%, respectively. Therefore, the shortcomings of multi-component concrete can be compensated by the collaborative optimization of several admixtures.

In recent years, the admixtures of sulfate corrosion resistance (AS) have been developed. The respective functions of expansion, excitation, filling, and water-reducing are provided by the components of AS to resist the salt erosion. The AS was prepared by Yang [21] using the anhydrite ($CaSO_4$), the ultra-fine slag powder, and the anhydrous calcium sulfoaluminate (CSA). The experimental results show that the AS can significantly improve the resistances of concrete to chloride penetration, sulfate corrosion, and carbonation. Meanwhile, other research also shows that AS can effectively promote the hydration of cement [22], improve the pore structure of concrete with reduced porosity [23], and increase the density of concrete [24]. This comprehensively improves the performance of concrete.

With the needs of corrosion determination, test methods are specified in related test codes [25,26]. Meanwhile, different methods were innovated to research the corrosion mechanisms of concrete. For instance, the thermal analysis method for attack products and the chemical titrating method for sulfate ion content were developed for the concrete in sulfate corrosion [27], and the non-destructive testing measurements, including ground penetrating radar, were used to detect the damage of corroded concrete [28]. This is an important research aspect dealing with the issues of test methodology.

Based on the above analyses, the research of this paper is concentrated on the preparation of high-durability concrete with supplementary cementitious materials. It aims at the economical and environment-friendly production of concrete by comprehensive utilization of industrial byproducts. Combined with the durability requirement of concrete structures built in severe environments, a high-performance concrete with multiple resistances to chloride penetration, sulfate corrosion, and frost was developed. The supplementary cementitious admixtures of FA, SF, GGBS, and AS were used in binary and ternary combinations. Fifteen mix proportions of concrete were designed and prepared for the experimental study. The results are discussed with the explanation of the admixtures' effects on internal structure of concrete, and the mix proportions of concrete are selected to produce high-durability concrete in engineering applications.

2. Materials and Test Methods

2.1. Materials

The cement was grade P·O 42.5 ordinary Portland cement produced by Xinhua Cement Co. Ltd. Hubei, China. The supplementary cementitious admixtures were the fly ash (FA), silicate fume (SF), ground granulated blast-furnace slag (GGBS), and admixture of sulfate corrosion-resistance (AS). Their chemical compositions are presented in Table 1, of which LOI is the loss on ignition. The physical and mechanical properties of cement are presented in Table 2, which meet the relevant specifications of China codes GB 175 [29]. The physical and mechanical properties of FA, SF, GGBS, and AS are presented in Table 3, which meet the relevant specifications of China codes GB/T 1596, GB/T 27690, GB/T 18046, and JC/T 1011 [30–33].

Table 1. Chemical compositions of cement, fly ash (FA), silica fume (SF), ground granulated blast-furnace slag (GGBS), and admixture of sulfate corrosion-resistance (AS).

Materials	Chemical Compositions (%)									
	SiO_2	Fe_2O_3	Al_2O_3	CaO	MgO	SO_3	f-CaO	K_2O	K_2O	LOI
cement	21.6	3.3	4.9	56.4	3.4	2.2	0.9	0.1	0.1	5.3
FA	55.9	5.9	17.3	6.6	3.8	1.9	0.3	1.9	1.9	2.6
SF	88.3	0.7	0.9	1.2	0.2	0.9	0.3	0.7	0.7	0.8
GGBS	25.9	2.6	8.4	41.4	4.5	0.1	0.1	0.5	0.5	4.0
AS	48.0	2.4	8.6	20.3	1.9	7.4	3.7	0.6	0.6	1.9

Table 2. Physical and mechanical properties of ordinary Portland cement.

Density (kg/m^3)	Fineness (m^2/kg)	Setting Time (min)		Flexural Strength (MPa)		Compressive Strength (MPa)	
		Initial	Final	3d	28d	3d	28d
3133	324	236	308	5.1	7.8	24.9	47.2

Table 3. Physical and mechanical performances of FA, SF, GGBS, and AS.

Materials	Density (kg/m^3)	Fineness (m^2/kg)	Water Demands Ratio(%)	Active Index (%)	
				7d	28d
FA	2342	406	84	—	73.3
SF	2149	—	101.7	97.8	—
GGBS	2998	438.8	—	76	97.6
AS	2703	380	—	95	102

The fine aggregate was river sand with a fineness modulus of 2.86 and an apparent density of 2640 kg/m^3. The coarse aggregate was crushed granite gravel with continuous grading and a maximum particle of 20 mm, and an apparent density of 2730 kg/m^3. As presented in Figure 1, the gradations of fine and coarse aggregates met the specifications of China codes GB/T 14684 and GB/T14685 [34,35]. The polycarboxylate-based superplasticizer (PS) was used with water-reducing rate of 27% and solid content of 23%. The mixing water was tap water.

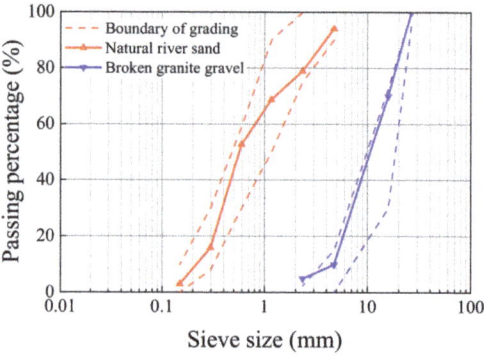

Figure 1. Grading of fine and coarse aggregates.

2.2. Mix Proportion

To study the effects of supplementary cementitious admixtures on the compressive strength and durability of concrete, fifteen mix proportions were designed by the absolute volume method in accordance with China code JGJ 55 [36]. FA and SF were mainly used to adjust the w/b, while GGBS and AS replaced cement to keep the constant of w/b. The content of PS in the percentage of the total mass of cementitious materials was adjusted by trial mixing to keep the slump of the fresh concrete at (70 ± 20) mm. Table 4 presents the confirmed mix proportions of test concretes, of which the first letters of SF, GGBS, FA, and AS are used respectively to identify the materials.

Table 4. Mix proportions of high-performance concrete.

Identifier of Mixture	w/b	Dosage of Raw Materials (kg/m³)							PS (%)
		Cement	FA	SF	GGBS	AS	Sand	Gravel	
S1-1	0.33	425	0	14	0	0	721.0	1227.6	2.6
C1 1	0.33	425	0	0	14	0	674.8	1253.2	1.6
S1G2-1	0.33	385	0	14	40	0	715.1	1217.5	2.4
F1S2-2	0.29	425	35	40	0	0	697.1	1187.0	2.2
F1G2-2	0.29	425	35	0	40	0	652.5	1211.8	1.8
F1S2G2-2	0.29	385	35	40	40	0	691.2	1177.0	2.2
F2S3-3	0.25	425	55	100	0	0	668.3	1138.0	2.2
F2G3-3	0.25	425	55	0	100	0	625.6	1161.8	1.8
F2S3G2-3	0.25	385	55	100	40	0	662.4	1127.9	2.4
F1S2A1-2	0.29	375	35	40	0	50	657.7	1221.5	3.0
F1G2A1-2	0.29	375	35	0	40	50	650.1	1207.4	2.0
F1S2G2A1-2	0.29	335	35	40	40	50	652.1	1211.1	2.8
F1S2A2-2	0.29	365	35	40	0	60	657.2	1220.6	2.8
F1G2A2-2	0.29	365	35	0	40	60	649.6	1206.5	2.2
F1S2G2A2-2	0.29	325	35	40	40	60	651.6	1210.2	2.8

2.3. Test Method

The mixtures of concrete were mixed with a horizontal-shaft mixer, and the specimens were fabricated in molds and compacted on a vibrating table. All specimens were demolded after 24 h and cured in a standard curing room with a temperature of (20 ± 2) °C and relative humidity of 95% for 56 days before testing.

The test for compressive strength was in accordance with China code GB/T 50081 [37]. The specimen was a cube with a dimension of 100 mm. As the cubic compressive strength (f_{cu}) is corresponding to the standard cube with a dimension of 150 mm, the test value was converted by multiplying a coefficient of 0.95. The loading rate was 0.8 MPa/s on the tested cube by a compression-testing machine.

Tests for chloride ion penetration, sulfate corrosion, and rapid freezing and thawing were in accordance with China code GB/T 50082 [25]. The first one is the electric flux test, which is similar to the test method specified in ASTM C1202 [38]. The samples of ϕ100 mm × 50 mm for chloride ion penetration test was cut from the specimen of ϕ100 mm × 200 mm. The test procedure was as follows: (1) The sample was put into a vacuum to be water saturated; (2) the samples were installed into the testing cell, ensuring the reliable sealing of the sample with the testing cell; (3) 0.3 M NaOH solution and 3% NaCl solution were placed in the testing cells on two sides of the sample, respectively, and connected to the positive and negative terminals of the power supply; (4) the DC power was switched on, and the total electric flux passed through the sample was automatically recorded for 6 h. Finally, the measured total electric flux (Q_{100}) was converted to become the value (Q_s) as a standard sample with diameter of 95 mm, that is,

$$Q_s = Q_{100} \times \left(\frac{95}{100}\right)^2 \tag{1}$$

Three groups of cubic specimens with a dimension of 100 mm were used for the sulfate resistance test of the same concrete. One group was used for the reference cured at the standard conditions. Another two were for the test of sulfate resistance as per the following procedure: (1) put the specimens into the oven to be dried at (80 ± 5) °C for 48 h and cool down to room temperature; (2) move the specimens into the sulfate solution with 5% Na_2SO_4 of automatic sulfate dry–wet cycle testing machine. One group went through 120 dry–wet cycles, another went through 150 dry–wet cycles. A cycle included the soaking in solution for 15 h, heating to 80 °C after draining the solution, maintaining at the temperature of 80 ± 5 °C for 5 h, and cooling after drying. Each cycle lasted for about (24 ± 2) h. The corrosion resistance coefficient (K_f) was computed by the compressive strength ratio of the sulfate corroded to reference specimens, as follows:

$$K_f = \frac{f_{cn}}{f_{c0}} \qquad (2)$$

where f_{cn} is the compressive strength of specimens after sulfate corrosion for n dry–wet cycles; f_{c0} is the compressive strength of reference specimens at the same time.

Before the rapid freezing and thawing test, the prismatic specimens of 100 mm × 100 mm × 400 mm were immersed in water with a temperature of (20 ± 2) °C for 4 days, and the initial value of fundamental frequency (f_0) and initial value of weight (W_0) of the specimens were measured. Then, the specimens were put into the freeze–thaw testing machine. Each cycle of freezing and thawing was completed within 5–6 h, in which the melting time was not less than one quarter of the total freeze–thaw time. The fundamental frequency (f_n) and the weight (W_n) of specimens were measured for each of 25 cycles of freezing and thawing. The relative dynamic elastic modulus (P_n) and the weight loss rate (ΔW_n) were calculated as follows:

$$P_n = \frac{f_n^2}{f_0^2} \times 100 \qquad (3)$$

$$\Delta W_n = \frac{W_0 - W_n}{W_0} \times 100 \qquad (4)$$

3. Results and Discussion

3.1. Compressive Strength

Test results of compressive strength are presented in Table 5. Comparing groups of S1-1 with G1-1, F1S2-2 with F1G2-2, and F2S3-3 with F2G3-3, the compressive strength ratios are 1.30, 1.23, and 1.47. This indicates that, due to the higher activity, as presented in Table 3, the SF has a higher strengthening effect than GGBS in the same conditions. As we know, the high activity of SF comes from the amorphous silica with high volcano ash activity that can react with $Ca(OH)_2$, a cement hydration product, to form a calcium–silicate–hydrate (C–S–H) gel [14–17].

As presented in Tables 1 and 3, the chemical composition and pozzolanic activity of GGBS were similar to cement, therefore, the hydration reaction of GGBS and cement was basically equivalent. Theoretically, replacing equal weight of cement by GGBS should not lead to an obvious decrease in the concrete strength. Comparing with group S1-1, the compressive strength of S1G2-1 reduced about 10%. However, this also depends on the joint admixing and contents of FA and SF. Comparing groups F1S2-2 with F1S2G2-2, and F2S3-3 with F2S3G2-3, the compressive strength ratios are 1.00 and 1.24. This indicates that with proper contents of SF and FA, the secondary hydration of SF and FA with $Ca(OH)_2$ released from the hydration of cement could come up to a sufficient status, and the strength of concrete could remain unchanged by replacing 40 kg/m³ cement with GGBS. When the content of SF was much higher than FA, the insufficient secondary hydration of SF and FA took place with a competition for $Ca(OH)_2$. This led to the obviously reduced strength of concrete with the replacing of 40 kg/m³ cement with GGBS. In the last condition of F2S3G2-3, the SF could not release the potential activity during the hardening of concrete.

Table 5. Test results of compressive strength and resistances to chloride penetration and sulfate attack.

Identifier of Mixtures	f_{cu} (MPa)	Q_s (C)	f_{cu} at 120 Times Cycle (MPa)		K_f (%)	f_{cu} at 150 Times Cycle (MPa)		K_f (%)
			Sulfate Attack	Ref.		Sulfate Attack	Ref.	
S1-1	87.2	589.8	90.0	89.5	100.6	88.2	89.1	99.0
G1-1	67.3	1756.3	76.4	81.5	93.7	77.7	86.0	90.3
S1G2-1	79.0	499.9	84.9	81.5	104.2	86.7	83.7	103.6
F1S2-2	89.4	41.8	92.4	85.7	107.8	89.1	82.5	108.0
F1G2-2	72.7	1193.3	82.0	84.4	97.2	85.3	85.9	99.3
F1S2G2-2	89.5	182.4	88.5	87.1	101.6	87.3	88.4	98.8
F2S3-3	113.7	48.3	120.0	110.3	108.8	125.0	108.0	115.7
F2G3-3	77.5	708.3	87.4	89.1	98.1	89.1	91.0	98.0
F2S3G2-3	91.5	164.6	101.4	88.8	114.2	100.6	89.4	112.5
F1S2A1-2	102.8	31.3	107.7	102.0	105.6	103.5	100.6	102.9
F1G2A1-2	70.2	221.0	76.6	72.5	105.7	79.2	77.1	102.7
F1S2G2A1-2	103.9	27.4	114.1	100.0	114.1	121.0	103.2	117.2
F1S2A2-2	101.5	27.3	95.9	99.9	96.0	103.6	96.3	107.6
F1G2A2-2	85.8	184.2	85.9	81.0	106.1	89.0	86.7	102.7
F1S2G2A2-2	105.9	24.9	113.1	102.9	109.9	103.1	98.7	104.5

Meanwhile, with the increasing contents of supplementary cementitious admixtures, the decreased water to binder ratio (w/b) led to an increased strength of concrete [39,40]. As presented in Figure 2, higher compressive strength of concrete was given out by the binary use of FA and SF or the ternary use of SF, FA, and GGBS. This is contributed mainly from the highest volcano ash activity of amorphous silica of SF. Due to the lower activity of FA, the effect of FA is to reduce the strength of concrete.

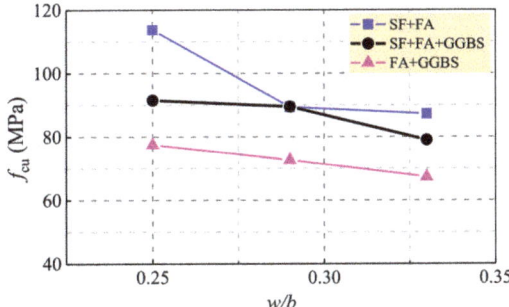

Figure 2. Compressive strength changed with water to binder ratio.

Replacing cement with 50 kg/m^3 of AS, the compressive strength of concrete with SF and FA increased by 15.0%, and that of concrete with SF, FA, and GGBS increased by 16.1%. A slight reduction of 3.5% took place on the compressive strength of concrete with FA and GGBS. Replacing cement with 60 kg/m^3 of AS, the compressive strength of the former two concretes increased by 15.9%, while the latter one increased by 18.0%. This means that, apart from the main chemical compositions of SiO_2 and CaO, the AS with higher content of Al_2O_3 and f-CaO provided good condition of secondary hydration for SF and FA, and played an expansion role in concrete due to the higher content of calcium sulfoaluminate [33,41].

3.2. Chloride Resistance

The test results of total electric flux passed through concrete samples are presented in Table 5. Overall, the total electric flux decreased with the increase of the compressive strength, as presented in Figure 3. This means that the resistance of concrete to chloride penetration was basically positive to the strength of concrete. At the same time, due to the difference of macro- and micro- structures of concrete with pores, the resistance changed within limits, with the changes of supplementary cementitious admixtures. According to the specification of China code JGJ/T 193 for durability assessment of concrete [26], for the concretes in this experiment, the grade of resistance to chloride penetration can be divided as Q-III, Q-IV, and Q-V, respectively, corresponded to the total electric flux within 2000–1000 C, 1000–500 C, and lower than 500 C. The concretes with GGBS (specimen G1-1) or with FA and GGBS (specimen F1G2-2) belong to grade Q-III; the concretes with a little amount of SF with and without GGBS (specimens S1-1 and S1G2-1), and the concretes with FA and a larger amount of GGBS (specimen F2G3-3) belong to grade Q-IV; other concretes belong to grade Q-V.

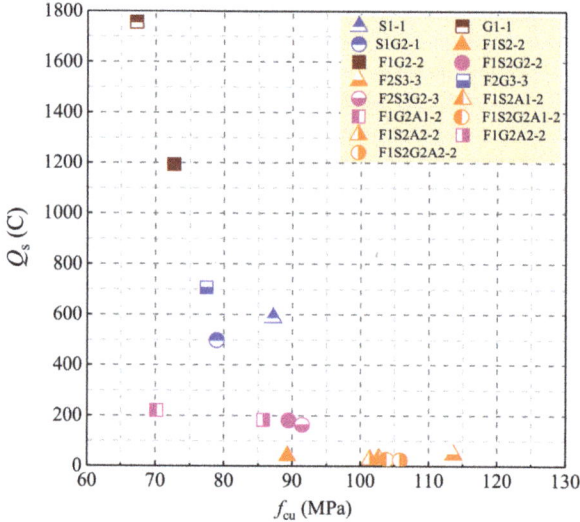

Figure 3. Total electric flux changed with compressive strength.

For the concretes at grades Q-III and Q-IV, due to the similar chemical composition of GGBS with cement, GGBS could not produce the efficiency to refine the pores pattern and distribution of concrete. Due to the filling effect and the secondary hydration of FA with the Ca(OH)$_2$ produced by the hydration of cement, FA refines the pores of concrete to benefit the resistance to chloride penetration [8,13]. Although only 14 kg/m^3 of SF, about 3.2% of total weight of binders, was used, the beneficial effect of SF raised the resistance of concrete to chloride penetration from grade Q-III to Q-IV.

The concretes at grade Q-V can be further divided into two levels of Q-V-1 and Q-V-2 with the total electric flux around 200 C and below 50 C, respectively. The concretes at Q-V-1 level were all prepared with FA and GGBS and admixed with SF (specimens F1S2G2-2 and F2S3G2-3) or AS (specimens F1G2A1-2 and F1G2A2-2). The concretes at Q-V-2 level were the concretes prepared with FA and SF (specimens F1S2-2 and F1S3-3), the concretes with FA, SF, and AS, and the concretes with all kinds of admixtures used in this test. This indicates that when the GGBS was admixed, the SF or AS was needed to get a good resistance to chloride penetration; the concretes with FA and SF could reach an ideal level of resisting chloride penetration. If the GGBS was used, the AS should be admixed additionally, apart from the FA and SF. This could also improve the concrete to become the ideal level. Generally, due to the soluble C–S–H of concrete reduced by the pozzolanic effect of SF, the performance of the interfacial transition zone is improved, and the pore structure of concrete is refined. Meanwhile, the pores of concrete filled by the fine particles of SF makes concrete more compacted to eliminate continuous pores [6,8,42].

3.3. Sulfate Resistance

The test results of sulfate resistance of concrete are presented in Table 5. The sulfate corrosion resistance coefficient is basically positive to the compressive strength of concrete, as presented in Figure 4 for the concretes after 150 dry–wet cycles in a sulfate solution. The same regularity exists for the concretes after 120 dry–wet cycles in a sulfate solution. According to the specification of China codes GB 50082 and JGJ/T 193 [25,26], the grade of resistance to sulfate corrosion is determined by the maximum number of dry–wet cycles when the corrosion resistance coefficient reaches 75%. All concretes in this experiment reached the highest grade, which is larger than KS150.

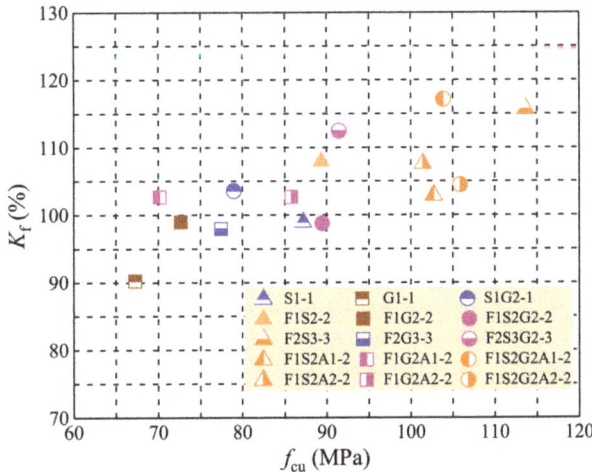

Figure 4. Sulfate corrosion resistance coefficient changed with compressive strength.

In general, the resistance of concrete to sulfate corrosion is similar to the resistance of concrete to chloride penetration, due to both of them relating to the macro- and micro structures of concrete. The grade Q-III and Q-IV concretes, including specimens G1-1, F1G2-2, S1-1, and F2G3-3, except S1G2-1, had a lower corrosion resistance coefficient than 100%. The concretes of grade Q-V had a larger corrosion resistance coefficient than 100%. This indicates that the compressive strength of concrete remained almost constant under the dry–wet recycles in a sulfate solution. In this condition, the filling effects and secondary hydration of supplementary cementitious admixtures improve the macro- and microstructures of concrete [16,18,43]; the concrete eroded layer by layer with sulfate corrosion is insignificant by the inhibited entrance of outside sulfate ions into concrete. As presented in Table 5, compared to the concrete tested at the curing age of 56 days, the compressive strength of reference concretes with GGBS and FA (specimens G1-1, F1G2-2, and F2G3-3) increased by 17.4%–27.8% at the curing age of 150 dry–wet cycles. Part of the increase of compressive strength comes directly from the strength development with the prolongation of curing age, due to the long-term increased strength of cement and the secondary hydration with pozzolanic activity of admixtures [44,45]. Others may come from the beneficial effects of filling original pores and increasing density of concrete, due the expansion of substances such as ettringite and gypsum produced by the reaction of sulfate with soluble C–S–H and layered $Ca(OH)_2$ [18,46,47].

By admixing the AS, the concretes appeared to have superior ability to resist the sulfate corrosion. This meets one of the expectations of this study, to produce a high resistance of concrete to sulfate corrosion.

3.4. Frost Resistance

The frost damage of concrete is induced by the freezing and thawing cyclic action of water in pores and defects of concrete. This appears as a looseness and peeling-off of surface materials and a frost heaving damage to pores and defects of concrete. Therefore, in addition to surface freezing and thawing of concrete, the decisive effect is the infiltration of external water into the interior of the concrete [48,49]. Based on the test results of compressive strength, chloride penetration, and sulfate corrosion, five groups of concrete without SF, including G1-1, F1G2-2, F2G3-3, F1G2A1-2, and F1G2A2-2, were selected to undertake the frost test.

Figure 5 presents the test results of the relative dynamic elastic modulus and the weight loss rate of concrete. After 300 cycles of freezing and thawing, all concretes had a decrease within 5% of relative dynamic elastic modulus, and an increase within 0.4% of weight loss rate. After 300 cycles of freezing and thawing, only the concrete with GGBS (specimen G1-1) had a rapid decrease in the relative dynamic elastic modulus and a sharp increase in the weight loss rate. This agrees with the resistances of concrete to chloride penetration and sulfate corrosion. By using the FA (specimens F1G2-2 and F2G3-3), the beneficial effect of FA on the refinement of pores of concrete recovered the negative effect of GGBS. Admixing the AS accompanied with the FA and GGBS (specimens F1G2A1-2 and F1G2A2-2), no obvious effect appeared on the frost resistance of concrete.

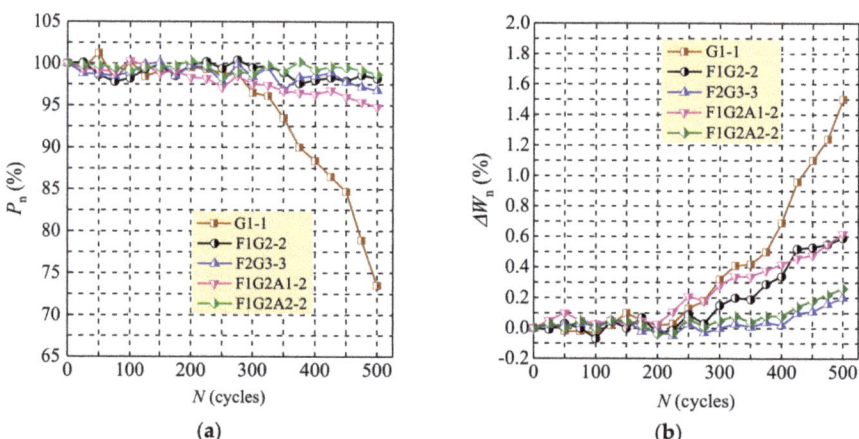

Figure 5. Changes of frost indexes of concrete: (a) relative dynamic elastic modulus; (b) weight loss.

The weight loss of specimens is mainly manifested in the damage of cementitious materials and aggregates on the surface of concrete. As per the surface morphology of specimens after 500 cycles of freezing and thawing presented in Figure 6, an obvious spalling phenomenon appeared on the surface of specimen G1-1, and some paste shedding phenomenon can be seen on surfaces of the other specimens. As seen from Figure 7, the explosion phenomena around the aggregates took place on the surface of specimen F1G2A1-2.

Figure 8 presents the changes of the relative dynamic elastic modulus and the weight loss rate with the compressive strength of concrete. High relative dynamic elastic modulus was kept with less weight loss rate for the concrete with higher compressive strength. This indicates that the high resistance of concrete to freezing and thawing was positive to the compressive strength of concrete. For concrete with at least binary cementitious admixtures including FA, SF, GGBS, and AS, the frost resistance can be comprehensively reflected by the compressive strength.

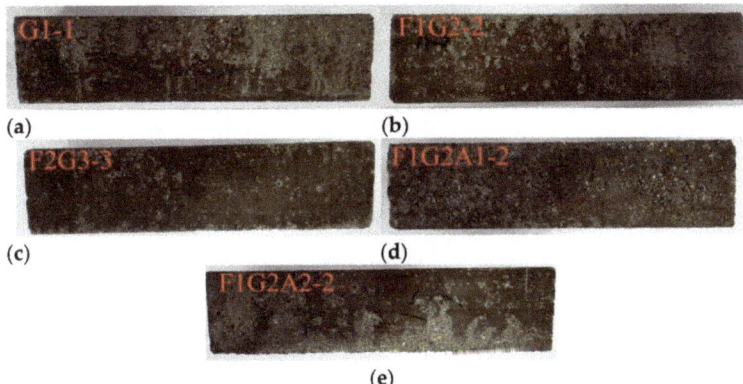

Figure 6. Appearance of specimens after 500 cycles of freezing–thawing: (**a**) obvious spalling on surface of G1-1; (**b**) paste shedding on surface of F1G2-2; (**c**) paste shedding on surface of F2G3-3; (**d**) paste shedding on surface of F1G2A1-2; (**e**) paste shedding on surface of F1G2A2-2.

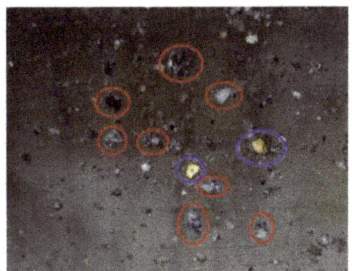

Figure 7. Typical photo for damage of aggregate in specimen F1G2A1-2 (red circle: coarse aggregate; blue circle: fine aggregate).

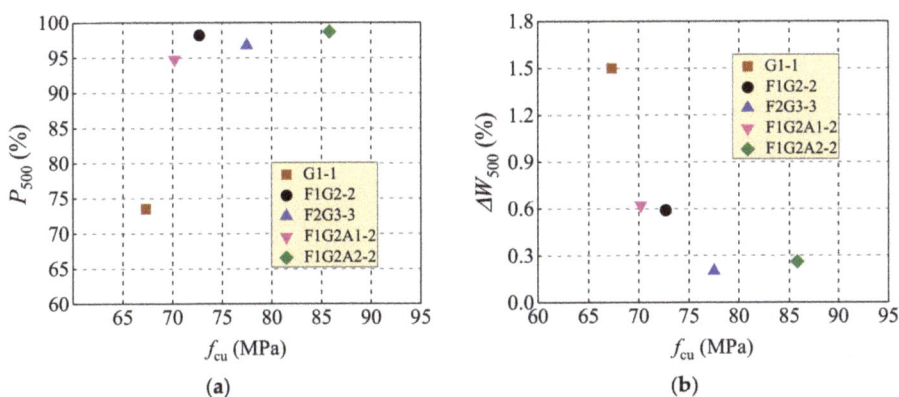

Figure 8. Frost resistance indexes of 500 cycles changed with compressive strength: (**a**) relative dynamic elastic modulus; (**b**) weight loss.

According to the specifications of China codes GB 50082 and JGJ/T 193 [25,26], the grade of the resistance of concrete to freezing and thawing is determined by the maximum number of freezing and thawing cycles, which corresponds to a relative dynamic elastic modulus no less than 60% and a weight loss rate of no more than 5%. Therefore, all concretes in this study are at the highest grade over F400.

4. Conclusions

Based on the experimental study of this paper, conclusions can be made as follows:

1. Due to the difference of chemical composition, fineness, and pozzolanic activity, the supplementary cementitious admixtures present different effects on the strength and durability of concrete. In general, SF and FA benefit the compressive strength and durability of concrete, and the negative effect of GGBS can be overcome with the hybrid use of SF, FA, and AS. AS benefits the compressive strength and the resistances of concrete to chloride penetration and sulfate corrosion, while showing no obvious effect on the resistance of concrete to frost.
2. With the different combinations of the supplementary cementitious admixtures, the concrete can be produced with the compressive strength ranging from 70 MPa to 110 MPa. With the binary use of FA and SF, FA and GGBS, or the ternary use of SF, FA, and GGBS, the compressive strength of concrete increases with the reduction of water to binder ratio. With the admixing of proper content of 7% FA and 8% SF, the concrete can be prepared with compressive strength of about 100 MPa, in which the cement can be replaced by equal weight of 8% GGBS and 10~12% AS.
3. Except for the concrete only mixed with GGBS or with GGBS and FA, other concrete can be produced with ideal resistance to chloride penetration. With proper content of FA and SF, the concretes present a superior resistance to chloride penetration of the highest grade over Q-V with the index of total electric flux lower than 500 C specified in the China code, in which the cement can be replaced by equal weight of 8% GGBS and 10~12% AS.
4. Except for the concrete only mixed with GGBS or with GGBS and FA, other concretes reaching the highest grade over KS150 withstood the 150 dry–wet cycles of sulfate corrosion specified in the China code. The concretes with the AS appeared to have a superior ability to resist the sulfate corrosion.
5. All concretes are at the highest grade over F400 for the resistance to frost, with the relative dynamic elastic modulus no less than 60% and the weight loss rate no larger than 5% specified in the China code, although the concrete admixed only with GGBS appears to have a relatively low resistance.
6. The durability is positively related to the compressive strength of concrete with hybrid admixtures. With water to binder ratio of 0.29 and total binders of 500 kg/m^3, the highest grades of resistances specified in the China codes to chloride penetration, sulfate corrosion, and frost can be reached for the concrete admixed with 7% FA + 8% SF + 8% GGBS or 7% FA + 8% SF + 8% GGBS + (10~12)% AS, while the compressive strength was about 100 MPa.

Author Contributions: Methodology: S.Z. and Z.R.; investigation and data curation: M.Z. (Miaomiao Zhu), X.D., and M.Z. (Mingshuang Zhao); writing: original draft preparation: M.Z. (Mingshuang Zhao), S.L., and J.D.; writing: review and funding acquisition: S.Z. and M.Z. (Miaomiao Zhu). All authors have read and agreed to the published version of the manuscript.

Funding: This research was funded by Key Scientific and Technological Research Project of University in Henan, China (19A560001, 20A560015), Postgraduate Innovation Project of NCWU (YK2019-29), State Key Research and Development Plan, China (2017YFC0703904), Innovative Sci-Tech Team of Eco-building Material, and Structural Engineering of Henan, China (YKRZ-6-066).

Institutional Review Board Statement: Not applicable.

Informed Consent Statement: Not applicable.

Data Availability Statement: The data presented in this study are contained in this article.

Conflicts of Interest: The authors declare no conflict of interest.

References

1. China Academy of Railway Sciences. *Code for Durability Design on Concrete Structure of Railway*; TB 10005-2010; China Railway Publishing House: Beijing, China, 2011.
2. Qu, F.L.; Liu, G.R.; Zhao, S.B.; Yan, L.Y. Mechanical properties of non-uniformly corroded steel bars based on length characterization. *J. Build. Mater.* **2016**, *19*, 566–570.
3. Kiesse, T.S.; Bonnet, S.; Amiri, Q.; Ventura, A. Analysis of corrosion risk due to chloride diffusion for concrete structures in marine environment. *Mar. Struct.* **2020**, *73*, 102804. [CrossRef]
4. Wolfram, M.; Robin, E.; Beddoe, D.H. Sulfate attack expansion mechanisms. *Cem. Concr. Res.* **2013**, *52*, 208–215.
5. Gao, R.D.; Zhao, S.B.; Li, Q.B.; Chen, J.H. Experimental study of the deterioration mechanism of concrete under sulfate attack in wet-dry cycles. *China Civ. Eng. J.* **2010**, *43*, 48–54.
6. Wang, D.Z.; Zhou, X.M.; Meng, Y.F.; Chen, Z. Durability of concrete containing fly ash and silica fume against combined freezing-thawing and sulfate attack. *Constr. Build. Mater.* **2017**, *147*, 398–406. [CrossRef]
7. Li, Y.; Zhai, Y.; Liang, W.B.; Li, Y.B.; Dong, Q.; Meng, F.D. Dynamic mechanical properties and visco-elastic damage constitutive model of freeze–thawed concrete. *Materials* **2020**, *13*, 4056. [CrossRef] [PubMed]
8. Adorján, B. Long term durability performance and mechanical properties of highperformance concretes with combined use of supplementary cementing materials. *Constr. Build. Mater.* **2016**, *112*, 307–324.
9. Duan, P.; Shui, Z.H.; Chen, W.; Shen, C.H. Efficiency of mineral admixtures in concrete: Microstructure, compressive strength and stability of hydrate phases. *Appl. Clay Sci.* **2013**, *83–84*, 115–121. [CrossRef]
10. Nehdi, M.L.; Sumner, J. Optimization of ternary cementitious mortar blends using factorial experimental plans. *Mater. Struct.* **2002**, *35*, 495–503. [CrossRef]
11. Elahi, A.; Basheer, P.A.M.; Nanukuttan, S.V.; Khan, Q.U.Z. Mechanical and durability properties of high performance concretes containing supplementary cementitious materials. *Constr. Build. Mater.* **2010**, *24*, 292–299. [CrossRef]
12. KojiKinomura, T. Enhanced hydration model of fly ash in blended cement and application of extensive modeling for continuous hydration to pozzolanic micro-pore structures. *Cem. Concr. Compos.* **2020**, *114*, 103733. [CrossRef]
13. Yan, P.Y. Mechanism of fly ash in hydration of composite cementitious materials. *J. China Ceram. Soc.* **2007**, *35(s1)*, 167–171.
14. Watcharapong, W.; Pailyn, T.; Athipong, N.; Arnon, C. Compressive strength and chloride resistance of self-compacting concrete containing high level fly ash and silica fume. *Mater. Des.* **2014**, *64*, 261–269.
15. Al-Dulaijan, S.U.; Maslehuddin, M.; Al-Zahrani, M.M.; Sharif, A.M.; Shameem, M.; Ibrahim, M. Sulfate resistance of plain and blended cements exposed tovarying concentrations of sodium sulfate. *Cem. Concr. Compos.* **2003**, *25*, 429–437. [CrossRef]
16. Erhan, G.; Mehmet, G.; Erdoğan, Ö. Strength and drying shrinkage properties of self-compacting concretes incorporating multi-system blended mineral admixtures. *Constr. Build. Mater.* **2010**, *24*, 1878–1887.
17. Mohammed, S.M.; Mohamed, A.I.; El-Gamal, S.; Fitriani, H. Performances evaluation of binary concrete designed with silica fumeand metakaolin. *Constr. Build. Mater.* **2018**, *166*, 400–412.
18. Liu, R.G.; Ding, S.D.; Yan, P.Y. Influence of hydration environment on the characteristics of ground granulated blast furnace slag hydration products. *B. China Ceram. Soc.* **2015**, *34*, 1594–1599.
19. Wu, M.Y.; Wang, Q.C.; Zhang, K.; Dong, Y.T.; Wang, X.L. Influence of mineral admixtures on impermeability and frost resistance of high performance concrete. *Highway* **2016**, *61*, 239–242.
20. Yan, X.L.; Gong, T.Z.; Chang, J.F.; Zheng, Y. Experimental research about sulfate resistance performance of high performance concrete. *J. Electr. Power* **2017**, *32*, 257–264.
21. Yang, B.J. Experimental Study on the Sulfate Corrosion-Resistance Admixturefor Concrete. Master's Thesis, Qingdao Technological University, Qingdao, China, 2011.
22. Xu, Y.L.; Zhang, P.Y. Effect and mechanism of sulfate corrosion-resistance admixture on concrete performance. *B. China Ceram. Soc.* **2016**, *35*, 2304–2308.
23. Chang, H.L.; Mu, S.; Liu, J.Z. Influence of anticorrosion agent and curing regimes on pore structure feature and moisture loss of concrete. *J. Southeast U Nat. Sci. Ed.* **2015**, *45*, 1155–1162.
24. Liang, R. Compounding of High Durability Concrete and Application in Strong Corrosive Environment. Master's Thesis, Harbin Institute of Technology, Harbin, China, 2018.
25. Ministry of Housing and Urban-Rural Construction of the People's Republic of China. *Standard for Test Methods of Long-Term Performance and Durability of Ordinary Concrete*; GB 50082-2009; China Building Industry Press: Beijing, China, 2009.
26. Ministry of Housing and Urban-Rural Construction of the People's Republic of China. *Standard for Inspection and Assessment of Concrete Durability*; JGJ/T 193-2009; China Building Industry Press: Beijing, China, 2009.
27. Gao, R.D.; Li, Q.B.; Zhao, S.B.; Yang, X.M. Deterioration mechanisms of sulfate attack on concrete under alternate action. *J. Wuhan Univ. Technol. Sci. Ed.* **2010**, *25*, 355–359. [CrossRef]
28. Sossa, V.; Pérez-Gracia, V.; González-Drigo, R.; Rasol, M.A. Lab non-destructive test to analyze the effect of corrosion on ground penetrating radar scans. *Remote Sens.* **2019**, *11*, 2814. [CrossRef]
29. General Administration of Quality Supervision, Inspection and Quarantine of the People's Republic of China. *Common Portland Cement*; GB 175-2007; China Standard Press: Beijing, China, 2007.
30. General Administration of Quality Supervision, Inspection and Quarantine of the People's Republic of China. *Fly Ash Used for Cement and Concrete*; GB/T 1596-2017; China Standard Press: Beijing, China, 2017.

31. General Administration of Quality Supervision, Inspection and Quarantine of the People's Republic of China. *Silica Fume Cement Mortar and Concrete*; GB/T 27690-2011; China Standard Press: Beijing, China, 2011.
32. General Administration of Quality Supervision, Inspection and Quarantine of the People's Republic of China. *Ground Granlated Blast Furnace Slag Used for Cement, Mortar and Concrete*; GB/T 18046-2017; China Standard Press: Beijing, China, 2008.
33. National Development and Reform Commission of the People's Republic of China. *Sulfate Corrosion-resistance Admixtures for Concrete*; JC/T 1011-2006; China Building Materials Industry Press: Beijing, China, 2006.
34. General Administration of Quality Supervision, Inspection and Quarantine of the People's Republic of China. *Sand for Construction*; GB/T 14684-2011; China Standard Press: Beijing, China, 2011.
35. General Administration of Quality Supervision, Inspection and Quarantine of the People's Republic of China. *Pebble and Crushed Stone for Construction*; GB/T 14685-2011; China Standard Press: Beijing, China, 2011.
36. Ministry of Housing and Urban Rural Development of the People's Republic of China. *Specification for Mix Proportion Design of Ordinary Concrete*; JGJ 55-2011; China Building Industry Press: Beijing, China, 2011.
37. Ministry of Construction of the People's Republic of China. *Standard for Test Methods on Mechanical Properties of Ordinary Concrete*; GB/T 50081-2002; China Building Industry Press: Beijing, China, 2002.
38. American Society for Testing and Materials. *Standard Test Method for Electrical Indication of Concrete's Ability to Resist Chloride Ion Penetration*; ASTM C1202-19; ASTM International: West Conshohocken, PA, USA, 2019.
39. Zhao, M.L.; Ding, X.X.; Li, J.; Law, D. Numerical analysis of mix proportion of self-compacting concrete compared to ordinary concrete. *Key Eng. Mater.* **2018**, *789*, 69–75. [CrossRef]
40. Ding, X.X.; Zhao, M.L.; Zhou, S.Y.; Fu, Y.; Li, C.Y. Statistical analysis and preliminary study on the mix proportion design of self-compacting steel fiber reinforced concrete. *Materials* **2019**, *12*, 637. [CrossRef] [PubMed]
41. Li, C.Y.; Shang, P.R.; Li, F.L.; Feng, M.; Zhao, S.B. Shrinkage and mechanical properties of self-compacting SFRC with calcium sulfoaluminate expansive agent. *Materials* **2020**, *13*, 588. [CrossRef]
42. Uysal, M.; Akyuncu, V. Durability performance of concrete incorporating Class F and Class C fly ashes. *Constr. Build. Mater.* **2012**, *34*, 170–178. [CrossRef]
43. Zhao, S.B.; Li, C.Y.; Zhao, M.S.; Zhang, X.Y. Experimental study on autogenous and drying shrinkage of lightweight-aggregate concrete reinforced by steel fibers. *Adv. Mater. Sci. Eng.* **2016**, *2016*, 2589383. [CrossRef]
44. Ding, X.X.; Li, C.Y.; Xu, Y.Y.; Li, F.L.; Zhao, S.B. Experimental study on long-term compressive strength of concrete with manufactured sand. *Constr. Build. Mater.* **2016**, *108*, 67–73. [CrossRef]
45. Li, C.Y.; Wang, F.; Deng, X.S.; Li, Y.Z.; Zhao, S.B. Testing and prediction of the strength development of recycled-aggregate concrete with large particle natural aggregate. *Materials* **2019**, *12*, 1891. [CrossRef] [PubMed]
46. Gao, R.D.; Li, Q.B.; Zhao, S.B. Concrete deterioration mechanisms under combined sulfate attack and flexural loading. *J. Mater. Civ. Eng.* **2013**, *25*, 39–44. [CrossRef]
47. Gao, R.D.; Zhao, S.B.; Li, Q.B. Deterioration mechanisms of sulfate attack on concrete under the action of compound factors. *J. Build. Mater.* **2009**, *12*, 41–46.
48. Mu, R. Durability and Service Life Prediction of Concrete Subjected to the Combined Action of Freezing-Thawing, Sustained External Flexural Stress and Salt Solution. Master's Thesis, Southeast University, Nanjing, China, 2000.
49. Zhao, M.S.; Zhang, X.Y.; Song, W.H.; Li, C.Y.; Zhao, S.B. Development of steel fiber-reinforced expanded-shale lightweight concrete with high freeze-thaw resistance. *Adv. Mater. Sci. Eng.* **2018**, *2018*, 9573849. [CrossRef]

Article

Tensile Behavior of Self-Compacting Steel Fiber Reinforced Concrete Evaluated by Different Test Methods

Xinxin Ding [1,*], Changyong Li [1], Minglei Zhao [2], Jie Li [2,*], Haibin Geng [1] and Lei Lian [3]

[1] International Joint Research Lab for Eco-building Materials and Engineering of Henan, School of Civil Engineering and Communications, North China University of Water Resources and Electric Power, Zhengzhou 450045, China; lichang@ncwu.edu.cn (C.L.); hbgeng@stu.ncwu.edu.cn (H.G.)

[2] School of Engineering, RMIT University, Melbourne, VIC 3001, Australia; s3339909@student.rmit.edu.au

[3] Changjiang Yichang Waterway Bureau, No. 273 Dongshan Avenue, Yichang 443500, China; lian_lei0125@163.com

* Correspondence: dingxinxin@ncwu.edu.cn (X.D.); jie.li@rmit.edu.au (J.L.)

Abstract: Due to the mechanical properties related closely to the distribution of steel fibers in concrete matrix, the assessment of tensile strength of self-compacting steel fiber reinforced concrete (SFRC) is significant for the engineering application. In this paper, seven groups of self-compacting SFRC were produced with the mix proportion designed by using the steel fiber-aggregates skeleton packing test method. The hooked-end steel fibers with length of 25.1 mm, 29.8 mm and 34.8 mm were used, and the volume fraction varied from 0.4% to 1.4%. The axial tensile test of notched sectional prism specimen and the splitting tensile test of cube specimen were carried out. Results show that the axial tensile strength was higher than the splitting tensile strength for the same self-compacting SFRC, the axial tensile work and toughness was not related to the length of steel fiber. Finally, the equations for the prediction of tensile strength of self-compacting SFRC are proposed considering the fiber distribution and fiber factor, and the adaptability of splitting tensile test for self-compacting SFRC is discussed.

Keywords: self-compacting SFRC; axial tensile strength; splitting tensile strength; notched section; axial tensile toughness; prediction

1. Introduction

The enhancement of steel fiber on the mechanical properties of concrete is affected by the mix proportion, fiber geometric characteristics and distribution pattern [1–3]. This directly affects the loading behaviors of reinforced concrete components including beams and slabs with the presence of steel fibers [4–8]. Compared with vibrated steel fiber reinforced concrete (SFRC), self-compacting SFRC is made with more contents of cementitious materials and fine aggregate, and smaller particle size of coarse aggregate [9–11]. This leads to the different fiber distribution and aggregate skeleton of self-compacting SFRC from those of vibrated SFRC, which further leads to the different tensile properties of self-compacting SFRC [1,12–14].

Splitting tensile test and axial tensile test are common test methods for the tensile strength of concrete. Meanwhile, the axial tensile stress-strain curve and the tensile toughness of concrete can be gotten from the axial tensile test [15]. However, restricted by the stiffness and clamping conditions of testing machine, the size and shape of axial tensile specimens are various, and the failure interface may appear in the non-tested cross-section [16–18]. Splitting tensile test is often used in research and practical engineering due to its easy operation, simple test process and low requirement for stiffness of testing machine [19]. This method is built upon the assumption that the uniform distribution of horizontal stress exists in the splitting section, except part of section closed to the splitting strips on top and bottom surfaces of the specimen. For conventional concrete close to

brittle material, this assumption is adaptable to get a relatively accurate result of tensile strength, which is always slightly higher than the axial tensile strength [20]. However, it is no longer fully applicable for SFRC due to the steel fibers involved in tensile work. With the development of a cracking section of splitting tensile specimen, the stress distribution becomes complex with the restrained local area of steel fiber near the splitting surface. The experimental studies on vibrated SFRC [17,21] indicated that, the enhancing coefficient of steel fiber on the axial tensile strength is less than that on the splitting tensile strength, and the ratio of axial tensile strength to splitting tensile strength decreases with the increase of fiber factor and the concrete compressive strength. The ductile behavior of the SFRC leads to a large compressive zone under the loading strip in the splitting tensile test, which makes the load distributed unevenly in the load direction [21].

At present, the splitting tensile test and axial tensile test are commonly used to determine the tensile strength of self-compacting SFRC. Ghanbarpour [13] reported that the ratio of splitting tensile strength to compressive strength for self-compacting SFRC is lower than that of vibrated SFRC, due to the characteristic of finer raw materials in self-compacting SFRC. Akcay and Tasdemir [1] presented that both the splitting tensile strength and flexural strength of self-compacting SFRC increased linearly with the volume fraction of steel fiber, while the increasing rate of flexural strength was higher due to the trends of steel fibers oriented along the bending direction; the effect of tensile strength of steel fiber on the splitting tensile strength was not obvious, but on the flexural strength was significant. Khala and Nazari [22] reported that the splitting tensile strength of self-compacting SFRC was improved with the randomly oriented steel fibers up to volume fraction of 1.2%. The study of You et al. [23] on the axial tensile strength of self-compacting SFRC by using notched prism specimen showed that, the axial tensile strength was enhanced obviously with longer fiber, and the growth tended to be stable or even decreased with the increasing volume fraction of steel fiber. Clifford et al. [24] reported that compared to the self-compacting SFRC with single hooked-end steel fiber and microfiber, the self-compacting SFRC with double hooked-end steel fiber and microfiber had higher axial tensile strength, lower fracture energy, smaller cracking strain and ultimate strain. Cunha et al. [25] studied the axial tensile properties of self-compacting SFRC by using notched cylinder specimen. Results indicated that the tensile properties after-cracking correlated positively with the effective number of bridged fibers in the failure section, pseudo strain hardening was observed in the descending portion of axial tensile stress-strain curve for self-compacting SFRC due to the pulling out of most bridged fibers, the residual tensile stress reduced significantly with the breakage of most bridged fibers. The study of Liao et al. [26] on the axial tensile properties of self-compacting SFRC by using large-end specimens presented that, with the hooked-end steel fiber in volume fraction from 0.38% to 1.96%, pseudo strain hardening appeared on the tensile stress-strain curve, and the failure was accompanied by multi-cracks.

Based on the above analyses, the comparative research is rare on splitting tensile strength and axial tensile strength of self-compacting SFRC, the axial tensile stress-strain curves of self-compacting SFRC are also short of experimental data. These efforts need to be enriched to facilitate the engineering application of self-compacting SFRC. In this paper, the self-compacting SFRC was produced based on the steel fiber-aggregates skeleton packing test method, in which the amount of cementitious material and reasonable sand ratio are related to the fiber factor. The axial tensile test was carried out using notched prism specimen to control the cracking section and to eliminate the wall effect of steel fiber on the tensile properties of self-compacting SFRC [16,18]. The splitting tensile test was carried out using cube specimen. With the comparison of test results, the influence of test methods on the tensile strength of self-compacting SFRC is studied. Finally, the practicability of the splitting tensile test method used for self-compacting SFRC is evaluated.

2. Materials and Methods

2.1. Raw Materials

All of the raw materials are the same as the former research [27]. The cement was grade P.O 42.5 ordinary silicate cement with density of 3085 kg/m^3, compressive strength at 28 d of 54.7 MPa and flexural strength at 28 d of 9.43 MPa. The class-II fly ash was used as the mineral admixture with density of 2349 kg/m^3 and activity index of 80.9%.

The coarse aggregate was crushed limestone in continuous grading with the particle sizes of 5–16 mm, the apparent density of 2736 kg/m^3, the bulk density of 1529 kg/m^3, and the crush index of 12.2%.

The fine aggregate was manufactured sand with the fineness modulus of 2.73, the apparent density of 2740 kg/m^3, the bulk density of 1620 kg/m^3, and the stone-powder content of 7.3%.

The water reducer was high-performance polycarboxylic acid type with measured water-reducing rate of 30% and solid content of 35%. The mix water was tap-water of Zhengzhou, China.

The hooked-end steel fibers with three different lengths were used, and identified as HFa, HFb and HFc, respectively. All fibers had a diameters d_f of 0.5 mm and tensile strength of 1150 MPa. The measured length l_f was 25.1 mm, 29.8 mm and 34.8 mm, and the fiber aspect ratio l_f/d_f was 50, 60 and 70, successively. The numbers of steel fibers per kilogram were 25,861, 21,637 and 18,546.

2.2. Preparation of Self-Compacting SFRC

The mix proportion of self-compacting SFRC was designed by using the steel fiber-aggregates skeleton packing test method [28,29]. Six groups of self-compacting SFRC were designed with the volume fraction of steel fiber v_f F 0.4%, 0.8%, 1.2%, 1.4%, 1.2% and 1.2%, and successively identified as HFb04, HFb08, HFb12, HFb14, HFa12 and HFc12. The length and the volume fraction of steel fiber were considered as main factors. The water to binder ratio w/b is constant of 0.31, the content of fly ash was 30% of the total mass of binders, the sand ratios changed rationally with the fiber factor λ_f. The fiber factor is the product of the volume fraction and the aspect ratio of steel fiber. That is, $\lambda_f = v_f \cdot l_f/d_f$. Detailed mix proportion of which, are presented in Table 1. One group of self-compacting concrete (SCC) without steel fiber was used as the reference concrete.

Table 1. Detailed mix proportion of self-compacting (steel fiber reinforced concrete) SFRC and reference (self-compacting concrete) SCC.

Concrete		SCC	HFb04	HFb08	HFb12	HFb14	HFa12	HFc12
w/b		0.31	0.31	0.31	0.31	0.31	0.31	0.31
Sand ratio β_s (%)		50	52	54	56	57	55	57
Fly ash content (%)		30	30	30	30	30	30	30
Steel fiber	Identifer	-	HFb	HFb	HFb	HFb	HFa	HFc
	v_f (%)	0	0.4	0.8	1.2	1.4	1.2	1.2
Water (kg/m^3)		192	201	210	219	223	214	223
Cement (kg/m^3)		433	454	474	494	504	484	504
Fly ash (kg/m^3)		186	194	203	214	216	207	216
Crushed stone (kg/m^3)		751	675	601	527	491	553	502
Sand (kg/m^3)		751	763	774	783	788	784	782
Water-reducer (kg/m^3)		5.57	5.51	5.42	5.30	5.40	5.19	5.40
Steel fiber (kg/m^3)		0	31.4	62.8	94.2	109.9	94.2	94.2

The mixture was mixed by a horizontal shaft forced-mixer. The cement, fly ash, manufactured sand and one-third of the water were mixed firstly for 1 min, and then the residual water mixed with water-reducer was added to mix for 3 min. After that, the crushed limestone and the steel fiber were added successively and mixed at least for 4 min.

The workability of fresh concrete was measured in accordance with the specification of China code JGJ/T 283 [29]. The slump flow of SCC, HFb04, HFb08, HFb12, HFb14, and HFa12 were ranged from 630 mm to 685 mm. The slump flow time T_{500} ranged from 4.33 s to 6.28 s, which is the time from the beginning of lifting the slump cone to the slump flow of fresh concrete reaching at 500 mm. It means that the six groups of fresh concretes are fit for constructing of the conventional reinforced concrete structures. HFc12 has slump flow of 560 mm and the flow time T_{500} of 6.71 s, which is fit for constructing of the concrete structures without reinforcement or with a minor amount of reinforcement.

2.3. Test Method

The splitting tensile strength of self-compacting SFRC was in accordance with the specification of China code GB50081 [19]. Three cube specimens with dimension of 150 mm were used as a group in the splitting test. A total of 21 cube specimens were cured in standard curing room for 90 days. Figure 1 presents the specimen with splitting strips. The test was carried out by a servo hydraulic universal testing machine made by SANS Co. Ltd. with maximum load of 600 kN. The loading direction was opposite to the gravity direction, and the loading speed was 2 kN/s. The test was finished with the splitting tensile fracture of the specimen.

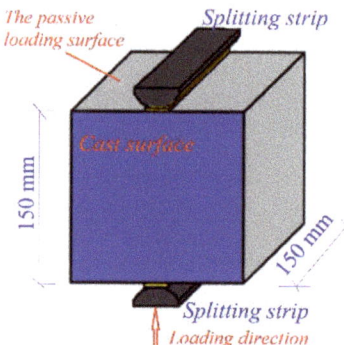

Figure 1. Specimens of splitting tensile test.

The axial tensile test was in accordance with the specification of China code CECS13 [15]. Three prism specimens as a group were used in the axial tensile test. As presented in Figure 2, the specimens were prisms with the centrally embedded tensile bars at the ends. The dimension of the prism was 150 mm × 150 mm × 550 mm. A total of 21 prism specimens were cured at standard curing room for 90 days. Referenced to the treatment method specified in RILEM TC-162 [30], the notch around cross-section of prism was cute before testing to eliminate the "wall effect" of steel fibers [15,31]. The notch was 25 mm deep and 10 mm wide. The notch section was square with side length of 100 mm.

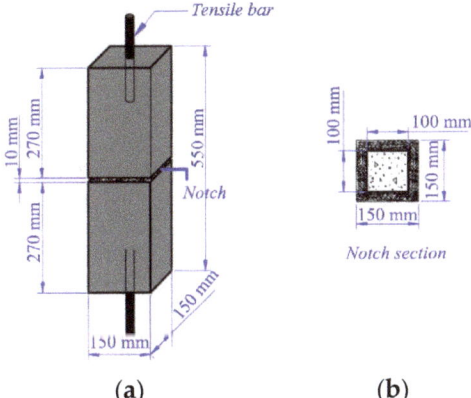

Figure 2. Specimens of axial tensile test: (**a**) notched specimen with tensile bars; (**b**) notch section of specimen.

The axial tensile test was carried out by a servo hydraulic universal testing machine with maximum load of 600 kN. As presented in Figure 3, the steel bars embedded in the ends of specimens were linked with the spherical universal hinge device of testing machine to eliminate the eccentricity effect [15,32]. Two LVDTs with measure scope of 5 mm, identified as LVDT1, and two LVDTs with measure scope of 20 mm, identified as LVDT2, were used to ensure the measuring precision and range simultaneously. After the specimen was fixed on the testing machine, four LVDTs were installed near the corner of specimen by using the fixing rings. The load and four displacements were collected synchronously and independently during the loading process. The average of four displacements was used as the measured deformation of the specimen. The tensile stress-strain curve was drawn based on the test results. The tensile stress was computed by the load on notch section, the tensile strain was computed with the gauge length of 150 mm. The loading speed was 0.1 mm/min, and the test was finished with the axial tensile fracture of the specimen.

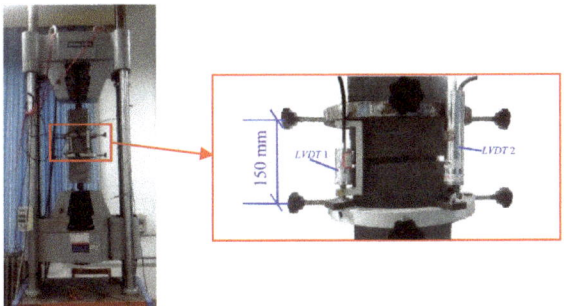

Figure 3. Testing machine and device for axial tensile strength of concrete.

3. Test Results and Discussion
3.1. Splitting Tensile Test
3.1.1. Test Phenomena

The specimen of the SCC broke into two parts due to the brittle cracking, the crack passed through both aggregates and hardened cement paste to form a flat failure surface. However, the different failure pattern appeared on the self-compacting SFRC specimens and the crushed zone are emphasized with red lines in Figure 4. Local splitting cracks first turned on the specimen near the splitting strips when the load reached a certain value, and the steel fibers bridging the cracks redistributed the tensile stress of the cracked SFRC matrix. The splitting section continuously cracked under the increased load, and the splitting failure presented a ductile feature with the development of the bond-slip of steel fiber [32]. The width of splitting crack decreased along with the direction from the passive loading surface to the active loading surface. With the increase of volume fraction of steel fiber, the crack width and the crushed zone of the specimen on the passive loading surface decreased significantly. This indicates that the more complex stress distribution took place on the splitting section of specimen of self-compacting SFRC.

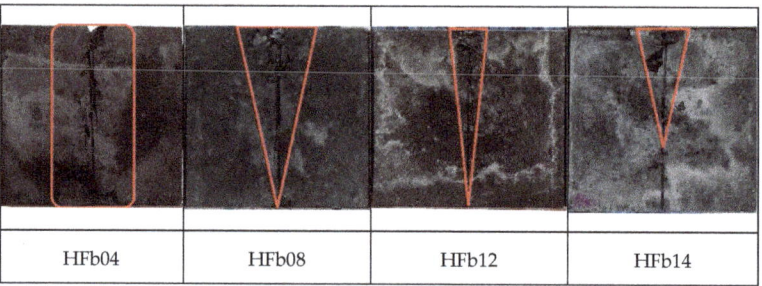

Figure 4. Failure patterns of splitting tensile specimens with the crushed zone emphasized by red lines.

3.1.2. Splitting Tensile Strength

The splitting tensile strength f_{st} could increase by 3.4~15.9% with the volume fraction of steel fiber HFb increased to be 1.2%. With the same $v_f = 1.2\%$, the f_{st} increased by 10.9% with the fiber length increased from 25.1 mm to 34.8 mm. The changes of the splitting tensile strength f_{st} of self-compacting SFRC with fiber factor λ_f is drawn in Figure 5. This indicates that a direct proportional linear relationship between f_{st} and λ_f can be fitted out, and presented as Equation (1) based on previous study [31],

$$f_{st} = f_{st,0}(1 + \alpha_{tb}\alpha_{te}\lambda_f) \tag{1}$$

where, α_{te} is a coefficient related to the fiber distribution, $\alpha_{te} = 0.441$ [31]; α_{tb} is the comprehensive coefficient of other factors affecting the bridging effect of steel fiber in the splitting tensile test; $f_{st,0}$ is the splitting tensile strength of SCC.

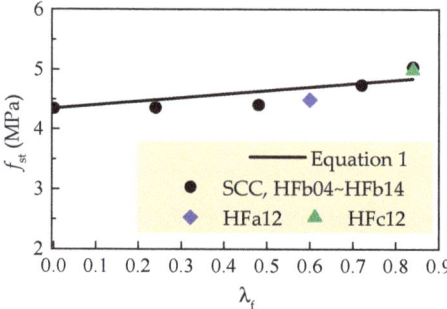

Figure 5. Changes of the splitting tensile strength with fiber factor.

By fitting the test data of f_{st} in this paper, the coefficient α_{tb} is 0.307. Figure 5 presents the comparison of test and calculation values of f_{st}. The average ratio is 1.008 with a dispersion coefficient of 0.035.

3.2. Axial Tensile Test

3.2.1. Test Phenomena

Typical failure patterns of axial tensile specimens of self-compacting SFRC are presented in Figure 6. Similar to the failure of notched cylinders [25], two failure modes were observed in this test: one mode was the breakage of bridged steel fibers, another was the pulling of bridged fiber out of the SFRC matrix. The former was observed in specimens of HFb04, and the latter was observed in specimens of HFb08, HFb12, HFb14, HFa12 and HFc12. This indicates that the bridged steel fibers were subjected to the tensile force which released from the cracked SFRC matrix. When the bridged steel fibers were insufficient to bear the tensile force, they would be broken with the suddenly overloading. Otherwise, the bridged steel fibers could continuously bear the tension to produce the bond-slip until pull out.

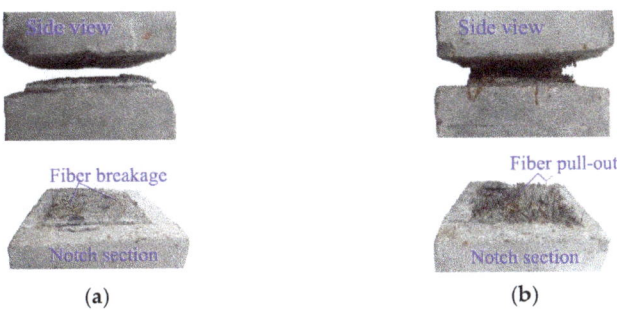

Figure 6. Typical failure patterns of axial tensile specimens: (**a**) fiber breakage of HFb04; (**b**) fiber pull-out of HFb12.

3.2.2. Axial Tensile Strength and Stress-strain Curve

The axial tensile stress-strain curves of self-compacting SFRC are presented in Figure 7. Due to difficult control in the post-peak region at the axial tensile test, specimens of SCC and HFb04 had no descending portion of axial tensile stress-strain curve. For specimen HFb08 with $v_f = 0.8\%$, a sharp drop existed on the stress-strain curve once the load reached the initial cracking resistance of the notch section, of which the initial cracking strength is the axial tensile strength. For specimens of HFb12, HFb14, HFa12 and HFc12 with $v_f \geq 1.2\%$, the descending portion of axial tensile stress-strain curves became gently with the gradually pulling out of steel fibers. Due to enough steel fibers bridged on notch section, the cracking load could be sustained or even increased slightly with the increase of tensile strain. This leads an increased axial tensile strength of self-compacting SFRC with the increasing volume fraction of steel fiber.

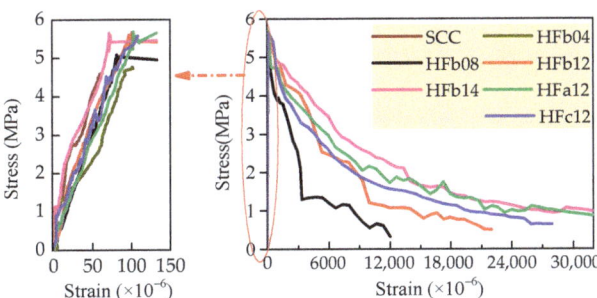

Figure 7. The axial tensile stress-strain curves of self-compacting SFRC.

With the increase of fiber length compared with HFa, the number of steel fiber per kilogram of HFb and HFc reduced by 19.5% and 39.4%. This led a reduction of the presence of steel fiber on notch section. As a result, the loading capacity of bridged steel fiber on notch section reduced after cracking. HFa has higher loading capacity with the same strain compared with HFb and HFc. This indicates the initial descending of the stress-strain curve dropped sharply with the increasing length of steel fiber. The stable presentation of stress-strain curves at descending portion are related to the better bond properties of hooked-end steel fiber in SFRC matrix [32].

The axial tensile test results of self-compacting SFRC are presented in Table 2. f_{fcr} is the initial cracking tensile strength of self-compacting SFRC, which is calculated with the initial cracking load divided by the notch section area of the specimen. f_{fa} is the peak tensile stress after the cracking of the notched specimen, which is calculated with the maximum load after cracking divided by the notch section area of the specimen. ε_{fcr} is the initial cracking tensile strain corresponding to f_{fcr}. ε_{fa} is the tensile strain corresponding to f_{fa}. According to China code CECS13 [15], the axial tensile strength f_{at} is larger one of f_{fcr} and f_{fa}, ε_{at} is the strain corresponded to the axial tensile strength.

Table 2. Test results of axial tensile strength and strain.

Item	SCC	HFb04	HFb08	HFb12	HFb14	HFa12	HFc12
Strengths							
f_{fcr} (MPa)	4.59	4.77	5.08	5.55	5.65	5.69	5.52
f_{fa} (MPa)	-	-	-	5.66	5.59	5.66	5.64
f_{at} (MPa)	4.59	4.77	5.08	5.66	5.65	5.69	5.64
Corresponding strains							
ε_{fcr} (10^{-6})	63.3	95.3	81.4	91.7	70.3	103.4	102.6
ε_{fa} (10^{-6})	-	-	-	101.1	73.5	133.3	160.3
ε_{at} (10^{-6})	63.3	95.3	81.4	101.1	70.3	103.4	160.3

Due to the brittle fracture after cracking, no values of f_{fa} were measured for the specimens with $v_f \leq 0.8\%$. For other specimens with $v_f \geq 1.2\%$, the values of f_{fcr} and f_{fa} are almost the same, while the strains ε_{fcr} and ε_{fa} increased slightly with the increasing volume fraction of steel fiber. This provides a good tension resistance of specimen with larger deformation.

The axial tensile strength f_{at} keeps a linear growth with the increase of fiber factor. f_{at} increases by 22.8–23.9% with the volume fraction of steel fiber HFb increased up to 1.2%. With the same $v_f = 1.2\%$, the axial tensile strength f_{at} decreases slightly with different fiber length. Figure 8 exhibits the changes of the axial tensile strength of self-compacting SFRC with the fiber factor λ_f. This indicates that a direct proportional linear relationship between f_{at} and λ_f can be fitted out. Therefore, the prediction of axial tensile strength of self-compacting SFRC considering fiber distribution can be done with Equation (2),

$$f_{at} = f_{at,0}(1 + \alpha_{ta}\alpha_{te}\lambda_f) \tag{2}$$

where, α_{te} is the influence coefficient of fiber distribution, α_{te} = 0.441 [31]; α_{ta} is the comprehensive coefficient of other factors affecting the bridging effect of steel fiber in the axial tensile test; $f_{at,0}$ is the axial tensile strength of SCC.

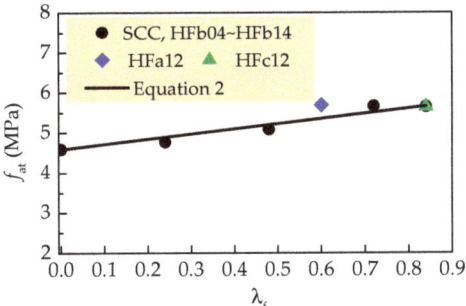

Figure 8. Axial tensile strength as a function of the fiber content.

By fitting the test data of f_{at} in this paper, α_{ta} is 0.631. Figure 8 presents the comparison of test and calculation values of f_{at}. The average ratio is 0.998 with a dispersion coefficient of 0.084.

3.2.3. Axial Tensile Toughness

According to the specification of China code CECS 13 [15], the tension work $W_{f,0.2}$ is the area under the axial tensile load-deformation curve. The maximum deformation of 0.3 mm is calculated by 0.2 times of gauge length $l_0 = 150$ mm. The axial tensile toughness ratio $R_{fe,0.2}$ is calculated by Equation (3).

$$R_{fe,0.2} = \frac{W_{f,0.2}}{0.2\% l_0 f_{at} A} \tag{3}$$

Where, A is the notch section area of the specimen.

From the axial tensile load-deformation curves presented in Figure 9, the calculation results of $W_{f,0.2}$ and $R_{fe,0.2}$ are shown in Figure 10. The positive correlations exist for $W_{f,0.2}$ and $R_{fe,0.2}$ with the fiber factor λ_f of HFb steel fiber, due to the more pull-out work needed to be exerted on the notch section of specimen with more steel fibers. However, no obvious relationships were observed in $W_{f,0.2}$ and $R_{fe,0.2}$ of self-compacting SFRC with the fiber length, as the hooked-end rather than fiber length plays a role controlling the bond performance in self-compacting SFRC [32].

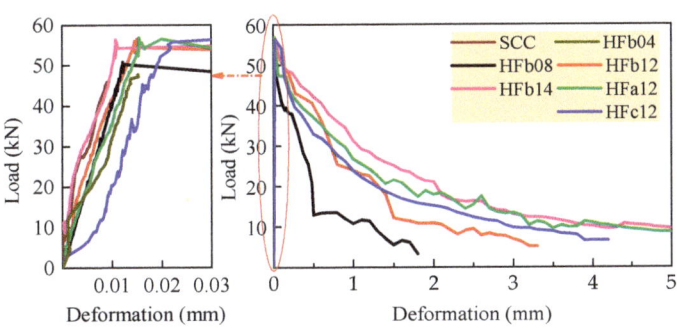

Figure 9. The axial tensile load-deformation curves of self-compacting SFRC.

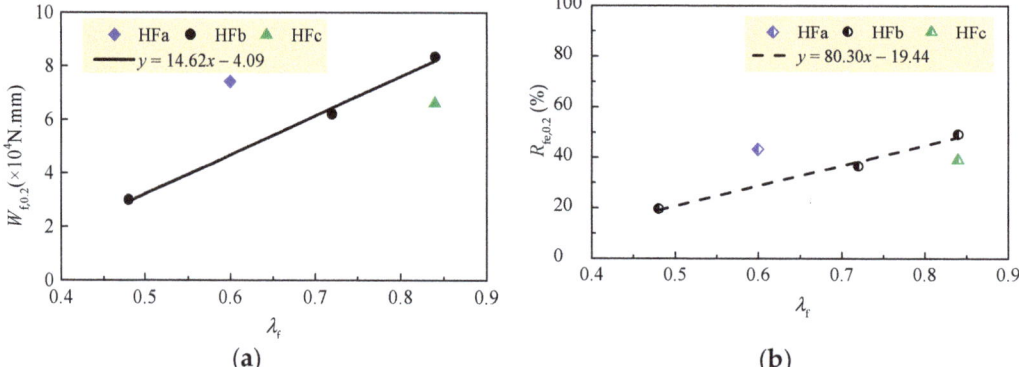

Figure 10. Axial tensile toughness changed with fiber factor: (**a**) tensile work; (**b**) toughness ratio.

3.3. Discussion on Tensile Strengths

The axial tensile strength of self-compacting SFRC f_{at} is higher than the splitting tensile strength f_{st}, the ratios vary from 1.09 and 1.27 with an average of 1.18. The strengthening effect of steel fiber on the axial tensile strength is greater than that of splitting tensile strength. This is significantly different from that of vibrated SFRC [17,33,34]. It has been reported that for vibrated SFRC, the axial tensile strength is less than splitting tensile strength, and the strengthening effect of splitting tensile strength is greater than that of axial tensile strength [17].

This difference comes mainly down to the significantly different distribution of steel fibers in the self-compacting SFRC. Due to the flat cast of prism specimens, the steel fiber trend to orientation along the flow direction of self-compacting SFRC [1,2,31]. This direction is consistent with the axial tensile direction which benefits to the axial tensile resistance of the self-compacting SFRC. At the same time, this orientation is perpendicular to the load direction of the splitting tensile test [2,31]. The strengthening effect of steel fiber on tensile strength could not be developed until the initiation of the first spitting cracks. In this condition, the complex stress exists in the splitting section with the coupling of shear stress and tensile stress [3,20]. This weakens the action of steel fiber along the splitting tensile direction. Therefore, the adaptability of splitting tensile test is questionable to be used for the self-compacting SFRC.

4. Conclusions

Based on the test results of the splitting tensile strength and axial tensile properties of self-compacting SFRC, the main conclusions can be drawn as follows:

(1) The splitting tensile strength of self-compacting SFRC is positively correlated with fiber factor, and the prediction model considering the influences of fiber distribution and volume fraction is proposed.

(2) The axial tensile strength of self-compacting SFRC is positively correlated with fiber factor, and the prediction model of which is proposed considering the influences of fiber distribution and volume fraction. The axial tensile work and tensile toughness ratio are positively correlated with the volume fraction of steel fiber, while no obvious relationship with the fiber length.

(3) The strengthening effect of steel fiber on axial tensile strength is greater than that of splitting tensile strength. This reflects the real distribution of steel fiber along the flow direction of self-compacting SFRC.

(4) With the improvement of notching techniques, the forming of notched specimens become easier. The axial tensile test with notched specimen is a feasible method to evaluate the tensile properties of self-compacting SFRC.

Author Contributions: Conceptualization, X.D. and M.Z.; methodology, C.L.; validation, J.L.; formal analysis, X.D.; investigation, H.G. and L.L.; data curation, X.D.; writting—original draft preparation, X.D.; writing—review and editing, M.Z. and C.L.; funding acquisition, X.D. and J.L. All authors have read and agreed to the published version of the manuscript.

Funding: This research was funded by National Key Research and Development Plan of China, grant number "2017YFC0703904"; Attracting Foreign Talents Fund of Henan, China, grant number "YWZ2018-6-HS2"; Natural Science Foundation of Henan, China, grant number "212300410192" and the Key Scientific and Technological Research Project of University in Henan, China, grant number "20A560015".

Data Availability Statement: The data presented in this study are contained in this article.

Conflicts of Interest: The authors declare no conflict of interest.

References

1. Akcay, B.; Tasdemir, M.A. Mechanical behaviour and fibre dispersion of hybrid steel fibre reinforced self-compacting concrete. *Constr. Build. Mater.* **2012**, *28*, 287–293. [CrossRef]
2. Zhao, M.L.; Li, J.; David, L. Effects of flowability on SFRC fibre distribution and properties. *Mag. Concr. Res.* **2017**, *69*, 1043–1054. [CrossRef]
3. Khaloo, A.; Raisi, E.M.; Hosseini, P.; Tahsiri, H. Mechanical performance of self-compacting concrete reinforced with steel fibers. *Constr. Build. Mater.* **2014**, *51*, 179–186. [CrossRef]
4. Li, C.Y.; Geng, H.B.; Deng, C.H.; Li, B.C.; Zhao, S.B. Experimental investigation on columns of steel fiber reinforced concrete with recycled aggregates under large eccentric compression load. *Materials* **2019**, *12*, 445. [CrossRef]
5. Li, X.K.; Pei, S.W.; Fan, K.P.; Geng, H.B.; Li, F.L. Bending performance of SFRC beams based on composite-recycled aggregate and matched with 500MPa rebars. *Materials* **2020**, *13*, 930. [CrossRef]
6. Zhao, M.S.; Li, C.Y.; Su, J.Z.; Shang, P.R.; Zhao, S.B. Experimental study and theoretical prediction of flexural behaviors of reinforced SFRELC beams. *Constr. Build. Mater.* **2019**, *208*, 454–463. [CrossRef]
7. Martinelli, P.; Colombo, M.; Fuente, A.; Cavalaro, S.; Pujadas, P.; Prisco, M. Characterization tests for predicting the mechanical performance of SFRC floors: Design considerations. *Mater. Struct.* **2021**, *54*, 2. [CrossRef]
8. Martinelli, P.; Colombo, M.; Pujadas, P.; Fuente, A.; Cavalaro, S.; Prisco, M. Characterization tests for predicting the mechanical performance of SFRC floors: Identification of fibre distribution and orientation effects. *Mater. Struct.* **2021**, *54*, 3. [CrossRef]
9. Pereira, E.N.B.; Barros, J.A.O.; Camões, A. Steel fiber-reinforced self-compacting concrete: Experimental research and numerical simulation. *J. Struct. Eng.* **2008**, *134*, 1310–1321. [CrossRef]
10. Ferrara, L.; Park, Y.D.; Shah, S.P. A method for mix-design of fiber-reinforced self-compacting concrete. *Cem. Concr. Res.* **2007**, *37*, 957–971. [CrossRef]
11. MertYücel, Y.; Bülent, B.; Mehmet, A.S. Effect of fine to coarse aggregate ratio on the rheology and fracture energy of steel fibre reinforced self-compacting concretes. *Sadhana* **2014**, *39*, 1447–1469.
12. Silva, M.A.D.; Pepe, M.; Pfeil, M.S.; Pfeila, M.S.; Filho, R.D.T. Rheological and mechanical behavior of high strength steel fiber-river gravel self-compacting concrete. *Constr. Build. Mater.* **2017**, *150*, 606–618. [CrossRef]
13. Ghanbarpour, S. The Effect of type and volume fraction (VF) of steel fiber on the mechanical properties of self-compacting concrete. *J. Eng. Des.* **2010**, *8*, 247–256. [CrossRef]
14. Tameemi, W. Correlations Between Compressive, Flexural, and Tensile Behavior of Self-Consolidating Fiber Reinforced Concrete. Master's Thesis, University of Kansas, Lawrence, KS, USA, October 2015.
15. China Association of Engineering Construction Standardization. Test methods for steel fiber reinforced concrete. In *CECS 13:2009*; China Planning Press: Beijing, China, 2010.
16. Cao, D.F.; Fu, L.Z.; Qin, X.C.; Yang, Z.W. Tensile constitutive characteristics of concrete under freeze-thaw cycles. *J. Jiangsu Univ.* **2011**, *32*, 359–363.
17. Han, R.; Zhao, S.B.; Qu, F.L. Experimental study on tensile properties of steel fiber reinforced concrete. *China Civ. Eng. J.* **2006**, *11*, 67–71.
18. Shen, C.M.; Cui, Y.Y. Experimental study on tensile strength of concrete with steel slag and nano-SiO_2. *Concrete* **2016**, *11*, 24–27.
19. Ministry of Housing and Urban-Rural Development of the People's Republic of China. Standard for test methods of concrete physical and mechanical properties. In *GB/T 50081-2019*; China Building Industry Press: Beijing, China, 2019.
20. Zhao, S.B. *Design Principle of Concrete Structures*; Tongji University Press: Shanghai, China, 2013.
21. Goaiz, H.A.; Farhan, N.A.; Sheikh, M.N.; Yu, T.; Hadi, M.N.S. Experimental evaluation of tensile strength test methods for steel fibre-reinforced concrete. *Mag. Concr. Res.* **2019**, *71*, 385–394. [CrossRef]
22. Khalaj, G.; Nazari, A. Modeling split tensile strength of high strength self compacting concrete incorporating randomly oriented steel fibers and SiO_2, Nanoparticles. *Compos. B Eng.* **2012**, *43*, 1887–1892. [CrossRef]
23. You, Z.G.; Tao, Z.Q.; Xu, G.Q.; Han, Y.T.; Zhou, Y.L. Experimental study on axial tensile strength of hybrid fiber reinforced self-compacting concrete. *Struct. Eng.* **2019**, *35*, 202–208.

24. Clifford, A.O.O.; David, W.B.; Stephanie, J.B.; Nikos, N. Behaviour of hybrid steel fibre reinforced self compacting concrete using innovative hooked-end steel fibres under tensile stress. *Constr. Build. Mater.* **2019**, *202*, 753–761.
25. Cunha, V.M.C.F.; Barros, J.A.O.; Sena-Cruz, J.M. An integrated approach for modelling the tensile behaviour of steel fibre reinforced self-compacting concrete. *Cem. Concr. Res.* **2011**, *41*, 64–76. [CrossRef]
26. Liao, W.C.; Chao, S.H.; Park, S.Y.; Naaman, A.E. Self-Consolidating High Performance Fiber Reinforced Concrete (SCHPFRC)–Preliminary Investigation. Report No. UMCEE 06. 2006. Available online: https://www.rilem.net/publication/publication/58?id_papier=9603 (accessed on 25 February 2021).
27. Ding, X.X.; Zhao, M.L.; Li, J.; Shang, P.R.; Li, C.Y. Mix proportion design of self-compacting SFRC with manufactured sand based on the steel fiber-aggregates skeleton packing test. *Materials* **2020**, *13*, 2833. [CrossRef]
28. Ding, X.X.; Zhao, M.L.; Zhou, S.Y.; Fu, Y.; Li, C.Y. Statistical analysis and preliminary study on the mix proportion design of self-compacting steel fiber reinforced concrete. *Materials* **2019**, *12*, 637. [CrossRef] [PubMed]
29. Ministry of Housing and Urban-Rural Development of the People's Republic of China. Technical specification for application of self-compacting concrete. In *JGJ/T 283-2012*; China Architecture & Building Press: Beijing, China, 2012.
30. RILEM TC 162-TDF. Test and design methods for steel fiber reinforced concrete, bending test. *Mater. Struct.* **2002**, *35*, 579–582. [CrossRef]
31. Ding, X.X.; Li, C.Y.; Han, B.; Lu, Y.Z.; Zhao, S.B. Effects of different deformed steel-fibers on preparation and fundamental properties of self-compacting SFRC. *Constr. Build. Mater.* **2018**, *168*, 471–481. [CrossRef]
32. Ding, X.X.; Zhao, M.L.; Li, C.Y.; Li, J.; Zhao, X.S. A multi-index synthetical evaluation of pull-out behaviors of hooked-end steel fiber embedded in mortars. *Constr. Build. Mater.* **2021**, *276*, 122219. [CrossRef]
33. Zhao, M.L.; Zhao, M.S.; Chen, M.H.; Li, J.; David, L. An experimental study on strength and toughness of steel fiber reinforced expanded-shale lightweight concrete. *Constr. Build. Mater.* **2018**, *183*, 493–501. [CrossRef]
34. Zhao, S.B.; Li, C.Y.; Du, H.; Qian, X.J. Experimental study on steel fiber reinforced high strength concrete with large size aggregate. *J. Build. Mater.* **2010**, *13*, 155–160.

Article

Research on the Mechanical Performance of Carbon Nanofiber Reinforced Concrete under Impact Load Based on Fractal Theory

Wei Xia [1,*], Jinyu Xu [1,2] and Liangxue Nie [1]

[1] School of Aeronautical Engineering, Air Force Engineering University, Xi'an 710038, China; xujinyuafeua@163.com (J.X.); nieliangxueafeu@163.com (L.N.)
[2] College of Mechanics and Civil Architecture, Northwest Polytechnic University, Xi'an 710072, China
* Correspondence: xiaweiafeu@163.com; Tel.: +86-18729263358

Abstract: The research is focused on the dynamic compressive strength, impact toughness and the distribution law of fragmentation size for the plain concrete and the carbon nanofiber reinforced concrete with four fiber volume contents (0.1%, 0.2%, 0.3% and 0.5%) under impact load by using the Φ100 mm split-Hopkinson pressure bar. Based on the fractal theory and considering the micropore structure characteristics of the specimen, the impact of the strain rate and the dosage of carbon nanofibers on the dynamic mechanical performance of concrete is analyzed. According to the results, both the dynamic compressive strength and the impact toughness increase continuously with the improvement of the strain rate level at the same dosage of fiber, showing strong strain rate strengthening effect; at the same strain rate level, the impact toughness increases gradually with the increase in the fiber dosage, while the dynamic compressive strength tends to increase at first and then decrease; the distribution of the fragmentation size of concrete is a fractal in statistical sense, in general, the higher the strain rate level, the higher the number of fragments, the lower the size, and the larger the fractal dimension; the optimal dosage of carbon nanofibers to improve the dynamic compressive strength of concrete is 0.3%, and the pore structure characteristics of carbon nanofiber reinforced concrete exhibit obvious fractal features.

Keywords: carbon nanofiber reinforced concrete; impact load; dynamic compressive strength; pore structure characteristics; fractal dimension

1. Introduction

After its continuous development for almost two centuries, concrete has become one of the most widely used materials for the construction in the fields of both civil infrastructure and military defensive projects. In recent years, sudden explosion accidents due to terrorist attacks, local wars or negligence during industrial production and daily life have happened repeatedly, putting many concrete structures under the threat of extreme external loads, including impact and explosion [1–5]. In order to promote the performance of concrete in terms of explosion and penetration resistance, and improve its mechanical properties under impact stress, various fiber-modified forms of concrete have been developed to meet the requirements [6–9].

As a kind of new multifunctional material with excellent performance, carbon nanofibers (CNFs) [10–12] are characterized by small self-dimension, large specific surface area, and strong cohesiveness within the concrete matrix compared with carbon fiber, steel fiber, etc. CNFs are a kind of discontinuous nanoscale graphite fibers, which have excellent characteristics of both carbon fibers and nanomaterials. CNFs enjoy wide application prospects in the field of the modification design of composite materials, specifically, they are able to provide excellent performance in terms of tensile-resistance, crack-resistance, electric conduction and fatigue-resistance to the concrete when being mixed into concrete. However, most of the research on carbon nanofiber reinforced concrete (CNFC) focuses on certain basic physical and mechanical performances [13–17]; in the meantime, the research on the characteristics of strength and energy, as well as the damage mode and distribution law of the fragments during impact breaking, is relatively rare. The essence of the failure of the concrete due to the impact load is the process that the internal damage cracks of the concrete are continuously expanded, extended and connected under the driving of energy, thereby resulting in material instability and failure [18–20]. The number and the size distribution of the broken fragments are exactly the macroscopic representation of this process. The particle size distribution and the pore structure characteristics inside the matrix of the destroyed products of the concrete under impact load show certain self-similarity and fractal characteristics [21]. Fractal theory can be used to describe fractal irregular features effectively. Fractal theory was founded by Mandelbrot in the 1970s. Its research object is the disordered but self-similar system widely existing in nature. At present, as a new method and concept, fractal theory has developed rapidly in many fields such as physics, biology, materials science, etc. Through the research on the changes of the strength, energy absorption and the breakup characteristics of the modified concrete under different dosages of CNFs and loading rates, the comprehensive analysis and assessment of the damage degradation degree, the energy consumption evolution mechanism and the ability to resist impact load can be implemented. Therefore, more in-depth research on the dynamic compression mechanical properties of CNFC is necessary, so as to understand their dynamic response law under impact load, thus ensuring the safety of the engineering structures in practice.

In view of this, the Φ100 mm split-Hopkinson pressure bar (SHPB) is used as the test device for the impact compression test on the plain concrete (PC) and the modified concrete with four volume dosages of carbon nanofiber, thereby carrying out screening statistics against the fragments of the impact failure and exploring the effects of the impact velocity of the bullet (strain rate) and fiber dosage on the dynamic compressive strength, impact toughness, failure mode and fractal dimension of the fragments. In the meantime, taking into consideration the mercury intrusion test, the analysis of microscopic mechanism for the pore structure characteristics of the specimen is implemented based on the fractal theory, so as to provide better guidance to the engineering practice.

2. Materials and Methods

2.1. Raw Materials and Specimen Preparation

The following raw materials are utilized, including the cement of P·O 42.5R of brand Qinling with the initial setting time of 2 h, and the final setting time of 5 h; the coarse aggregate of limestone gravel, with the particle size ranging from 5~10 mm (accounting for 15%), and 10~20 mm (accounting for 85%); fine aggregate of medium sand which is used after washing and drying, with the apparent density of about 2630 kg/m^3; clean tap water; water reducing agent of FDN high-efficiency water reducing agent with water reducing rate of 20%; as well as the fiber material of CNFs from Beijing Deke Daojin Science and Technology Co., Ltd., with its physical performance index shown in Table 1. Table 2 shows the mix proportion of the plain concrete with the strength grade of C40 and carbon nanofiber reinforced concrete, in which PC represents the plain concrete, CNFC01, CNFC02, CNFC03, and CNFC05 refer to the volume dosages of carbon nanofiber of 0.1%, 0.2%, 0.3% and 0.5%, respectively. The pouring of concrete specimens was based on the "method of sand and rubbles enveloped with cement". CNFs are prepared into dispersion solution and then uniformly mixed into the concrete mixture [22]. The test results of four-probe method show that the resistance of the sample is obviously reduced and the conductive effect is excellent. From this, it can be judged that CNFs achieves the purpose of dispersion. The preparation process of CNFC group mixture is shown in Figure 1. Sand and gravel are added to the mixer in turn, and part of the mixed liquid is added while stirring, and then cement was added, stirring for 120 s. After that, the remaining mixed liquid is added and stirred for 120 s to prepare the concrete mixture. The mixture is stirred evenly, and then put into the cylinder for die test and molding. The mold is removed after standing indoors for 24 h, and moved into the curing room for standard curing. After 28 days, the mixture is taken out for polishing, thereby obtaining the short cylinder specimen with the geometric size of Φ98 mm × 48 mm for the impact compression test (as shown in Figure 2).

Table 1. Main physical performance index of CNTs.

Purity/%	Diameter/nm	Resistivity/$\Omega \cdot cm$	Thermal Expansion Coefficient/$°C^{-1}$	Specific Surface Area/$m^2 \cdot g^{-1}$	Density/$g \cdot cm^{-3}$
99.9	100~200	<0.012	1	300	0.18

Table 2. Mix proportions of concrete (kg/m^3).

Specimen No.	Cement	Coarse Aggregate	Fine Aggregate [1]	Water	Defoaming Agent [2]	Water Reducing Agent	CNTs
PC	495	1008	672	100	0	0	0
CNFC01	495	1008	672	180	0.30	5.0	0.18
CNFC02	495	1008	672	180	0.45	7.5	0.36
CNFC03	495	1008	672	180	0.60	10.0	0.54
CNFC05	495	1008	672	180	0.90	15.0	0.90

[1] River sand of 2.8 fineness modulus. [2] Aqueous defoamer of tributyl phosphate.

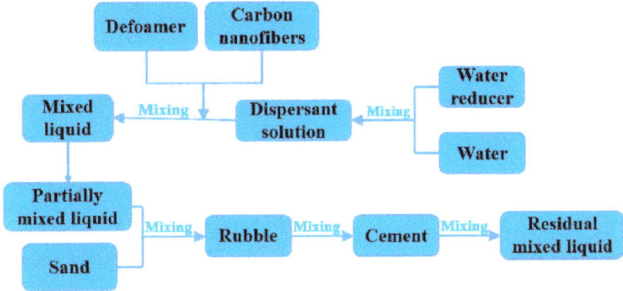

Figure 1. The preparation process of carbon nanofiber reinforced concrete (CNFC) group mixture.

Figure 2. The processed cylinder specimens.

2.2. Test Equipment and Method

The Φ100 mm SHPB test device is used for the impact compression test (as shown in Figure 3). Figure 4 shows the propagation process of the stress wave in the test. A pneumatic gun is used in this device to drive the bullet, and makes it collide with the incident rod in a high-speed coaxial manner, thereby producing the incident wave $\varepsilon_I(t)$. The specimen placed between the incident bar and the transmission bar generates high-speed deformation under the loading of the incident wave. In the meantime, it transmits the reflected wave $\varepsilon_R(t)$ and the transmitted wave $\varepsilon_T(t)$ to the incident bar and the transmission bar, respectively. These required waveform information is measured and recorded by the high dynamic strain indicator, waveform memory, etc., and then the data are processed with the "three-wave method" (as shown in Formula (1)) [23,24], thereby obtaining the relevant parameters reflecting the dynamic compression mechanical properties of the specimen. The impact velocity of the bullet is jointly decided by the air pressure and its action distance applied. During the test, the action distance of air pressure is kept fixed, and the bullet velocity is controlled by the adjustment of the pressure of the input air. In the meantime, the impact velocity of bullet is also affected by the test environment, in this case, although the input pressure and its action distance can be kept constant each time, the bullet velocity may be different. The strain rate can also be deemed as the reflection of bullet impact velocity, and there is an approximate linear correlation between them [25,26]. The typical strain rate time history curve of the concrete specimen under dynamic compression is shown in Figure 5, where Point A refers to the inflection point of the rising section of the curve, and Point B represents the inflection point of the corresponding falling section. The average strain rate of the middle platform section is selected as the representative value of the strain rate of the specimen under the current impact velocity of bullet [27,28]. A total of five strain rate levels are set for the test, and the corresponding input pressures are 0.3 MPa, 0.35 MPa, 0.4 MPa, 0.45 MPa and 0.5 MPa, respectively. H62 circular brass sheet with the thickness of 1 mm is selected to shape the initial stress wave, and the typical waveforms of the shaped incident wave, transmission wave and reflection wave are shown in Figure 6. To ensure the effectiveness and reliability of the test results, the test should be repeated at least three times under each input pressure, then the average value of the obtained test data should be calculated and taken as the representative value of the test data under this working condition.

$$\begin{cases} \varepsilon_s(t) = \frac{C_e}{L}\int_0^\tau [\varepsilon_I(t) - \varepsilon_R(t) - \varepsilon_T(t)]dt \\ \dot{\varepsilon}_s(t) = \frac{C_e}{L}[\varepsilon_I(t) - \varepsilon_R(t) - \varepsilon_T(t)] \\ \sigma_s(t) = \frac{E_e A_e}{2A_s}[\varepsilon_I(t) + \varepsilon_R(t) + \varepsilon_T(t)] \end{cases} \quad (1)$$

where C_e, E_e and A_e refer to the wave velocity, elastic modulus and cross-sectional area of the compression bar; $\varepsilon_s(t)$, $\dot{\varepsilon}_s(t)$, $\sigma_s(t)$, A_s and L represent the strain, strain rate, stress, end area and length of the specimen, respectively; and τ denotes the propagation time of stress wave in the bar.

Figure 3. Φ100 mm SHPB test system.

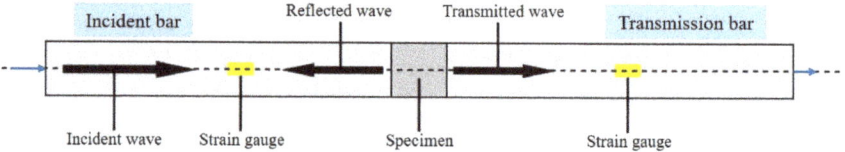

Figure 4. Schematic diagram of stress wave propagation.

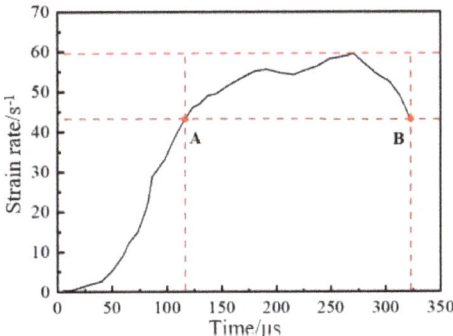

Figure 5. Typical time history curve of strain rate.

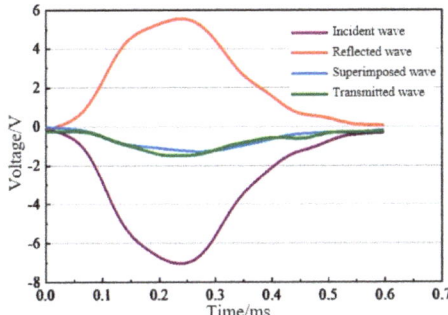

Figure 6. Typical waveform diagram after shaping.

The screening test adopts the bushing screen with the screen size of 50 mm, 40 mm, 31.5 mm, 26.5 mm, 20 mm, 16 mm, 10 mm, 5 mm, and 2.36 mm, respectively. In the meantime, the mass of fragments retained on each screen is measured by the electronic balance. Mercury intrusion porosimetry (MIP) is used to study the structural characteristics of the microscopic pore of concrete [29]. The experimental equipment refers to PoreMaster33 automatic mercury intrusion instrument produced by Quantachrome Instruments from the United States of America. The fragments of the specimen with appropriate size after impact compression test are selected as the sample for analysis. Remove the residues and dust on the surface of the sample before placing them in a constant-temperature drying box for complete dehydration, and then carry out the determination.

3. Mechanical Properties of Dynamic Compression

3.1. Dynamic Compressive Strength

The dynamic compressive strength ($f_{c,d}$) represents the peak stress in the stress–strain curve of the specimen, indicating the strength characteristics of the concrete under impact load. Figure 7 illustrates the variation law of dynamic compressive strength of concrete specimen under different strain rates. It can be seen from the analysis that: (1) with the increase in strain rate level, the dynamic compressive strength of both PC group and CNFC group increases continuously, showing a significant strain rate strengthening effect. (2) It can be seen from the change in the dynamic compressive strength with the strain rate that there is an approximate linear correlation between them. When carrying out linear fitting, it is found from Formula (2) that the fitting effect of them is relatively good. (3) In general, compared with PC, the addition of appropriate amount of CNFs can significantly improve the dynamic compressive strength of concrete. At the same strain rate level, the dynamic compressive strength increases first and then decreases with the increase in the addition amount of fiber. When the addition amount of fiber content reaches 0.3%, the improvement effect is the best; in contrast, when the addition amount of fiber content is 0.5%, the improvement effect is relatively small. This may be due to the excessive addition of fiber, i.e., 0.5%, which means that CNFs cannot be evenly dispersed in the concrete matrix, resulting in the phenomenon of "agglomeration". The excessive CNFs intensify the internal defects of concrete matrix structure, resulting in stress concentration in local areas under impact load, which is not conducive to further improving the concrete strength characteristics.

$$\begin{cases} \text{PC}: & f_{c,d} = 22.0878 + 0.5216\dot{\varepsilon} & (R^2 = 0.9874) \\ \text{CNFC01}: & f_{c,d} = 21.4446 + 0.5448\dot{\varepsilon} & (R^2 = 0.9793) \\ \text{CNFC02}: & f_{c,d} = 22.3441 + 0.5472\dot{\varepsilon} & (R^2 = 0.9899) \\ \text{CNFC03}: & f_{c,d} = 26.0754 + 0.5202\dot{\varepsilon} & (R^2 = 0.9934) \\ \text{CNFC05}: & f_{c,d} = 28.1470 + 0.4573\dot{\varepsilon} & (R^2 = 0.9702) \end{cases} \quad (2)$$

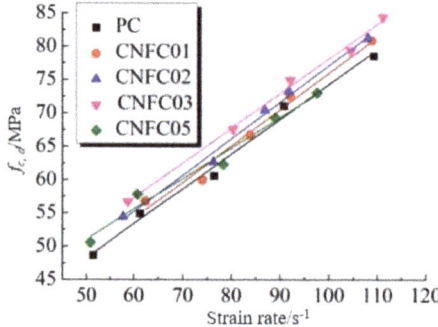

Figure 7. Dynamic compressive strength of concrete under different strain rates.

3.2. Impact Toughness

The impact failure process of concrete is bound to be accompanied by the changes in energy, especially under the impact load; the transformation and dissipation of energy are extremely fast and active. The impact toughness (IT) can be used to characterize the entire stress–strain development process of the specimen under impact load until the energy for the failure of the specimen is absorbed. The physical meaning of impact toughness refers to the area surrounded by stress–strain curve and transverse axis [28], expressed as follows:

$$IT = \int_0^{\varepsilon_u} f d\varepsilon \quad (3)$$

where f refers to the dynamic stress–strain curve of the specimen; and ε_u represents the dynamic ultimate strain of the curve.

The variation relationship between the impact toughness and the strain rate of each group of specimen is shown in Figure 8. The analysis shows that: (1) the impact toughness is also highly sensitive to the strain rate, specifically, the impact toughness increases gradually with the increase in the strain rate level. (2) At a relatively low strain rate level, when the volume dosage of CNFs reaches 0.2%, the improvement effect on the impact toughness of concrete is weaker than that of 0.3%; in contrast, the improvement effect is opposite at high strain rate level. (3) It can be seen from the change in the impact toughness with the strain rate that there is an approximate linear correlation between them. Formula (4) shows the result of linear fitting between them, and it is found that the fitting effect is good. (4) Compared with PC, the addition of CNFs can improve the impact toughness of concrete to a certain extent. In general, with the increase in the dosage of CNFs, the impact toughness of concrete at the same strain rate level is increased to certain degree. The possible reason for the results of this experiment may lie in the fact that the impact toughness is jointly decided by the dynamic compressive strength and the corresponding impact compression deformation. After the addition of CNFs, both the strength and deformation of concrete will receive certain enhancement, and the combined effect of the two leads to the improvement of the impact toughness of concrete under all dosages of fiber.

$$\begin{cases} \text{PC}: & IT = 3.2639 + 10.1479\dot{\varepsilon} & (R^2 = 0.9648) \\ \text{CNFC01}: & IT = -271.9173 + 13.7023\dot{\varepsilon} & (R^2 = 0.9709) \\ \text{CNFC02}: & IT = -448.3160 + 17.2420\dot{\varepsilon} & (R^2 = 0.9315) \\ \text{CNFC03}: & IT = -205.5699 + 14.4042\dot{\varepsilon} & (R^2 = 0.9055) \\ \text{CNFC05}: & IT = 10.9772 + 12.2302\dot{\varepsilon} & (R^2 = 0.9303) \end{cases} \quad (4)$$

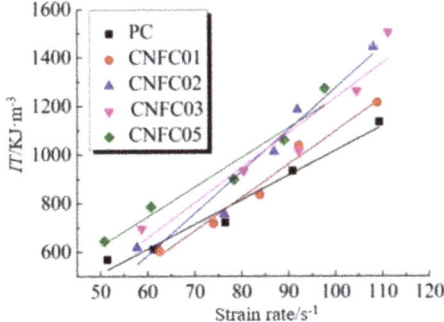

Figure 8. Impact toughness of concrete under different strain rates.

3.3. Mechanism Analysis

Both the dynamic compressive strength and the impact toughness of PC and CNFC increase with the increase in impact velocity of bullet, showing strong strengthening effect of strain rate, besides, both the dynamic compressive strength and impact toughness of CNFC are higher than their counterparts of PC. It can be seen from the microscopic mechanism of concrete failure that the initiation and propagation of internal microcracks are the main causes of the failure of the specimen. The greater the impact velocity of bullet, the greater the deformation rate of the specimen, the more the number of cracks generated, and the more energy absorbed. Under high-speed impact load, the action time of bullet on the specimen is rather short, and the material deformation buffering is small. Therefore, most of the energy accumulation of the specimen is achieved by increasing the stress instead of the strain, leading to the increase in the dynamic compressive strength of the material with the increase in the loading rate. In addition, as per the microstructure test results of CNFC impact fracture specimen, the matrix compactness has been significantly improved (the specific mechanism is shown in Section 5.2), and the failure of specimen is mainly due to the pull-out or fracture of the fiber (as shown in Figure 9a). The reason is that, for CNFC, CNFs can play a role of crack resistance and bridging adsorption [30]. The deformation released after the initiation of microcracks can first result in fiber debonding, rather than supporting the propagation of microcracks, thereby delaying the fracture process, and enhancing the toughness of concrete specimen. However, excessive addition of CNFs may result in "agglomeration" (as shown in Figure 9b), and CNFs may be intertwined to form new weak areas within the concrete matrix, which is not conducive to further improve the concrete strength.

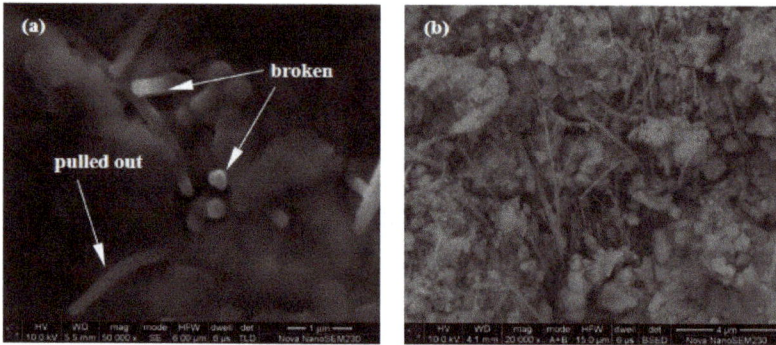

Figure 9. Microstructure and morphology of CNFC (**a**) CNFs pull-out and fracture; (**b**) CNFs aggregation and winding.

4. Fractal Characteristics of Impact Fragmentation Size

4.1. Impact Failure Mode and Fragmentation Size Distribution

The instability and failure of concrete under the impact load refer to the process of continuous inoculation, development and aggregation of internal microdamage cracks under external load, which eventually leads to macroscopic breaking. Besides, different impact velocities and fiber dosages will inevitably result in the change in the breaking morphology. Figure 10 shows the typical failure modes of the specimen under different strain rate levels. Due to the space limitation of this paper, only PC group and CNFC02 group are used as the representative specimens for analysis. The comparative analysis shows that the failure modes of the specimen can be basically classified into four types, i.e., edge failure, core-retaining failure, fragment failure and crushing failure. With the increase in strain rate level, the number of broken concrete fragments increases, the size decreases and tends to be uniform, and the degree of fragmentation increases continuously. In addition, at the same strain rate level, the particle size of the PC group specimen after failure is smaller, while the breaking morphology of the concrete specimen modified by CNFs is greatly improved, and the particle size of the fragment is relatively large, indicating that the addition of CNFs has a significant improvement effect on the impact resistance of concrete.

Figure 10. Typical failure mode of concrete under different strain rates (**a**) PC; (**b**) CNFC02.

To further describe the distribution law and the dimensional characteristics of concrete fragmentation size, the mass screening statistics of the specimen fragments under the impact load is carried out [21,31]. Additionally, based on the statistical theory, the average size d_{ave} of the fragments of the specimen is calculated, namely:

$$d_{ave} = \sum d_i \eta_i / \sum \eta_i \tag{5}$$

where d_i refers to the average particle size of the remaining fragments on each sieve screen, taking the average pore diameter of the primary sieve and the upper sieve; η_i represents the percentage of the mass of the retained fragments in the total mass of each sieve.

The variation law between d_{ave} and strain rate is shown in Figure 11. According to the analysis, both of them meet the requirements of relationship distribution of $y = A - B\ln(x + C)$. We carried out the nonlinear fitting based on the strain rate as the transverse axis, and d_{ave} as the longitudinal axis, with the fitting results shown in Formula (6). It was found that d_{ave} of each group of specimens decrease with the increase in strain rate level. At low strain rate level, d_{ave} decreased sharply, while at high strain rate level, d_{ave} decreased slightly. The reason is that the size of the fragments changes greatly when the mode of the specimen evolves from the "edge failure" to the "core-retaining failure", while the size change in the fragments is relatively small when the mode of the specimen evolves from the "core-retaining failure" to the "fragment failure", and the size change in the fragments is smaller when the mode of the fragment evolves from the "fragment failure" to the "crushing failure". Therefore, under the condition of high strain rate, even if the strain rate level increases greatly, the change amplitude of impact crushing degree of concrete is still relatively moderate.

$$\begin{cases} \text{PC}: & d_{ave} = 26.2690 - 3.5176\ln(\dot{\varepsilon} - 51.4599) & (R^2 = 0.9990) \\ \text{CNFC01}: & d_{ave} = 51.9066 - 10.6044\ln(\dot{\varepsilon} - 61.4724) & (R^2 = 0.9612) \\ \text{CNFC02}: & d_{ave} = 90.3682 - 20.7881\ln(\dot{\varepsilon} - 53.3802) & (R^2 = 0.9295) \\ \text{CNFC03}: & d_{ave} = 156.8900 - 35.8741\ln(\dot{\varepsilon} - 49.0569) & (R^2 = 0.9249) \\ \text{CNFC05}: & d_{ave} = 20.5064 - 1.4370\ln(\dot{\varepsilon} - 50.8900) & (R^2 = 0.9996) \end{cases} \tag{6}$$

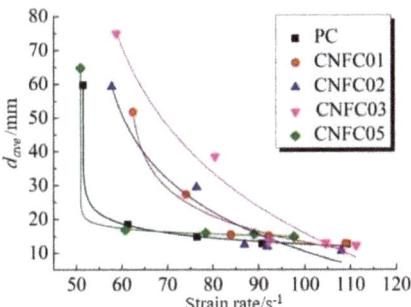

Figure 11. Relationship between d_{ave} and strain rate.

4.2. Fractal Dimension

When the concrete specimen is broken under impact load, the particle size distribution of the fragments is in accordance with the distribution law of Gate-Gaudin-Schuhmann [21,32]. The distribution equation based on mass-frequency is:

$$y = \frac{M(r)}{M_T} = \left(\frac{r}{r_m}\right)^q \tag{7}$$

where r refers to the particle size of the fragments of the broken specimen; r_m represents the maximum particle size of fragments; $M(r)$ denotes the sum of mass of all fragments whose particle size is less than r; M_T is the total mass of fragments when the specimen is broken; q refers to the fragment mass distribution parameter, taking the slope of $\lg[M(r)/M_T]$-$\lg r$ linear fitting curve.

According to the definition of fractal dimension, that is, $N = r^{-D_b}$ (N refers to the number of fragments with particle size greater than r, and D_b represents the fractal dimension of fragments); in the meantime, considering the relationship between the increment of fragment number and the increment of fragment mass, that is, $dM \propto r^3 dN$, the fractal dimension D_b of fragments can be calculated with the mass-particle size method, namely:

$$D_b = 3 - q \tag{8}$$

According to the above analysis, the slope of the linear fitting curve between $\lg[M(r)/M_T]$ and $\lg r$ is $3-D_b$. Taking $\lg r$ as the abscissa and $\lg[M(r)/M_T]$ as the ordinate, the scatter diagram of the two is drawn, and the linear fitting is carried out, as shown in Figure 12.

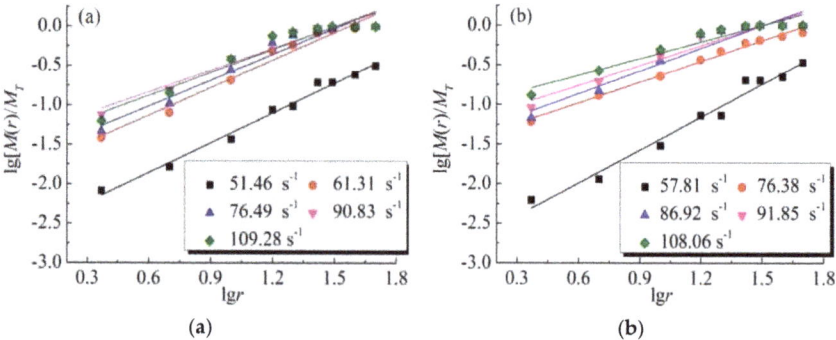

Figure 12. $\lg[M(r)/M_T]$-$\lg r$ linear fitting curve (a) PC; (b) CNFC02.

The data points in Figure 12 show good linear correlation in the double logarithmic coordinate system, indicating that the distribution of concrete fragments after impact failure has fractal characteristics. This is because the microscopic cracks and pores in the concrete show self-similarity at different scales; besides, the breaking process and the shape of fragments are the direct results of crack propagation, thereby resulting in the power-law distribution of fragments, which is a fractal in the statistical sense. D_b describes the distribution characteristics of the concrete fragment size after breaking. The larger D_b is, the smaller the average particle size of the concrete fragments is, and the higher the degree of crushing of the specimen is. According to the analysis, the strain rate has a significant influence on the crushing morphology of the concrete under impact load. The relationship between the strain rate and D_b is shown in Figure 13. It is found that there is no strict positive correlation between the strain rate and D_b in this test. However, on the whole, the higher the strain rate level, the larger the value of D_b. The test results can reflect the distribution characteristics of the particle size of the broken specimen to a certain extent. Moreover, in this paper, the same series of screening apertures (the same size) are used under different working conditions; therefore, the fractal dimension values obtained are comparable.

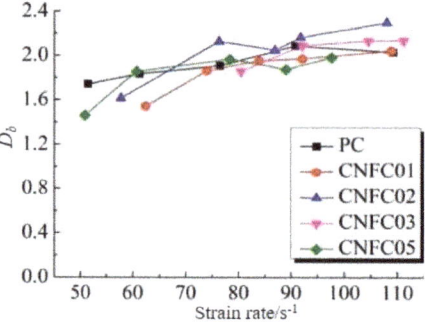

Figure 13. Relationship between D_b and strain rate.

5. Fractal Characteristics of Microscopic Pore Structure

During the hardening and forming of concrete, due to its own drying shrinkage, external curing conditions, internal hydration reaction and other factors, a certain number of initial defects, such as pores and cracks, will be generated in the matrix. The damage of concrete is caused by the gradual development and evolution of these initial defects, and the gradual intensification of new damages. Therefore, it can be concluded that the microscopic pore structure characteristics, such as pore size distribution composition, porosity characteristics in concrete, determine its macroscopic mechanical properties.

5.1. Fractal Model Based on Thermodynamic Relationship

The pore structure parameters of each group of concrete samples are shown in Table 3. Figure 14 shows the differential curve of pore size distribution of PC group and CNFC02 group samples. The pore distribution characteristics measured by mercury intrusion method can be used to divide the pore into gel pores (<10 nm), transition pores (10~100 nm), fine pores (100~1000 nm) and large pores (>1000 nm). The pore volume distribution of the four types of pores and their percentage distribution of total pores in the samples of each group are shown in Figures 15 and 16, respectively. The analysis shows that with the increase in CNFs addition, the content of large pores and fine pores in the concrete decreases significantly, while the proportion of gel pores and transition pores increases to a certain extent.

Table 3. Pore characteristic parameters of each group of samples.

Sample No.	Most Probable Pore Size [1]/nm	Medium Pore Diameter [2]/nm	Total Pore Volume/mL·g^{-1}	Average Pore Size [3]/nm	Pore Proportion/%			
					<10 nm	10~100 nm	100~1000 nm	<1000 nm
PC	9463	309.10	0.0445	111.58	4.27	33.93	22.25	39.55
CNFC01	33.02	87.66	0.0403	72.18	4.08	47.70	15.26	32.96
CNFC02	13.92	31.87	0.0384	54.76	12.23	52.13	10.62	25.02
CNFC03	36.92	55.91	0.0372	45.90	4.29	59.68	14.52	21.51
CNFC05	53.19	91.15	0.0395	83.21	2.12	48.68	18.78	30.42

[1] The aperture corresponding to the peak value on the differential distribution curve of aperture. [2] The corresponding pore size when half of the mercury is injected. [3] Ratio of total pore volume to pore surface area.

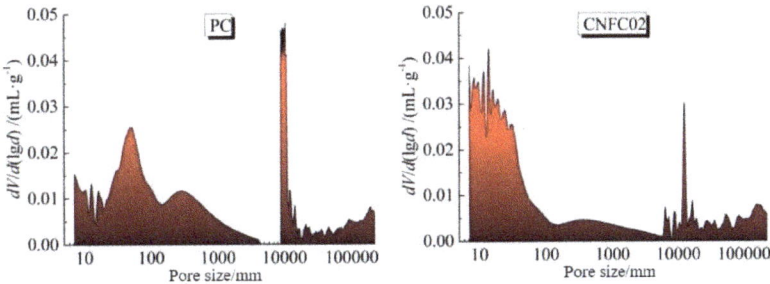

Figure 14. Differential curves of typical pore size distribution of samples.

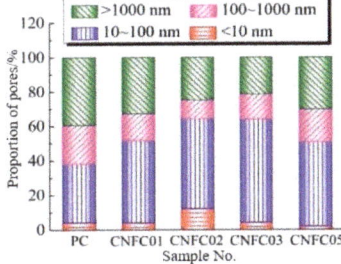

Figure 15. Proportion of various pores.

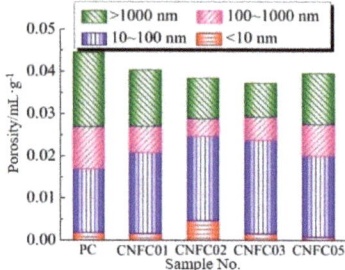

Figure 16. Pore volume distribution.

Under the vacuum state, mercury is pressed into the pores of concrete samples. According to thermodynamic theory, the work done by external force on mercury is equal to the increase in surface energy pressed into mercury [33], that is:

$$\int_0^V p\,dV = -\int_0^S \sigma \cos\theta\,dS \tag{9}$$

where p refers to the external pressure on mercury; σ represents the surface tension pressed into mercury; V denotes the volume pressed into mercury; S is the pore surface area of the sample; and θ refers to the mercury infiltration angle.

According to dimensional analysis, the pore surface area S of the tested sample can be expressed by the amount of mercury intake V and the pore size r. In the mercury-intake stage, Formula (9) can be obtained after discretization:

$$\sum_{i=1}^{n} \overline{p}_i \Delta V_i = C r_n^2 \left(V_n^{1/3}/r_n \right)^{D_p} \tag{10}$$

where n refers to the interval number of pressure applied throughout the mercury-intake stage; $\overline{p}$ and ΔV_i represent the average pressure and mercury intake corresponding to the ith mercury intake, respectively; r_n and V_n denote the corresponding pore size and cumulative mercury intake at the nth mercury intake, respectively; D_p is the fractal dimension of pore surface area based on thermodynamic relationship.

Let $W_n = \sum_{i=1}^{n} \overline{p}_i \Delta V_i$, $Q_n = V_n^{1/3}/r_n$, and substitute it into Formula (10), and take the logarithm:

$$\ln\left(W_n/r_n^2\right) = D_p \ln Q_n + \ln C \tag{11}$$

5.2. Fractal Characteristics of Pore Structure

The fractal dimension is the characterization of the randomness and irregularity of the internal pores in the concrete, reflecting the distribution characteristics of the internal pores of concrete. The larger the fractal dimension, the more complex the pore structure inside the concrete, that is, the higher the distribution characteristics of pores and the complexity of their composition, specifically, the higher the content of large-aperture pores, the lower the content of small-aperture pores. According to Formula (11) and considering the mercury intrusion test results, the $\ln(W_n/r_n^2)$ and $\ln Q_n$ values of each group of samples are obtained, and the linear fitting of the two is performed. The x and y in the fitting equation shown in Figure 17 represent $\ln Q_n$ and $\ln(W_n/r_n^2)$, respectively. The corresponding fitting correlation coefficients R^2 are all above 0.998; therefore, the slope after curve fitting of each group could be used as the fractal dimension D_p of the pore surface area of this group of specimens. The relationship between the samples of each group D_p and CNFs dosage is shown in Figure 18. The analysis shows that the relationship between CNFs dosage and D_p is opposite to the strength characteristics of the concrete, that is, with the increase in CNFs dosage, D_p decreases first and then increases, and reaches the minimum at the dosage of 0.3%, which also proves the improvement law of CNFs on the internal pore structure of the concrete.

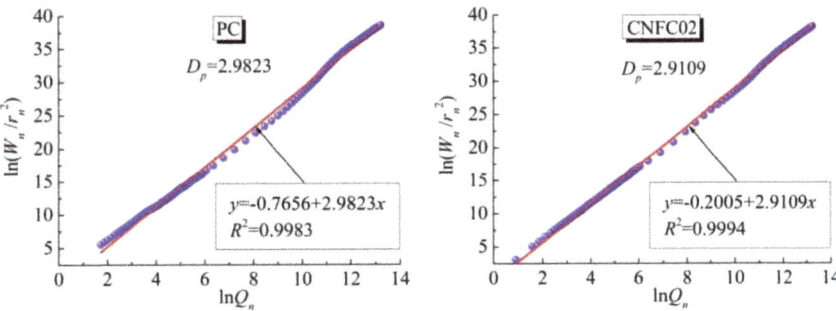

Figure 17. $\ln(W_{ii}/r_{ii}^2)$-$\ln Q_n$ linear fitting curve.

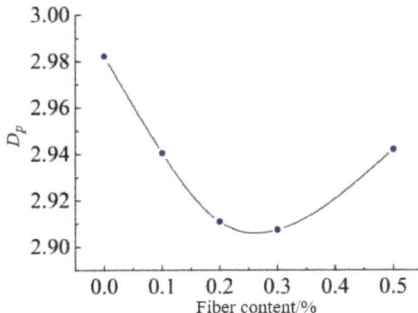

Figure 18. The relationship between D_p and CNFs dosage.

In summary, the internal pore structure characteristics of CNFC show significant fractal characteristics. The addition of CNFs into concrete can mitigate the defects, such as filling the micro voids in concrete, etc., thus improving the original pore structure of the concrete, and effectively promoting its dynamic mechanical properties. On the one hand, the addition of CNFs has a significant improvement effect on the internal pore content of concrete, which is manifested by a sharp decrease in the pore content of large aperture. Although the pore content of small aperture increases slightly, the total pore content in the matrix decreases significantly, indicating that the compactness of the material has been effectively improved. The diameter of single filament of CNFs is only 100~200 nm, and the addition of CNFs can have a good filling effect on the dry shrinkage cracks produced by concrete molding. For the micropores generated after evaporation and consumption of water, the content of such pores is generally small, and the pore size is small. However, CNFs can be dispersed within such pores, thus further reducing the content of such pores. Therefore, CNFs can effectively reduce the pore content in concrete, and thus improve the compactness of the concrete. On the other hand, the addition of CNFs can improve the internal pore structure of concrete, reducing the proportion of large-aperture pores and increasing the proportion of small-aperture pores, indicating that the internal pore structure of concrete is effectively refined. For the pores with pore size greater than 100 nm, the addition of an appropriate amount of CNFs can effectively fill them, resulting in significantly decrease in such pores. At the same time, due to the addition of CNFs, a small amount of pores will be generated on the contact interface between CNFs and concrete matrix. Such pores are generally micropores with pore size less than 100 nm, therefore, to a certain extent, the content of such pores will increase.

6. Conclusions

The impact compression tests of PC and CNFC were carried out using Φ100 mm SHPB test device. The dynamic compressive strength, impact toughness and fragmentation size distribution law of PC and CNFC were analyzed, respectively. Considering the macroscopic failure mode and microscopic pore structure characteristics of the specimens, the change mechanism of the dynamic compressive mechanical properties of the concrete was explained based on fractal theory. The main conclusions are as follows:

(1) Both the dynamic compressive strength and the impact toughness of PC and CNFC increase continuously with the increase in strain rate level at the same dosage of CNFs, showing a strong strengthening effect of the strain rate—besides, both the dynamic compressive strength and impact toughness of CNFC are higher than their counterparts of PC.

(2) At the same strain rate level, the impact toughness of both PC and CNFC gradually increases with the increase in CNFs dosage, while the dynamic compressive strength first increases and then decreases with the increase in CNFs dosage.

(3) The fragmentation size distribution of PC and CNFC after impact failure shows self-similarity, which is a fractal in a statistical sense. In general, the higher the strain rate level, the larger the number of fragments, the smaller the size, and the larger the fractal dimension of particle size.

(4) CNFs can improve the internal pore structure of concrete, play the role of crack resistance, and effectively enhance the macroscopic dynamic mechanical properties of concrete. The optimal dosage of CNFs to improve the dynamic compressive strength of concrete is 0.3%, and the microscopic pore structure characteristics of CNFC show significant fractal features.

Author Contributions: Conceptualization, W.X. and J.X.; methodology, W.X.; software, L.N.; validation, J.X.; formal analysis, W.X.; investigation, J.X.; resources, W.X. and J.X.; data curation, W.X.; writing—original draft preparation, W.X.; writing—review and editing, L.N.; visualization, J.X. and L.N.; project administration, J.X. All authors have read and agreed to the published version of the manuscript.

Funding: This research was funded by the National Natural Science Foundation of China, grant number 51208507 and 51378497.

Institutional Review Board Statement: The study did not involve humans or animals.

Informed Consent Statement: The study did not involve humans.

Data Availability Statement: The data used to support the findings of this study are available from the corresponding author upon reasonable request.

Acknowledgments: The authors express their thanks to all members of the laboratory team for their help with the technical support.

Conflicts of Interest: The authors declare no conflict of interest.

References

1. Torok, R. The 9/11 commission report and the reframing of the 'war on terror' as a new type of warfare. *J. Polic. Intell. Count. Terror.* **2011**, *6*, 137–150. [CrossRef]
2. Bamonte, P.; Gambarova, P.G. A study on the mechanical properties of self-compacting concrete at high temperature and after cooling. *Mater. Struct. Constr.* **2012**, *45*, 1375–1387. [CrossRef]
3. Akca, A.H.; Özyurt, N. Effects of re-curing on residual mechanical properties of concrete after high temperature exposure. *Constr. Build. Mater.* **2018**, *159*, 540–552. [CrossRef]
4. Yongliang, L.; Xiangming, K.; Yanrong, Z.; Peiyu, Y. Static and Dynamic Mechanical Properties of Cement-Asphalt Compsites. *J. Mater. Civ. Eng.* **2013**, *25*, 1489–1497. [CrossRef]
5. Ossa, A.; Gaxiola, A. Effect of water on the triaxial behavior under dynamic loading of asphalt concrete used in impervious barriers. *Constr. Build. Mater.* **2019**, *208*, 333–342. [CrossRef]
6. Sun, X.; Zhao, K.; Li, Y.; Huang, R.; Ye, Z.; Zhang, Y.; Ma, J. A study of strain-rate effect and fiber reinforcement effect on dynamic behavior of steel fiber-reinforced concrete. *Constr. Build. Mater.* **2018**, *158*, 657–669. [CrossRef]
7. Afroughsabet, V.; Biolzi, L.; Ozbakkaloglu, T. High-Performance Fiber-Reinforced Concrete: A Review. *J. Mater. Sci.* **2016**, *51*, 6517–6551. [CrossRef]
8. Hassan, M.; Wille, K. Comparative experimental investigations on the compressive impact behavior of fiber-reinforced ultra high-performance concretes using split Hopkinson pressure bar. *Constr. Build. Mater.* **2018**, *191*, 398–410. [CrossRef]

9. Daghash, S.M.; Soliman, E.M.; Kandil, U.F.; Reda Taha, M.M. Improving Impact Resistance of Polymer Concrete Using CNTs. *Int. J. Concr. Struct. Mater.* **2016**, *10*, 539–553. [CrossRef]
10. Li, W.; Li, M.; Adair, K.R.; Sun, X.; Yu, Y. Carbon nanofiber-based nanostructures for lithium-ion and sodium-ion batteries. *J. Mater. Chem. A* **2017**, *5*, 13882–13906. [CrossRef]
11. Gao, Y.; Zhu, X.; Corr, D.J.; Konsta-Gdoutos, M.S.; Shah, S.P. Characterization of the interfacial transition zone of CNF-Reinforced cementitious composites. *Cem. Concr. Compos.* **2019**, *99*, 130–139. [CrossRef]
12. Galao, O.; Baeza, F.J.; Zornoza, E.; Garcés, P. Strain and damage sensing properties on multifunctional cement composites with CNF admixture. *Cem. Concr. Compos.* **2014**, *46*, 90–98. [CrossRef]
13. Gao, D.; Sturm, M.; Mo, Y.L. Erratum: Electrical resistance of carbon-nanofiber concrete. *Smart Mater. Struct.* **2009**, *18*, 095039. [CrossRef]
14. Konsta-Gdoutos, M.S.; Aza, C.A. Self sensing carbon nanotube (CNT) and nanofiber (CNF) cementitious composites for real time damage assessment in smart structures. *Cem. Concr. Compos.* **2014**, *53*, 162–169. [CrossRef]
15. Meng, W.; Khayat, K.H. Effect of graphite nanoplatelets and carbon nanofibers on rheology, hydration, shrinkage, mechanical properties, and microstructure of UHPC. *Cem. Concr. Res.* **2018**, *105*, 64–71. [CrossRef]
16. Sanchez, F.; Ince, C. Microstructure and macroscopic properties of hybrid carbon nanofiber/silica fume cement composites. *Compos. Sci. Technol.* **2009**, *69*, 1310–1318. [CrossRef]
17. Wang, B.M.; Zhang, Y. Synthesis and properties of carbon nanofibers filled cement-based composites combined with new surfactant methylcellulose. *Mater. Express* **2014**, *4*, 177–182. [CrossRef]
18. Cao, S.; Hou, X.; Rong, Q.; Zheng, W.; Abid, M.; Li, G. Effect of specimen size on dynamic compressive properties of fiber-reinforced reactive powder concrete at high strain rates. *Constr. Build. Mater.* **2019**, *194*, 71–82. [CrossRef]
19. Ren, W.; Xu, J.; Su, H. Dynamic compressive behavior of basalt fiber reinforced concrete after exposure to elevated temperatures. *Fire Mater.* **2016**, *40*, 738–755. [CrossRef]
20. Cao, S.; Hou, X.; Rong, Q. Dynamic compressive properties of reactive powder concrete at high temperature: A review. *Cem. Concr. Compos.* **2020**, *110*, 103568. [CrossRef]
21. Ren, W.; Xu, J. Fractal Characteristics of Concrete Fragmentation under Impact Loading. *J. Mater. Civ. Eng.* **2017**, *29*, 04016244. [CrossRef]
22. Gay, C.; Sanchez, F. Performance of carbon nanofiber-cement composites with a high-range water reducer. *Transp. Res. Rec.* **2010**, 109–113. [CrossRef]
23. Bagher Shemirani, A.; Naghdabadi, R.; Ashrafi, M.J. Experimental and numerical study on choosing proper pulse shapers for testing concrete specimens by split Hopkinson pressure bar apparatus. *Constr. Build. Mater.* **2016**, *125*, 326–336. [CrossRef]
24. Nie, L.; Xu, J.; Bai, E. Dynamic stress-strain relationship of concrete subjected to chloride and sulfate attack. *Constr. Build. Mater.* **2018**, *165*, 232–240. [CrossRef]
25. Lv, T.H.; Chen, X.W.; Chen, G. Analysis on the waveform features of the split Hopkinson pressure bar tests of plain concrete specimen. *Int. J. Impact Eng.* **2017**, *103*, 107–123. [CrossRef]
26. Hassan, M.; Wille, K. Experimental impact analysis on ultra-high performance concrete (UHPC) for achieving stress equilibrium (SE) and constant strain rate (CSR) in Split Hopkinson pressure bar (SHPB) using pulse shaping technique. *Constr. Build. Mater.* **2017**, *144*, 747–757. [CrossRef]
27. Hao, Y.; Hao, H. Dynamic compressive behaviour of spiral steel fibre reinforced concrete in split Hopkinson pressure bar tests. *Constr. Build. Mater.* **2013**, *48*, 521–532. [CrossRef]
28. Hou, X.; Cao, S.; Zheng, W.; Rong, Q.; Li, G. Experimental study on dynamic compressive properties of fiber-reinforced reactive powder concrete at high strain rates. *Eng. Struct.* **2018**, *169*, 119–130. [CrossRef]
29. Zhang, B.; Liu, W.; Liu, X. Scale-dependent nature of the surface fractal dimension for bi- and multi-disperse porous solids by mercury porosimetry. *Appl. Surf. Sci.* **2006**, *253*, 1349–1355. [CrossRef]
30. Wang, B.M.; Zhang, Y.; Liu, S. Influence of carbon nanofibers on the mechanical performance and microstructure of cement-based materials. *Nanosci. Nanotechnol. Lett.* **2013**, *5*, 1112–1118. [CrossRef]
31. Xie, H.; Zhou, H.W. Application of fractal theory to top-coal caving. *Chaos Solitons Fractals* **2008**, *36*, 797–807. [CrossRef]
32. Rieu, M.; Sposito, G. Fractal Fragmentation, Soil Porosity, and Soil Water Properties: I. Theory. *Soil Sci. Soc. Am. J.* **1991**, *55*, 1231–1238. [CrossRef]
33. Wang, F.; Li, S. Determination of the Surface Fractal Dimension for Porous Media by Capillary Condensation. *Ind. Eng. Chem. Res.* **1997**, *36*, 1598–1602. [CrossRef]

Article

Effect of $K_yAl_4(Si_{8-y}) O_{20}(OH)_4$ Calcined Based-Clay on the Microstructure and Mechanical Performances of High-Performance Concrete

David O. Nduka [1,*], Babatunde J. Olawuyi [2], Olabosipo I. Fagbenle [1] and Belén G. Fonteboa [3]

[1] Department of Building Technology, College of Science and Technology, Km 10 Idiroko Road, Covenant University, Ota 112233, Nigeria; olabosipo.fagbenle@covenantunivesity.edu.ng
[2] Department of Building, School of Environmental Technology, Federal University of Technology, Minna 920211, Nigeria; babatunde@futminna.edu.ng
[3] Department of Civil Engineering, School of Civil Engineering, University of A Coruña, 15001 A Coruña, Spain; belen.gonzalez.fonteboa@udc.es
* Correspondence: david.nduka@covenantuniversity.edu.ng

Abstract: The work described in this paper has been performed to determine the potential use of meta-illite ($K_yAl_4(Si_{8-y}) O_{20}(OH)_4$) calcined clay (MCC) as a supplementary cementitious material (SCM) in a binary Portland cement (PC) for high-performance concrete (HPC) production. To obtain the properties of the cementitious materials, the chemical composition, mineral phases, morphology, calcination efficiency and physical properties were quantitatively analysed using the advanced techniques of X-ray fluorescence (XRF), scanning electron microscopy/energy dispersive X-ray (SEM/EDX), X-ray diffraction (XRD), Fourier transform infrared/attenuated total reflection (FTIR/ATR), thermogravimetric analysis (TGA), laser particle sizing and Brunauer–Emmett–Teller (BET) nitrogen absorption method. The MCC's effect on the workability and mechanical properties (compressive, splitting tensile and flexural strengths) and microstructure (morphology and crystalline phases) of hardened MCC-based HPCs were determined. The XRF result shows that the oxide composition of MCC confirmed the pozzolanic material requirements with recorded high useful oxides content. At the same time, the SEM image presents particles of broad, solid masses with a wider surface area of irregular shape. The XRD results show that the MCC was majorly an illite-based clay mineral calcined at a maximum temperature of 650 °C, as revealed by the TGA. The MCC addition increases the slump flow of HPCs at 5–15% cement replacement. The MCC incorporation at 10% cement replacement best improved the porosity of HPCs at a later age resulting in increased mechanical and microstructural properties of tested samples. Therefore, it is recommended that MCC addition within 10% cement replacement be adopted for low W/B Class I HPC at no deleterious results on mechanical and microstructural properties of the concrete.

Keywords: dehydroxylation; high-performance concrete; superabsorbent polymers; superplaticiser; meta-illite calcined clay; supplementary cementitious materials

1. Introduction

The past two decades have seen the rapid use of high-performance concrete (HPC) in constructing critical structural elements of super-tall buildings and other complex architecture, engineering, and construction (AEC) structures worldwide. HPC is an innovative high quality, and cost-efficient concrete compared to normal-strength concrete that meets the new generation's desire for complex engineering structures [1,2]. Such a fact is not surprising because HPC had provided a pleasant living environment and safety in high-rise buildings and other AEC facilities. Specifically, HPC is essentially a concrete with a low-water-to-binder (W/B) ranging from 0.2–0.38 [3] that meets the performance challenges of structural elements regarding increasing heights, span length, and load. Moreover, many demonstrated AEC projects had been accomplished using HPC in many countries [4,5]. Therefore, the use of HPC in a developing country such as Nigeria would improve infrastructure projects' future performances.

Among the various constituents of HPC, supplementary cementitious materials (SCMs) of pozzolanic nature play significant roles in meeting HPC requirements. SCMs are mainly siliceous/aluminous finely divided solid minerals used as partial/whole substitutes for cement in concrete and mortar production. SCM in the cement matrix will react chemically to deplete calcium hydroxide to form a more cementitious product later [6]. Significance of SCMs includes the refinement and improvement of pore size distribution; capillary pores and interfacial transition zone of hardened concrete; reduced large pores; improved density of cementitious products; improved workability in the fresh state; and lowering of W/B in HPC mix [1,7]. In addition, SCMs have also been documented to have the potentials to reduce carbon emissions generated during cement production and promising resources in lowering clinker content [8].

In this respect, thermally transformed illite-based clay presents a viable option as SCM in HPC due to the global availability, economic and circular economy attributes [9] and little effects on water demand and 28-day compressive strength [10], among others. Illite is a 2:1 structured aluminosilicate clay mineral sandwich of silica tetrahedron (T)—alumina octahedral (O)—silica tetrahedron (T) layers in the mica family [11]. Song et al. [11] gave the chemical formula as $K_y Al_4(Si_{8-y}) O_{20}(OH)_4$, here y is estimated to be 1.5. The dehydroxylation of illite commences at about 350 °C and terminates at 800 °C with loss in crystallinity [12]. In a pozzolanic reactivity test, Avet et al. [12] showed that low-grade kaolinitic calcined clays' compressive strength compared favourably with plain Portland cement. Zhuo et al. [6] demonstrated that abandoned London clay calcined up to 900 °C exhibited similar compressive strength properties with slag and pulverised fuel ash in concrete application. Trümer et al. [13] developed a binary cementitious material that consists of 30 wt.% of calcined bentonite clay and 70 wt.% of PC in concrete. They suggested that their calcined clay could apply to the majority of concrete works. Ferreiro et al. [14] indicated that 2:1 structured illite clay thermally activated at a temperature of 930 °C is appropriate for developing ternary calcined clay-limestone blended cement, which corresponds to improved workability and strength performance in concrete.

Also, Irassar et al. [15] experimentally studied the thermal transformation of raw illite-chlorite-based clay to understand the material production's best calcination conditions for concrete use. Similarly, Laidani et al. [16] used bentonite-rich calcined clay as an SCM to improve self-compacting concrete's fresh and hardened (i.e., mechanical and durability) properties. The authors reported improved compressive strength at 15–20% content of the calcined bentonite clay. Finally, Marchetti et al. [10] demonstrated the applicability and performance of illite calcined clay thermally activated at a temperature of 950 °C on a low-energy mortar. The authors' findings revealed an enhanced packing density and compressive strength of mortar at a later age. Therefore, using illite calcined clay as a cement replacement could be regarded as a sustainable approach in HPC production due to the improved mechanical, durability and microstructural properties and minimisation of PC consumption.

With many studies conducted on HPC in many regions, information on HPC materials and structural properties produced with calcined clay is still scarce in Nigeria. The present study attempts to fill the gap in the literature by investigating the influence of meta-illite calcined clay (MCC) on Class 1 (50–75 MPa) HPC internally cured with superabsorbent polymers (SAP). The justification of this research is that for the first time, a Nigeria manufactured Pozzolan (NBRRI cement) named MCC in this study, which is readily available, is to be investigated as a binder component for use in HPC. This study will further provide direction to construction industry stakeholders on new materials that can revolutionise high-rise buildings and other heavy civil engineering infrastructure projects in developing countries such as Nigeria.

2. Materials and Methods

2.1. Materials

Portland-limestone cement (CEM II B-L, 42.5 N)—"3X" produced by Dangote cement PLC., Ibese Plant Ogun, State, Nigeria conforming to BS EN 197-1 [17], and NIS 444-1 [18] was used as the main binder. A commercially available Pozzolan (MCC) manufactured by the Pozzolan Cement Plant of the Nigeria Building and Road Research Institute (NBRRI), Ota, Ogun State Nigeria, served as the SCM. The SCM was incorporated in powdered form for the various HPC mixtures required by the mix design. Masterglenium Sky 504—a polycarboxylic ether (PCE) polymer-based superplasticiser supplied by BASF Limited (West Africa) was used to improve the workability of the HPC mixtures and administered within the manufacturer's optimum specification of $\leq$2% by weight of binder (b_{wob}). The specific gravity of the superplaticiser was 1.115, and it is chlorine-free. Superabsorbent polymers (SAP) tagged "FLOSET 27CC" $\geq$ 300 µm as described in an earlier publication of Olawuyi and Boshoff [19] at a constant content of 0.3% by weight of binder (b_{wob}) was used as an internal curing agent. As specified by BS EN 1008 [20], potable water available within the concrete laboratory of the Department of Building Technology, Covenant University, Ota, was used for mixing.

The river sand used as fine aggregate was at the air-dry condition with a minimum particle size of 300 µm (i.e., all the particles smaller than 300 µm were removed using the sieving method) in compliance with the requirement of fine aggregate specification for HPC production [21–23]. Crushed granite stone passing through 13.50 mm sieve size and retained on 9.50 mm sieve size was used as coarse aggregate in compliance with typical HPC mixes found in the literature [4,22–24]. The crushed granite was used in saturated surface dry conditions after it has been washed to eliminate fine content that will likely increase water demand. Results of the physical characteristics of the materials are presented in Section 3.1. For a more scientific explanation of the binders, laser diffraction PSD, Brunauer–Emmett–Teller (BET) specific surface area (SSA), specific gravity, initial and final setting times, and soundness CEM II and MCC were determined. The binders' PSD was performed using a Malvern Mastersizer 3000. The specific surface area of each binder was measured by nitrogen adsorption for the BET model using Nova Station B Quantachrome Instrument.

The powdery binder samples' chemical compositions were investigated using a wavelength dispersive X-ray fluorescence (XRF) [Bruker AXS. S4, explorer]. The scanning electron microscopy/energy-dispersive X-ray(SEM/EDX) was performed to examine the binder's samples' morphology and microstructure using SEM (Model: Phenom ProX, PhenomWorld Eindhoven, Netherlands). The mineralogical phases of the binders were measured using a Rigaku Miniflex 600, Japan X-ray diffraction (XRD) technique adopting the reflection-transmission spinner stage of Theta-Theta settings. The two-Theta starting position was 2° and ended at 75° with a two-theta step of 0.026 at 8.67 s per step.

The MCC in powdery forms was analysed for functional groups using Cary 630 Fourier-transform infrared spectroscopy/Attenuated total reflectance (FTIR/ATR) spectrometer made by Agilent Technologies, Malaysia. The powdery sample was placed on the sample stage while the optimal spectra in the range of 500–4000 cm^{-1} were obtained at a high speed greater than 110 spectra/s. The calcination efficiency of MCC was analysed using the PerkinElmer thermogravimetry (TGA), the Netherlands, operating at a maximum temperature of 1200 °C and a maximum heat rate of 20 °C.

2.2. HPC Production

2.2.1. Mix Proportions

Table 1 shows the mix compositions of the seven HPC mixtures designed for 28-day Class 1, HPC having characteristic cube strength of 67 MPa (i.e., C55/67) following the margin for mix design. The equation can be written as: $f_m = f_c + ks$; where f_m = target mean strength; f_c = the specified characteristic strength; ks = the margin, which is a product of s = the standard deviation and k = a constant. The binder Type 1 (i.e., CEM II only) was adopted as an HPC control mixture and denoted as control. Binder Type 2 (i.e., CEM II + MCC) was adopted for MCCC-5 to MCCC-30 HPC mixtures. The MCCC-5 to MCCC-30 refers to the blend of CEM II with MCC from 5% to 30% at 5% intervals of SCM content adopted for the HPC mixtures. All the mixtures were prepared at a fixed W/B of 0.3%, fixed SAP content of 0.3% (b_{wob}), and superplasticiser content (1.5% b_{wob}). Additional water of 12.5 g/g of SAP was provided on the ground of the SAP absorbency determined by the work of Olawuyi [25]. Every constituent of HPC was measured by weight (kg/m^3), and hence, the MCC addition was taken to be by weight of the binder. This same measurement was also adopted for SAP contents, as reflected in Table 1 following the British method of HPC design. Several studies have adopted the weight (%) method of cement replacement to arrive at their desired mix [6,26–30]. The HPC groups were designated with the name MCCC with MCC content reflected. For instance, the HPC containing MCC with 5% MCC content was coded as MCCC-5.

Table 1. Mix constituents of HPC with MCC.

Constituents	Mix Blends (kg/m^3)						
	Control	MCCC-5	MCCC-10	MCCC-15	MCCC-20	MCCC-25	MCCC-30
Water	156	156	156	156	156	156	156
Cement (CEM II)	540	513	486	459	432	405	378
MCC	0	27	54	81	108	135	162
Coarse aggregate	1050	1050	1050	1050	1050	1050	1050
Sand ($\geq$300 μm)	700	700	700	700	700	700	700
SAP (0.3% b_{wob})	1.62	1.62	1.62	1.62	1.62	1.62	1.62
Superplastiser (1.5% b_{wob})	8.10	8.10	8.10	8.10	8.10	8.10	8.10
Water/binder (W/B) *	0.3	0.3	0.3	0.3	0.3	0.3	0.3
Additional water	20.30	20.30	20.30	20.30	20.30	20.30	20.30

* W/B = ((water + liquid content of superplasticiser)/(cement + MCC).

2.2.2. Batching Procedure and Curing Conditions

The fine aggregate was first poured into the 50-litre capacity pan-mixer, followed by the binders, which had first been thoroughly hand-mixed to enhance the MCC's dispersion until a uniform colour was observed. After mixing for about 30 s, the dry SAP particles were then poured in, and all the fine contents were mixed for another 30 s. The coarse aggregate was then added, and mixing continued for another 1 min before water already mixed with superplasticiser (Masterglenium Sky 504—a PCE) was added. The mixing continued for another 3 min, as recommended in the literature [21–23,25]. Only about half of the mixer volume was maintained as the maximum volume of concrete produced per batch, noting that the mix is very stiff for these low W/B concrete and becomes difficult if 50% volume of the mixer is exceeded. Once the mixture met the required workability and consistency for the specified design mix, a specimen for the various tests was cast in two layers on a vibrating table into the previously oiled moulds. The cast specimen was then covered in the laboratory with thick vapour barriers (jute bags) and allowed to harden for 24 h, before demoulding and placed to cure immersed in water in a curing tank at 20 $\pm$ 3 °C till the requisite curing ages (7, 28, 56 and 90 days respectively) before testing in accordance to standards [31].

2.3. Test Methods

2.3.1. Workability

Slump flow measurement was conducted using the flow table test described in BS EN 12350-5 [32] to measure the workability of HPC mixtures.

2.3.2. Setting Times Test

The setting times (initial and final) of the HPC mixtures were determined using a penetration resistance method under ASTM C403 [33]. A standard 4.75 mm sieve was used to obtain mortar samples from the fresh HPC mixtures which were cast in two layers into 150 mm cube moulds to about 10 mm below the height edge. The specimens were then kept in a climate control room set at 24 ± 2 °C temperature and 70 ± 5% humidity. The pocket type H-4134 concrete mortar penetrometer was then used to measure the concrete's resistance by forcing the penetrometer's shaft into the mortar to a depth of 25 mm at a constant rate and time intervals. The resistance in MPa is shown on the penetrometer's direct-reading scale. The initial set of concrete is reached when the penetration resistance is 3.50 MPa. The initial set is the semi-hardened, partially hydrated condition of the concrete beyond which it can no longer be worked or consolidated by vibration. The plot of penetration resistance on (Y-axis) against time in minutes (X-axis) gives the initial (3.5 MPa resistance) and final setting (27.6 MPa resistance) times, respectively.

2.3.3. Mechanical Properties

Compressive and splitting tensile strengths were determined in line with BS EN 12390-3 and 6 [34] and RILEM Technical Recommendation TC14-CPC 4 [35], respectively. 210 triplicates of 100 mm cubes and 126 Ø100 × 200 mm cylinders were used to investigate the compressive and splitting tensile strengths using the digitised materials testing machine (Model YES-2000, Eccles Technical Engineering Ltd., Eccles, England) with 2000 kN maximum loading capacity. In addition, 126 prismatic beams of 100 × 100 × 500 mm were tested for flexural strength test using a manually operated three-points contact 50 kN Impact AO 320 flexural machine following BS EN 12390-5 [36].

2.3.4. Non-Destructive Tests on HPC Mixtures

Selected HPC cube specimens of the respective MCC contents (control, 10% and 20%), at constant SAP cured for 90 days, were subjected to SEM-EDX and XRD examinations. Thus, this procedure enables the understanding of the effect of MCC contents on the mechanical behaviours of the HPC. SEM data was collected on thin sections of about 3 mm thickness samples after hydration stoppage. The hydration of the specimen was truncated by soaking the crushed HPC samples in acetone. XRD analysis of HPCs samples was carried out on remains of crushed hardened specimen passing through 75 μm standard sieve.

3. Results and Discussions

3.1. Material Characterisation

3.1.1. Physical Properties

The physical characteristics of the sand analysed through the sieve particle size distribution (PSD) were fineness modulus (FM) = 2.87; coefficient of uniformity (C_u) = 2.39; coefficient of curvature (C_c) = 0.94; dust content = 0.45%, specific gravity (SG) value of 2.65 and water absorption of 1.2% was recorded in the physical properties' tests and presented in Figure 1 and Table 2, respectively. The granite analysed for the study had a specific gravity of 2.7, water absorption of 1.05%, aggregate crushing and impact values of 28% and 11%, respectively.

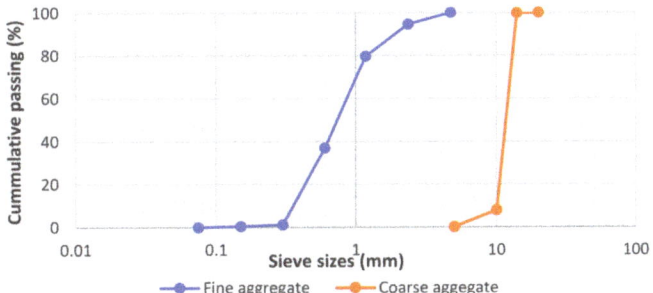

Figure 1. Particle size distribution of aggregates.

Table 2. Physical and mechanical properties of aggregates.

Properties	Sand	Granite
Fineness modulus	2.87	-
Specific gravity	2.65	2.7
Water absorption, %	1.44	1.26
Aggregate crushing value, %	-	28
Aggregate impact value, %	-	11
D_{10}	360	10,000
D_{30}	540	11,000
D_{60}	860	13,000
C_u	2.39	1.30
C_c	0.94	0.93

The PSD plot for the MCC and CEM II is presented in Figure 2, which shows that 90% (D_{90}) of CEM II and MCC particles are smaller than 4000 and 950 nm, respectively. The median particle size, D_{50} of CEM II and MCC are 48.8 and 450 nm, respectively. Furthermore, the particle size below 10% (D_{10}) falls within 4.88 and 275 nm, respectively, for the same samples. From Table 3, the SSA measured via the SinglePoint and MultiPoint BET model showed 5.590×10^2 and 3.026×10^2 m^2/g and 8.182×10^2 and 4.649×10^2 m^2/g for CEM II and MCC, respectively. The pore diameter of the binders analysed through the DA BET mode indicated corresponding values of 2.92 and 2.88 nm for CEM II and MCC samples. Based on these values, CEM II is finer than MCC. The differences in particle size seen in the binders may be attributed to the different production methods used. Comparing the two binders DA BET analysis, the MCC sample has a lower pore size diameter, conforming to macro-mesoporous material [37].

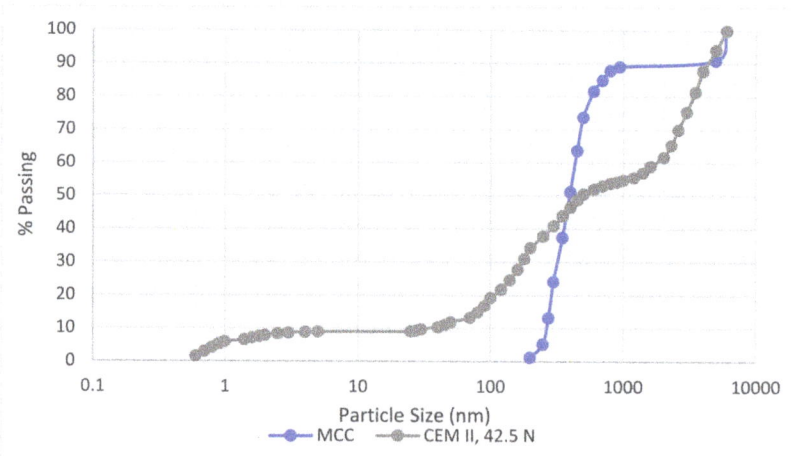

Figure 2. Particle size distribution of MCC and CEM II.

Table 3. Physical properties of CEM II and MCC.

Binders	BET SSA (m²/g)		Specific Gravity	Initial and Final Setting Time (min)	Soundness (%)	D_{90}	D_{50}	D_{10}	Pore Diameter Mode—DA (nm)
	SinglePoint	MultiPoint							
CEM II	5.590×10^2	8.182×10^2	3.12	90 and 205	0.75	4000	48.8	4.88	2.92
MCC	3.026×10^2	4.649×10^2	2.81			950	450	275	2.88

3.1.2. Chemical and Microstructure Analyses of Binders

Figure 3 depicts the morphology of the dark brown MCC powder at 200 μm XRF magnifications, revealing that MCC powder particles are broad, solid masses of the wider surface area and irregularly shaped. Table 4 further reveals that MCC mainly comprises SiO_2, Al_2O_3 and Fe_2O_3 while CEM II oxide components are CaO, SiO_2 and Al_2O_3.

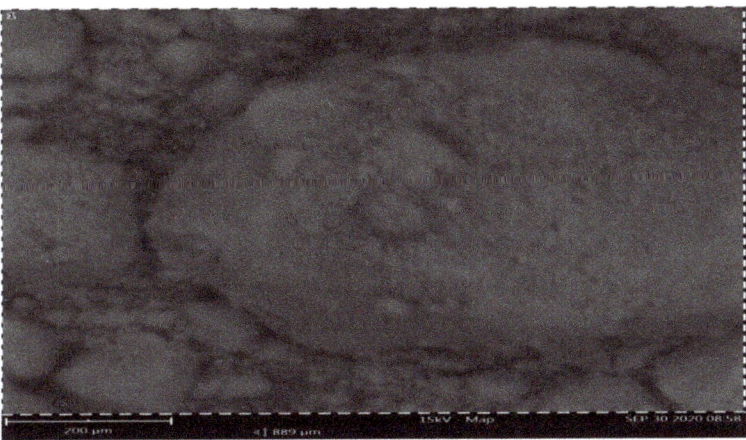

Figure 3. SEM Image of MCC.

Table 4. Oxide Composition of Binder Constituents.

Oxides	MCC (%)	CEM II (%)
SiO_2	60.92	15.38
Al_2O_3	20.92	4.14
Fe_2O_3	3.43	3.19
CaO	1.38	56.92
MgO	0.24	2.44
SO_3	0.00	1.59
K_2O	0.35	0.21
Na_2O	0.00	0.04
M_2O_5	0.12	0.04
P_2O_5	0.18	0.28
TiO_2	3.24	0.21
LOI	7.68	15.59
$SiO_2 + Al_2O_3 + Fe_2O_3$	85.27	22.71
Total	98.46	100.00

The XRD phase spectra presented in Figure 4 shows that illite and kalicinite minerals dominate with 13% mineral contents of the MCC, followed by quartz with 8% mineral representation. Calcite and garnet recorded 6% mineral contents, respectively. Rutile recorded the least mineral content of 5% mineral.

The FTIR/ATR absorption spectra for MCC are shown in Figure 5. The spectra depict the intensities of the OH stretching, Si–O stretching and bending and Al–OH bending bands, indicating the development of kaolinite and illite minerals sensitive to cation exchange [38]. The OH stretching of the inner surface and the outer hydroxyl groups are observed at 3683, 3623 and 3534, 3414 cm^{-1} for sandwiched octahedral sheets between two layers for kaolinites tetrahedral-octahedral-tetrahedral structure of illite. The high region with a band at 3623 cm^{-1} resulted from the low frequency of kaolinites' inner surface hydroxyls, indicating kaolinite and illite minerals sensitive to cation exchange [15]. The bands at 1651 cm^{-1} are due to the deformation of water molecules. Functional Al–OH bending bands at 1033–913 cm^{-1} typical for all smectite mineral clay groups are observed, thus validating the previous work on kaolin's inner hydroxyls [39]. The vibrational modes of the octahedral aluminium ions of kaolinite and the Si–O stretching of quartz are recorded for the MCC sample at 693 cm^{-1} validating the XRD result.

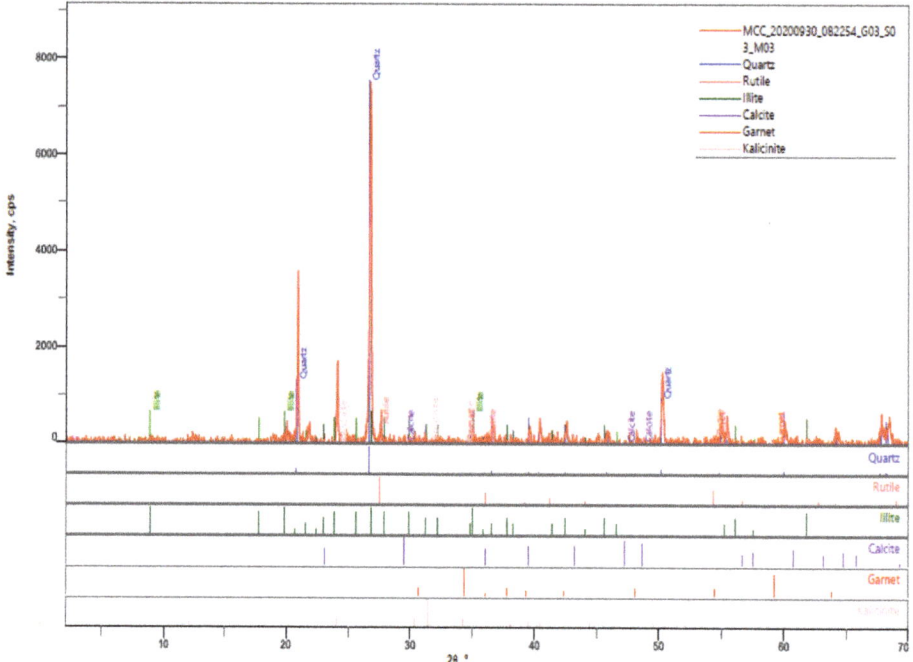

Figure 4. XRD patterns of MCC.

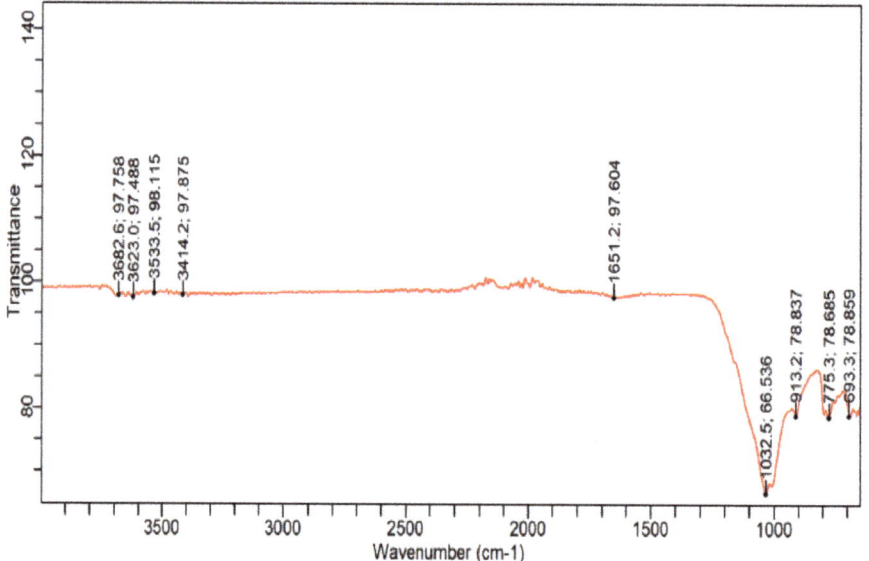

Figure 5. FTIR/ATR patterns of MCC.

Characterising the MCC for TGA (as shown in Figure 6) was borne out of knowing the dehydroxylation trend of the samples. The studied samples' carbonation effects were controlled by maintaining the nitrogen environment with a 50 mL/min flow rate within the heating chamber. As shown in Figure 6, after subjecting the MCC sample to about 270 °C, the absorbed water located in the clay sample's interlayer space got dehydrated. Further measurement to about 350 °C to 450 °C, brought about dehydroxylation of possible kaolinite. At 500 °C, it shows complete dehydroxylation of the remaining kaolinite. At about 650 °C to 888 °C, the TGA curve remains flattened, showing the maximum calcination temperature to which MCC manufacturers subjected the raw clay material. Zhou [39] averred that at 800 °C, a compete dehydroxylation of illite and montmorillonite clay minerals appears. Garg and Skibsted [40] pointed out that illite/smectite mineral-based clay attains dehydration, dehydroxylation, amorphisation and recrystallisation at the temperature ranges of 25 to 200 °C; 600 to 800 °C; 800 to 900 °C; 950 °C and above, respectively. Thus, the MCC showed an amorphous phase considering the flattering of the TGA curve from 650 °C to 888 °C and supports the postulations of Garg and Skibsted [40].

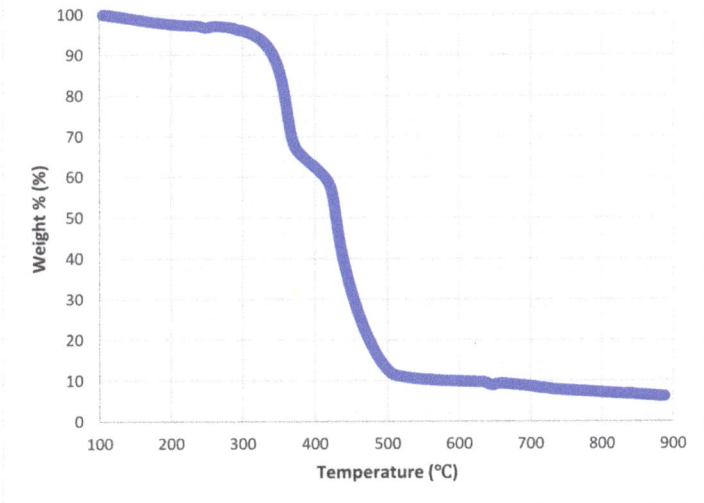

Figure 6. TGA Curve of MCC.

3.2. Fresh Properties

3.2.1. Slump Flow Test

Before casting into moulds, fresh HPC mixtures were examined for workability by slump flow test. The slump flow test results on the HPC mixtures containing MCC is presented in Figure 7.

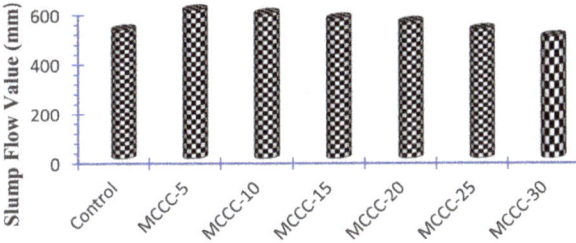

Figure 7. Slump flow of HPC made with different contents of MCC.

The result demonstrates an average flow value of 520 mm for the control mixture. In contrast, the mixtures containing MCC show an improvement inflow up to 20% MCC content. Specifically, there is an improved flow of 15.39%, 12.50%, 8.65%, and 5.77% for MCCC-5 to MCCC-20 compared with the control. MCCC-25 recorded the same flow value as the control, while MCCC-30 had a 5.77% decrease in flow value. These findings suggest the MCC's ability to absorb the Masterglenium Sky 504—a PCE superplasticiser used in the study, causing the dispersion and water reduction tendencies affecting MCC-based HPC [41–43]. The lowest flow value observed by the control mixture may be linked to the more angular shape of the CEM II particle size as compared with the broad irregular shape of clay particle size, as corroborated by SEM characterisation. However, the data obtained from slump flow tests is consistent with HPC mixtures in literature, with a slump flow range of 450–600 mm [23].

3.2.2. Initial and Final Setting Times

Figure 8 portrays the setting times test results for the various binary blend of HPC mortar mixtures using a concrete penetrometer under ASTM C403 [33]. From Figure 8, MCC-based HPC mixtures show that MCC's addition resulted in a gradual increase in the HPC's initial and final setting times. The initial setting time for control, MCCC-5, MCCC-10, MCCC-15, MCCC-20, MCCC-25 and MCCC-30, is 300, 420, 490, 540, 600, and 660 min, respectively and observed to be higher than the control. From the same Figure, the final setting times for control, MCCC-5, MCCC-10, MCCC-15, MCCC-20, MCCC-25 and MCCC-30 are 840, 840, 900, 960, 1020, and 1080 min, respectively. As observed from Figure 9, a relatively lower final setting time occurred in MCC modified HPCs than the initial setting time. The higher initial and final setting values seen for the HPC mixtures are linked to MCC's gradual addition into the HPCs. Cia et al. [44] inferred that the retardation of setting conditions for Portland cement pastes containing mineral clay is connected with the concentration influence of mineral clay and the partial performance of clay mineral's pozzolanic reactivity at an early age.

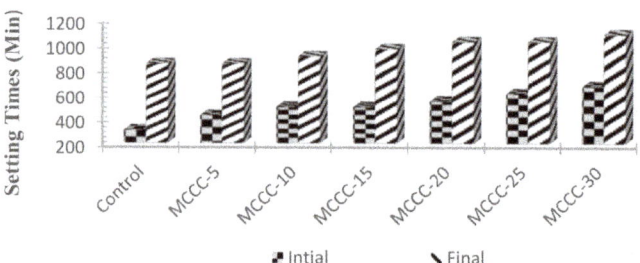

Figure 8. Initial and finals setting times of MCC-based HPC mortar.

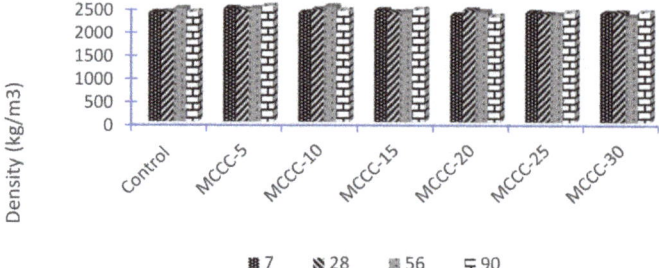

Figure 9. Density of HPC made with different contents of MCC.

3.3. Mechanical Properties

3.3.1. Density of HPC

The average dry densities of the control and various MCC-based HPCs at 7, 28, 56 and 90 hydration days are presented in Figure 9. The figure shows that the average density of the activated HPC samples containing MCC and the control varies from 2390 kg/m^3 to 2387 kg/m^3 for all ages. It also showed that the densities slightly increased by 0.04% as the curing age increases. Comparing the individual HPCs containing MCC and the CEM II, MCCC-5, MCCC-10, and MCCC-15 showed an average increase in dry densities of 2.92%, 1.80% and 0.46% over the control, respectively.

MCCC-20, MCCC-25, and MCCC-30 showed decreased dry densities corresponding to 1.35%, 1.51%, and 1.62% compared to the control over the ninety-day observation. The initial increases in density with lower MCC contents (MCCC-5 to MCCC-15) may be attributed to the binary combination of the cementitious materials filling the voids between the fine aggregates, thereby achieving a denser assembly. The fact that the dry densities of MCCC-5, MCCC-10 and MCCC-15 were improved, which was not the case for MCCC-20, MCCC-25 and MCCC-30, could be the reason that MCC was added as CEM II, replacement by weight. When incorporated to substitute cement to a higher mass, MCC has a lower density, the total volume of powder (CEM II+ MCC) was increased. The addition of MCC's higher weight with PSD closer to CEM II may have reduced the HPC density. This result agrees well with Zhou's [39] reported lower density with the higher replacement of London calcined clay in concrete.

3.3.2. Compressive Strength of HPC with MCC

HPC specimens' compressive strength data having MCC tested at 7, 28, 56 and 90 days are presented in Figure 10. HPC specimen with 5% cement replacement (MCCC-5) had the highest compressive strength at 7 days curing age, followed by MCCC-15, control, MCCC-20 and MCCC-25. There are comparable strength values between the control and the MCC-based HPCs. The control specimen showed a superior strength value at 28 days among other HPCs of the same age. On the other hand, the MCCC-10 HPC specimen had the highest compressive strength at 56- and 90-days curing ages, followed by MCCC-5 and MCCC-15 mixture types. MCCC-25 and MCCC-30 had a reduced strength at 56 and 90 days compared to control. The control and specimen with a higher MCC content had lower strength at later age following dilution effects and high content of calcined clay with low pozzolanic reactivity [6].

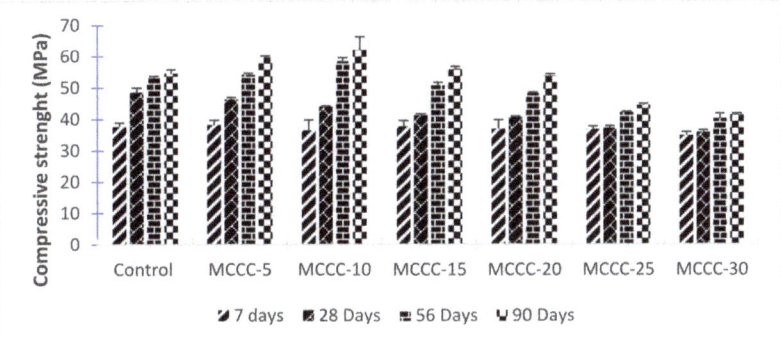

Figure 10. Compressive strength development of MCC-based HPC.

The HPC samples with higher MCC contents produce lower C$_3$S, b-C$_2$S phases, and high-water demand for cement hydration, giving room for decreased compressive strength [16]. The 56 and 90 days MCC-based HPC specimens gained strength at a greater rate than the control. Strength development continues beyond 28 days, and this result demonstrated later age minor pozzolanic reactions, filler effect, MCC fineness and cement hydration accomplishment [45]. It can be inferred that the insignificant content of decomposed kaolinite found in the MCC also had a marginal compressive strength impact on the HPCs design mix.

As NBBRI Pozzolan (MCC) was never used for HPC development before, the comparison with other researchers' results had to be done with similar compositions. Vejmelková et al. [46] prepared a ternary HPC mix containing 10–60% CEM I, 52.5 R cement class replaced with an industry prepared calcined Czech claystone calcined at a maximum temperature of 700 °C and silica fume with a target design strength of 120 MPa. The XRD analysis revealed that the mineralogical compositions of the clay were mostly kaolinite and illite. The authors achieved over 120 MPa compressive strength with 10, 20 and 30% replacements at 28, 90, 180- and 365-days age curing. The best result of Vejmelková et al. [46] was observed at 30% of the clay with compressive strength of ~20% higher than the control. Thus, the attainment of over 120 MPa may be linked to the introduction of silica fume and 1:1 clay structure in the HPC matrix, leading to their samples' improved microstructure. Trümer et al. [13] obtained a C40/50 concrete class with a 30/70 ratio with a CEM I, 42.5 R and calcined montmorillonite-based clay. The raw bentonite's calcination up to 900 °C brings about the total decomposition of montmorillonite mineral culminating in the clay's amorphousness and attainment of the designed strength. Schulze and Rickert [45] reached a strength class of 42.5 N using calcined clay, irrespective of the clay minerals (kaolinite, montmorillonite and muscovite/illite) content investigated. Laidani et al. [16] reported compressive strength of 74 and 70 MPa as their most successful strength, with 5% to 30% of calcined illite and quartz mineral-based clay in the cement-based blend. They found that their best mix showed an improved compressive strength of 20% and 15% over the control.

Strong evidence emanating from the entire result showed that blended CEM II with MCC could not produce the target designed strength of Class 1 HPC (50–75 MPa) at 28 days, while later age curing showed a promising result. Garg and Skibsted [40] pointed out that when illite/smectite clays are blended with Portland cement in mortar and concrete; there appears to be a little form of clay reaction at an early age while there is usually a considerable amount of reaction in a clay-based mortar or concrete mixtures at a later age. A further shortfall in the rate of strength development at an early age may be linked to the MCC's lower calcination temperature below 700 °C.

3.3.3. Splitting Tensile Strength of HPCs with MCC

Splitting tensile strength results of HPCs with varied MCC contents is presented in Figure 11.

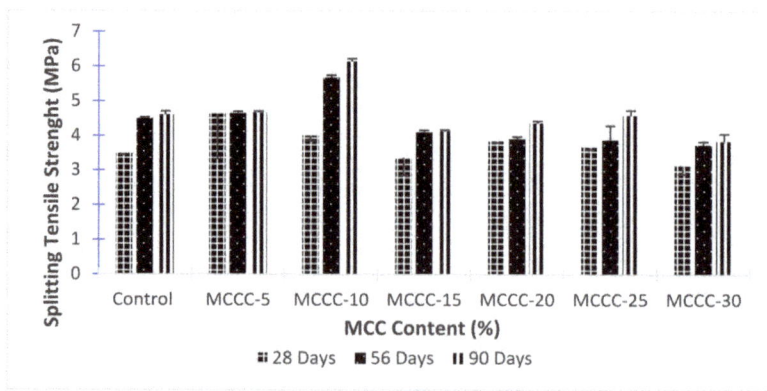

Figure 11. Splitting tensile strength results of HPCs at different treatments with MCC.

From Figure 11, the 28 days of hydration period of MCCC-5, MCCC-10, MCCC-20 and MCCC-25 enhances good splitting tensile strength value of the HPCs than the control. Only the splitting strength value of MCCC-10 was higher by 20.64% than the control specimen at 56 days. At the same age, MCCC-5, MCCC-15, MCCC-20, MCCC-25 and MCCC-30 decreased in strength with ~7%, 9%, 13%, 14% and 17%, respectively compared with control mix. Comparing 90 days splitting tensile strength values of MCCCs with control (4.61 MPa), MCCC-10 performed best with 6.15 MPa, followed by MCCC-5 with a slight margin of 4.67 MPa. Other MCCCs (MCCC-15, MCCC-20 and MCCC-25) recorded close splitting strengths values of 4.14, 4.37 and 4.59 MPa with control. Only MCCC-30 had a decreased value of about 16% compared with the control. The addition of MCC has a moderately positive effect on the splitting tensile strength of the HPC, especially at later hydration ages. These results indicate that 10% cement replacement with MCC (MCCC-10) is sufficient to accelerate 28, 56 and 90 days splitting tensile strength of the HPCs. These results were consistent with the splitting tensile strength result found in the literature [40,46].

3.3.4. Flexural Strength of HPCs with MCC

Flexural strength (modulus of rupture) results of HPCs as influenced by the MCC content is presented in Figure 12.

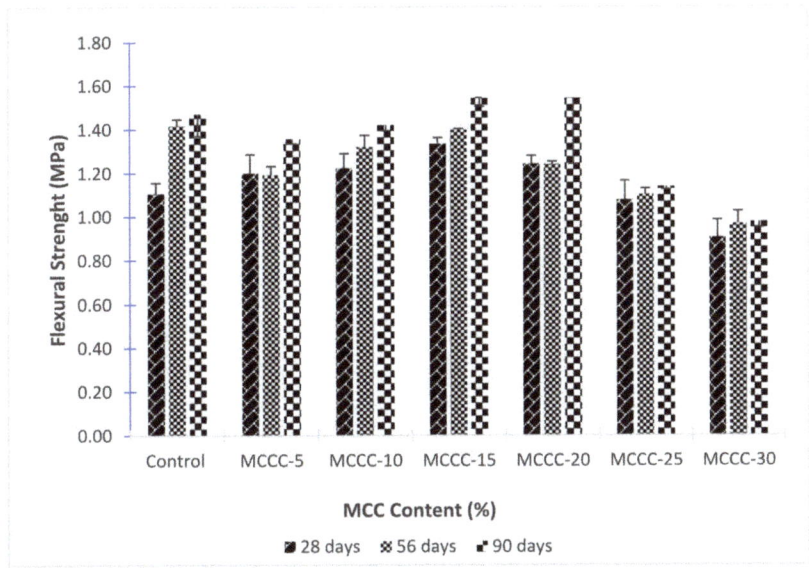

Figure 12. Flexural strength of HPCs at different treatment conditions with RHA.

At 28 days, for Figure 12, the HPC mixes with MCC had higher flexural strength than the reference mix, which can be attributed to the pozzolanic reactivity of MCC compared to the control mix at 28 days. The MCCC-15 with 15% cement replacement with MCC had the highest flexural strength (1.40 MPa). As the hydration progressed for 56 and 90 days, the strength increased faster with MCC mixes than the reference, especially for mix MCCC-15 and MCCC-10, indicating that MCC's pore structure refinement was more efficient in these mixes than others. Possible explanations for the strength improvement are contained in the studies of Garg and Skibsted [40], Zhou et al. [6], Vejmelková et al. [46] and Trümer et al. [13].

3.4. Non-destructive Tests on Hardened HPC Mixtures

SEM/EDX was conducted on selected HPC samples (Control, 10% and 20%) in furtherance to quantitatively assess the hydration products, molecular structure and the bond between the cement paste and aggregate at the interfacial transition zone (ITZ). The SEM images taken at 100 and 200 μm and oxides atomic concentrations for the control, MCCC-10 and MCCC-20 specimen at 90 days, are shown in Figure 13, Figure 14, Figure 15 and Table 5, respectively.

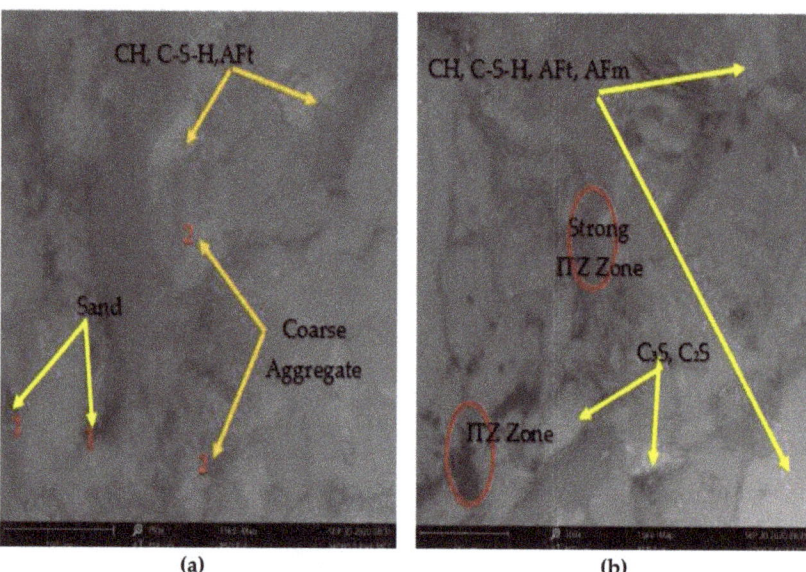

Figure 13. (a,b) SEM images of control HPC at 90 days.

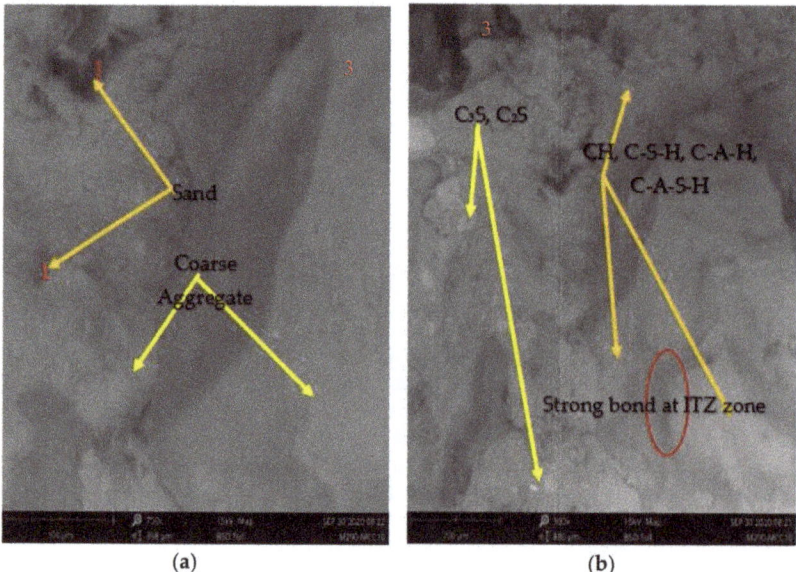

Figure 14. (a,b) SEM images of MCCC-10 HPC at 90 days.

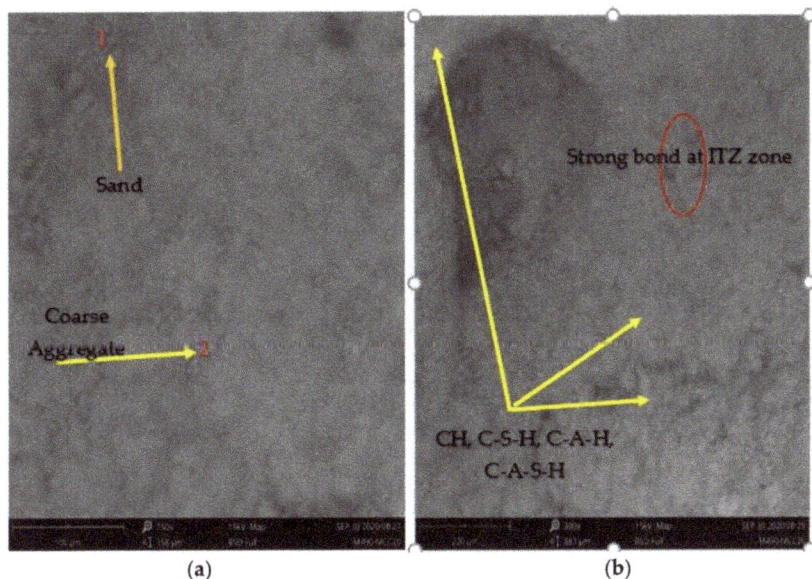

Figure 15. (a,b) SEM images of MCCC-20 HPC at 90 days.

Table 5. Oxides Atomic Concentration of the HPCs from EDX.

Element Symbol	Element Name	Atomic Concentration		
		Control	MCCC-10	MCCC-20
Si	Silicon	34.35	28.72	5.08
Ca	Calcium	18.78	38.99	70.90
O	Oxygen	22.24	22.92	15.37
Fe	Iron	5.11	2.69	1.20
Al	Aluminium	10.20	1.43	1.18
K	Potassium	2.94	1.30	1.19
Na	Sodium	2.59	0.00	0.37
Ti	Titanium	1.13	0.00	0.51
Ag	Silver	0.50	0.58	0.46
S	Sulfur	0.51	0.95	0.60
P	Phosphorus	0.52	0.93	0.60
Mg	Magnesium	0.62	0.25	0.59
'C	Carbon	0.53	0.33	0.65
Cl	Chlorine	0.00	0.90	0.76
Zn	Zinc	0.00	0.00	0.54
Ti	Titanium	0.00	0.00	0.51

3.4.1. SEM/EDX Analysis

Figure 13a,b depicts the SEM images of the reference HPC. The images revealed the general morphology and crystalline structure of the internal surface of the HPCs. The greyscale generally assists in identifying and analysing the specific elements for a well-defined chemical composition's evaluation. As shown in the SEM images taken at 100 μm, the part labelled 1 represents the sand of angular shape, darker in colour surrounding the coarse aggregate. The item labelled 2 represents the coarse aggregate, dark grey in colour, large, and irregular in shape. The cement paste matrix contains the light grey background portion, labelled 3 being spread all over the surface, which is assigned hydration products (C_3S, C2S, C_3A, C_4AF, CH, C-S-H, AFt and AFm). The 200 μm image also revealed the presence of a consolidated bond at the ITZ and unified surface. There appears to be a good bond between the aggregate and the cement paste leading to a dense interface with less porosity. Mindess, Young and Darwin [47] linked this phenomenon to factors such as chemical interaction between the aggregate and cement paste, surface roughness of aggregate, and micro filler inclusion. A good bonding system between the paste and aggregate can come from the chemical interaction between the concrete and the cement paste for the control mixture. The coarse aggregate's angularity is another important factor in forming the denser bond at the ITZ zone. Furthermore, the compositions of the oxide (Table 5) revealed specific atomic concentrations (%) by each constituent, giving silicon (34.35%), calcium (18.78%), oxygen (22.24%) and aluminium (10.20%) prominence in the phases. On the other hand, potassium traces (2.94%) and sodium (2.59%) were found in the constituents' phases.

Figures 14a,b and 15a,b show the SEM micrograph of MCCC-10 and MCCC-20 HPC specimen cured for 90 days. The SEM images indicated bright grey colour irregular shaped coarse aggregate surrounded by darker angular shaped sand. The cement paste matrix highlights the light grey background fragment spread all over the surface, which is assigned hydration products (C_3S, C_2S, CH, C-S-H, C-A-H, C-A-S-H, AFt and AFm). As can be seen from the images, there is evidence of a more dense and uniform transition zone between the cement paste and aggregates than the control. The addition of 10% and 20% of MCC resulted in a more compact and cohesive paste than the control sample; this can be correlated with these mixtures' high strength. This fact can be related to MCC's pozzolanic activity, which generated a close grid of hydration products with fewer portlandite residues. As shown in Figure 15a,b, the MCCC-20 mixture produced a few spot dark pores that interfered with the HPC's strength compared with the MCCC-10 mixture. EDX result (Table 5) of MCCC-10 and MCCC-20 mixtures showed the dominance of calcium (38.99%; 70.90%), silicon (28.78%; 5.08%) and oxygen (22.22%; 15.37%) in the elemental atomic weight % compositions, respectively. This result points to the formation of C-S-H and portlandite in the HPC mixtures. The elevated calcium formed in the tested MCC blended HPCs at 90 days of curing indicates portlandite formation compared to the reference mix.

3.4.2. XRD Analysis

The XRD diffraction pattern was used to determine the crystalline phase in the hardened HPC pastes, and the crystalline phase's amount at 90 days of hydration. XRD analysis from Figures 16–18 present the XRD patterns of control and MCC blended HPCs.

From Figure 16 quartz (SiO_2), calcite ($CaCO_3$), portlandite ($Ca(OH)_2$, phlogopite ($KMg_3AlSi_3O_{10}(F,OH)_2$), biotite ($K(Mg,Fe)_3(AlSi_3O_{10})(F,OH)_2$), andradite ($Ca_3Fe_2Si_3O_{12}$) are the major mineral phases of the hardened pastes. Calcite recorded the highest mineral phase (16%), possibly due to the constituent materials' carbonation during the production and samples preparation process. Quartz, phlogopite and biotite were 8%, 7%, and 5% of mineral phase contents. Portlandite and andradite recorded the lowest crystalline phase contents of 3% and 2%, respectively. Thus, these mineral compositions detected via XRD analysis are typical for a calcined clay blended cement [6,13,27,46]. XRD patterns of MCCC-10 and MCCC-20 mixtures at 90 days of hydration are demonstrated in Figures 17 and 18, respectively. As can be seen from the Figures, the percentages intensities of portlandite peaks reduced to 3% for both MCCC-10 and MCCC-20 mixtures, indicating depletion of portlandite in the MCC blended cement with the control sample. This phenomenon is consistent with the improved mechanical properties results of this study.

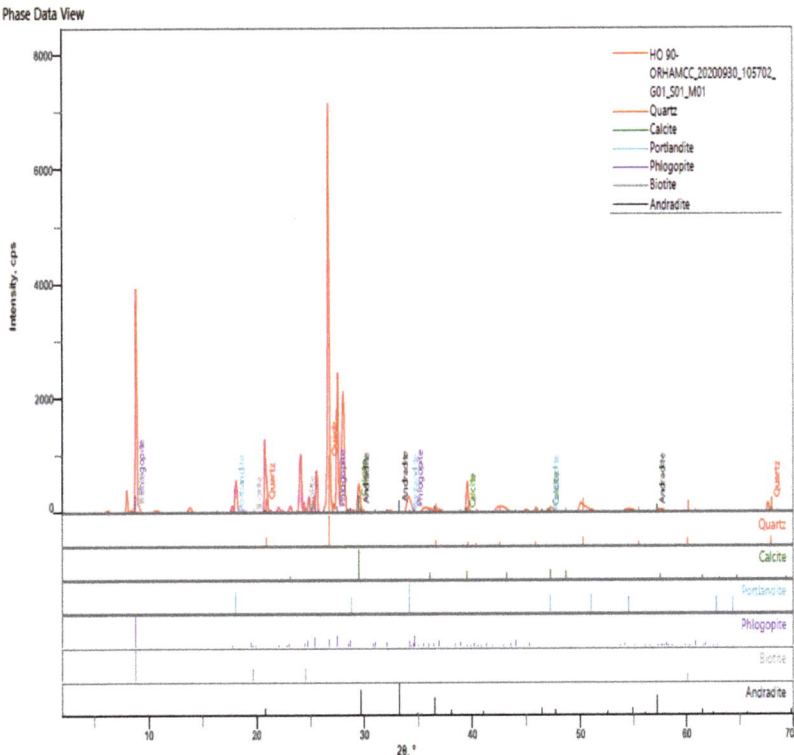

Figure 16. XRD pattern of the hardened control sample at 90 days.

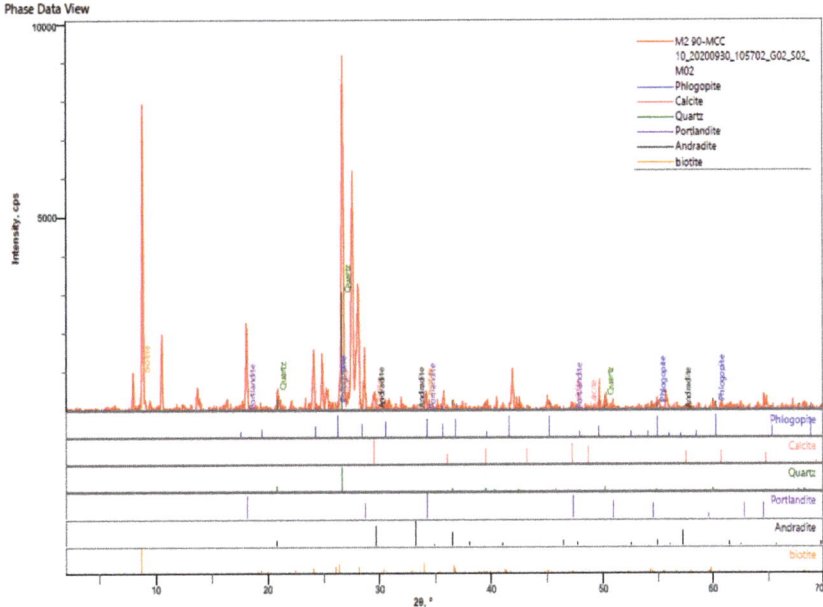

Figure 17. XRD pattern of the hardened MCCC-10 sample at 90 days.

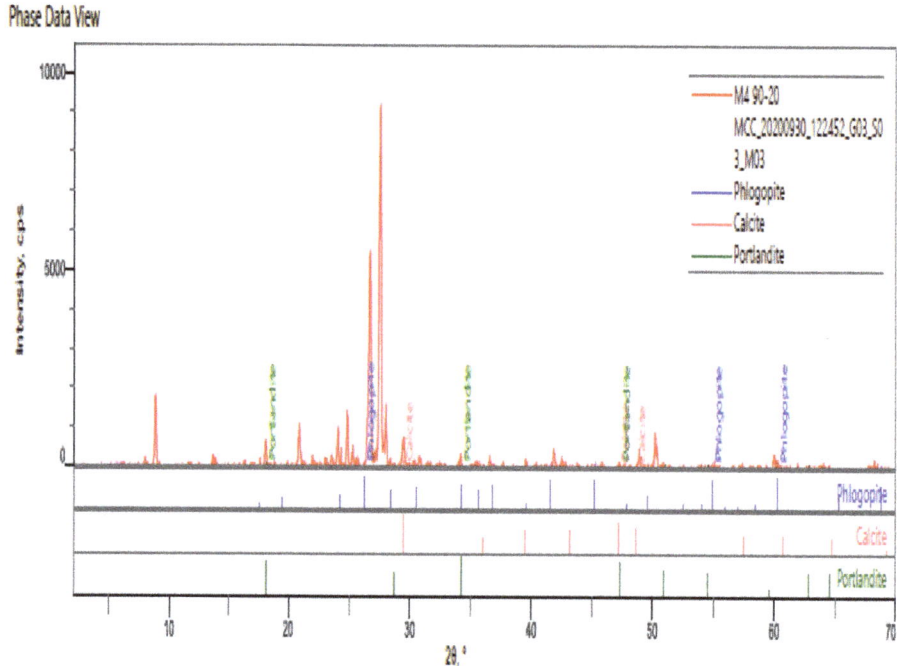

Figure 18. XRD pattern of the hardened MCCC-20 sample at 90 days.

4. Conclusions

In this paper, the potential use of MCC as an SCM in a binary blended cement for HPC production was investigated for the underlying mechanisms. The binder consists of MCC and PC, and the W/B was 0.3. SAP was introduced at 0.3 b_{wob} as an internal curing agent to increase internal moisture availability and prevent autogenous shrinkage. MCC was substituted at 5%, 10%, 15%, 20%, 25% and 30% of PC. The fresh properties were determined by slump flow test and setting times techniques, and the findings were compared with the recommendations of Neville [23] and ASTM C403 [33] for HPC mixtures. The hardened HPC samples were compared by compressive, splitting tensile and flexural strengths. The microstructural and mineralogical phases of the selected hardened HPC samples were analysed via SEM/EDX and XRD advanced techniques. The following inferences are drawn from the study.

- The MCCC slump flow is consistent with HPC mixtures having a slump flow range of 450 mm to 600 mm;
- MCC based HPC mixtures showed a gradual increase in the initial and final setting time;
- The average density of the activated HPC samples containing MCC and the control varies from 2390 kg/m^3 to 2387 kg/m^3 for all ages, conforming to the minimum density of normal weight concrete of 2400 kg/m^3;
- The blended CEM II with MCC HPCs could not produce the target designed strength of Class 1 HPC (50–75 MPa) at 28 days, while later age curing showed a promising result. HPC specimen type MCCC-10 had the highest compressive strength at 56- and 90-days curing ages;
- The addition of MCC has a moderate effect on the splitting tensile strength of the HPC, especially at later hydration days. The 10% cement replacement with MCC (MCCC-10) is sufficient to accelerate 28, 56 and 90 days splitting tensile strength of the HPCs;
- The MCC-based cement paste matrix highlights the light grey background fragment spread over the surface, assigned hydration products (C$_3$S, CH, C-S-H, AFm and AFt).

EDX result of MCCC-10 and MCCC-20 mixtures showed the dominance of calcium, silicon and oxygen in the elemental atomic weight% compositions, respectively;
- The hardened HPC control sample at 90 days revealed quartz, calcite, portlandite, phlogopite, biotite, andradite as the major mineral phases of the hardened paste. The percentages intensities of portlandite peaks reduced to 3% for both MCCC-10 and MCCC-20 mixtures, indicating depletion of portlandite in the MCC blended cement with the control sample.

Author Contributions: Conceptualization, D.O.N. and B.J.O.; methodology, D.O.N., B.J.O. and B.G.F.; investigation, D.O.N., B.J.O. and O.I.F.; writing—original draft preparation, D.O.N.; writing—review and editing, B.J.O., O.I.F. and B.G.F. All authors have read and agreed to the published version of the manuscript.

Funding: This research received no external funding, and the APC was funded by Covenant University Center for Research, Innovation and Discovery (CUCRID).

Acknowledgments: The authors extend their appreciation to the suppliers of the superplasticiser-Masterglenium Sky 504—BASF Limited, West Africa; the Superabsorbent Polymers (SAP)—SNF Floerger-ZAC de Milieux, France and 100 mm cube metal moulds—the Nigerian Building and Road Research Institute (NBRRI), Ota.

Conflicts of Interest: The authors declare no conflict of interest.

Abbreviations

AEC	Architecture, engineering, and construction
AFm	Monosulfate
AFt	Ettringite
BET	Brunauer-Emmett-Teller
C_2S	Dicalcium silicate
C_3S	Tricalcium silicate
Cc	Coefficient of curvature
CEM II	Portland-limestone cement
CH	Calcium hydroxide
C-S-H	Calcium silicate hydrate
Cu	Coefficient of uniformity
FM	Fineness modulus
FTIR-ATR	Fourier transform infrared-Attenuated total reflection
HPC	High-performance concrete
ITZ	Interfacial transition zone
$K_y Al_4 (Si_{8-y}) O_{20}(OH)_4$	Illite clay-based mineral
MCC	Meta-illite calcined clay
NBRRI	Nigeria Building and Road Research Institute
PC	Portland cement
PCE	Polycarboxylic ether
PSD	Particle size distribution
SAP	Superabsorbent polymers
SCM	Supplementary cementitious material
SEM-EDX	Scanning electron microscopy-energy dispersive X-ray
SG	Specific gravity
SSA	Specific surface area
TGA	Thermogravimetric analysis
W/B	Water-to-binder ratio
XRD	X-ray diffraction
XRF	X-ray fluorescence

References

1. Han, C.; Shen, W.; Ji, X.; Wang, Z.; Ding, Q.; Xu, G.; Lv, Z.; Tang, X. Behavior of high performance concrete pastes with different mineral admixtures in simulated seawater environment. *Constr. Build. Mater.* **2018**, *187*, 426–438. [CrossRef]
2. Nduka, D.O.; Ameh, J.O.; Joshua, O.; Ojelabi, R. Awareness and Benefits of Self-Curing Concrete in Construction Projects: Builders and Civil Engineers Perceptions. *Buildings* **2018**, *8*, 109. [CrossRef]
3. Di Bella, C.; Griffa, M.; Ulrich, T.; Lura, P. Early-age elastic properties of cement-based materials as a function of decreasing moisture content. *Cem. Concr. Res.* **2016**, *89*, 87–96. [CrossRef]
4. Aïtcin, P.C. *High-Performance Concrete*; Taylor & Francis e-Library: New York, NY, USA, 2004.
5. Abbas, S.; Nehdi, M.L.; Saleem, M.A. Ultra-high-performance concrete: Mechanical performance, durability, sustainability and implementation challenges. *Int. J. Concr. Struct. Mater.* **2016**, *10*, 271–295. [CrossRef]
6. Zhou, D.; Wang, R.; Tyrer, M.; Wong, H.; Cheeseman, C. Sustainable infrastructure development through use of calcined excavated waste clay as a supplementary cementitious material. *J. Clean. Prod.* **2017**, *168*, 1180–1192. [CrossRef]
7. Nwankwo, C.O.; Bamigboye, G.O.; Davies, I.E.; Michaels, T.A. High volume Portland cement replacement: A review. *Constr. Build. Mater.* **2020**, *260*, 120445. [CrossRef]
8. Fapohunda, C.; Akinbile, B.; Shittu, A. Structure and properties of mortar and concrete with rice husk ash as partial replacement of ordinary Portland cement—A review. *Int. J. Sustain. Built Environ.* **2017**, *6*, 675–692. [CrossRef]
9. Chen, Y.; Rodriguez, C.R.; Li, Z.; Chen, B.; Çopuroğlu, O.; Schlangen, E. Effect of different grade levels of calcined clays on fresh and hardened properties of ternary-blended cementitious materials for 3D printing. *Cem. Concr. Compos.* **2020**, *114*, 103708. [CrossRef]
10. Marchetti, G.; Rahhal, V.; Pavlík, Z.; Pavlíková, M.; Irassar, E.F. Assessment of packing, flowability, hydration kinetics, and strength of blended cements with illitic calcined shale. *Constr. Build. Mater.* **2020**, *254*, 119042. [CrossRef]
11. Song, T.; Ren, Z.; Li, H.; Sun, X.; Xue, M.; Yan, S. Modification of illite with calcium pimelate and its in-fluence on the crystallisation and mechanical property of isotactic polypropylene. *Compos. Part A Appl. Sci. Manuf.* **2019**, *123*, 200–207. [CrossRef]
12. Avet, F.; Snellings, R.; Diaz, A.A.; Ben Haha, M.; Scrivener, K. Development of a new rapid, relevant and reliable (R3) test method to evaluate the pozzolanic reactivity of calcined kaolinitic clays. *Cem. Concr. Res.* **2016**, *85*, 1–11. [CrossRef]
13. Trümer, A.; Ludwig, H.-M.; Schellhorn, M.; Diedel, R. Effect of a calcined Westerwald bentonite as supplementary cementitious material on the long-term performance of concrete. *Appl. Clay Sci.* **2019**, *168*, 36–42. [CrossRef]
14. Ferreiro, S.; Herfort, D.; Damtoft, J. Effect of raw clay type, fineness, water-to-cement ratio and fly ash addition on workability and strength performance of calcined clay–Limestone Portland cements. *Cem. Concr. Res.* **2017**, *101*, 1–12. [CrossRef]
15. Irassar, E.F.; Bonavetti, V.L.; Castellano, C.C.; Trezza, M.A.; Rahhal, V.F.; Cordoba, G.; Lemma, R. Calcined illite-chlorite shale as supplementary cementing material: Thermal treatment, grinding, color and pozzolanic activity. *Appl. Clay Sci.* **2019**, *179*, 105143. [CrossRef]
16. Laidani, Z.E.-A.; Benabed, B.; Abousnina, R.; Gueddouda, M.K.; Kadri, E.-H. Experimental investigation on effects of calcined bentonite on fresh, strength and durability properties of sustainable self-compacting concrete. *Constr. Build. Mater.* **2020**, *230*, 117062. [CrossRef]
17. European Committee for Standardization. *197-1: 2011. Cement, Composition, Specifications and Conformity Criteria for Common Cements*; British Standard Institution (BSI): London, UK, 2011.
18. *Nigeria Industrial Standard [NIS] 444-1. Composition, Specification and Conformity Criteria for Common Cements*; Standards Organization of Nigeria: Abuja, Nigeria, 2004.
19. Olawuyi, B.J.; Boshoff, W. Influence of SAP content and curing age on air void distribution of high performance concrete using 3D volume analysis. *Constr. Build. Mater.* **2017**, *135*, 580–589. [CrossRef]
20. European Committee for Standardization. *1008, 2002, Methods of Test for Water for Making Concrete (Including Notes on the Suitability of the Water)*; British Standards Institutions: London, UK, 2002.
21. Aïtcin, P.C. *High Performance Concrete*; CRC Press: Boca Raton, FL, USA, 1998.
22. Beushausen, H.; Dehn, F. *High-Performance Concrete. Fulton's Concrete Technology*, 9th ed.; Cement and Concrete Institute: Midrand, South Africa, 2009; pp. 297–304.
23. Neville, A.M. *Properties of Concrete*, 5th ed.; Pearson Educational Limited: London, UK, 2012.
24. Olawuyi, B.J.; Boshoff, W.P. Influence of superabsorbent polymer on the splitting tensile strength and fracture energy of high-performance concrete. In Proceedings of the MATEC Web of Conferences; EDP Sciences: Paris, France, 2018; Volume 199, p. 11004.
25. Olawuyi, B.J. The Mechanical Behaviour of High-Performance Concrete with Superabsorbent Polymers (SAP). Ph.D. Thesis, University of Stellenbosch, Stellenbosch, South Africa, 2016.
26. Mermerdaş, K.; Gesoğlu, M.; Güneyisi, E.; Özturan, T. Strength development of concretes incorporated with metakaolin and different types of calcined kaolins. *Constr. Build. Mater.* **2012**, *37*, 766–774. [CrossRef]
27. Shafiq, N.; Nuruddin, M.F.; Khan, S.U.; Ayub, T. Calcined kaolin as cement replacing material and its use in high strength concrete. *Constr. Build. Mater.* **2015**, *81*, 313–323. [CrossRef]
28. Rodriguez, C.; Tobon, J.I. Influence of calcined clay/limestone, sulfate and clinker proportions on cement performance. *Constr. Build. Mater.* **2020**, *251*, 119050. [CrossRef]

29. Du, H.; Dai Pang, S. High-performance concrete incorporating calcined kaolin clay and limestone as cement substitute. *Constr. Build. Mater.* **2020**, *264*, 120152. [CrossRef]
30. Cordoba, G.; Irassar, E.F. Sulfate performance of calcined illitic shales. *Constr. Build. Mater.* **2021**, *291*, 123215. [CrossRef]
31. BSI. *BS EN 12390-3, 5 & 6: Testing Hardened Concrete. Making and Curing Specimens for Strength Tests*; BSI: London, UK, 2009.
32. European Committee for Standardization. *12350-5. 2009. Testing Fresh Concrete-Part 5: Flow Table Test*; European Committee for Standardization: Brussels, Belgium, 2009.
33. ASTM. *ASTM C 403: Standard Test Method for Time of Setting of Concrete Mixtures by Penetration Resistance*; ASTM; BSI: West Conshohocken, PA, USA, 2008.
34. *BS EN 12390-3: 2019: Testing Hardened Concrete. Compressive Strength of Test Specimens*; BSI: London, UK, 2019.
35. International Union of Testing and Research Laboratories for Materials and Structures (RILEM). *CPC4—Compressive Strength of Concrete 1975, TC14-CPC, RILEM Technical Recommendations for the Testing and Use of Construction Materials*; E & FN, Spon: London, UK, 1994; pp. 17–18.
36. *BS EN 12390-5: 2019: Testing Hardened Concrete. Flexural Strength of Test Specimens*; BSI: London, UK, 2019.
37. Hollanders, S.; Adriaens, R.; Skibsted, J.; Cizer, Ö.; Elsen, J. Pozzolanic reactivity of pure calcined clays. *Appl. Clay Sci.* **2016**, *132*, 552–560. [CrossRef]
38. Ihekweme, G.O.; Shondo, J.N.; Orisekeh, K.I.; Kalu-Uka, G.M.; Nwuzor, I.C.; Onwualu, A.P. Characterisation of certain Nigerian clay minerals for water purification and other industrial applications. *Heliyon* **2020**, *6*, e03783. [CrossRef]
39. Zhou, D. Developing Supplementary Cementitious Materials from Waste London Clay. Ph.D. Thesis, Imperial College of London, London, UK, 2016.
40. Garg, N.; Skibsted, J. Pozzolanic reactivity of a calcined interstratified illite/smectite (70/30) clay. *Cem. Concr. Res.* **2016**, *79*, 101–111. [CrossRef]
41. Mehta, P.K.; Monteiro, P.J. *Concrete: Microstructure, Properties, and Materials*; McGraw-Hill Education: New York, NY, USA, 2014.
42. Elgamouz, A.; Tijani, N.; Shehadi, I.; Hasan, K.; Kawam MA, F. Characterisation of the firing behaviour of an illite-kaolinite clay mineral and its potential use as membrane support. *Heliyon* **2019**, *5*, e02281.
43. Ma, Y.; Shi, C.; Lei, L.; Sha, S.; Zhou, B.; Liu, Y.; Xiao, Y. Research progress on polycarboxylate based superplasticisers with tolerance to clays—A review. *Constr. Build. Mater.* **2020**, *255*, 119386. [CrossRef]
44. Cai, R.; He, Z.; Tang, S.; Wu, T.; Chen, E. The early hydration of metakaolin blended cements by non-contact impedance measurement. *Cem. Concr. Compos.* **2018**, *92*, 70–81. [CrossRef]
45. Schulze, S.E.; Rickert, J. Suitability of natural calcined clays as supplementary cementitious material. *Cem. Concr. Compos.* **2019**, *95*, 92–97. [CrossRef]
46. Vejmelková, E.; Koňáková, D.; Doleželová, M.; Scheinherrová, L.; Svora, P.; Keppert, M.; Černý, R. Effect of cal-cined Czech claystone on the properties of high performance concrete: Microstructure, strength and durability. *Constr. Build. Mater.* **2018**, *168*, 966–974. [CrossRef]
47. Mindess, S.; Young, F.J.; Darwin, D. *Concrete*, 2nd ed.; Prentile Hall: Prentile Hall, NJ, USA, 2003; p. 644.

Article

Service Life Modeling of Concrete with SCMs Using Effective Diffusion Coefficient and a New Binding Model

Mukhtar Oluwaseun Azeez [1] and Ahmed Abd El Fattah [2],*

[1] Department of Civil & Environmental Engineering, King Fahd University of Petroleum and Minerals, Dhahran 31261, Saudi Arabia; g201203220@kfupm.edu.sa
[2] Department of Architecture, King Fahd University of Petroleum and Minerals, Dhahran 31261, Saudi Arabia
* Correspondence: ahmedmohsen@kfupm.edu.sa; Tel.: +966-13-8603874

Received: 22 September 2020; Accepted: 24 October 2020; Published: 26 October 2020

Abstract: This paper presents a new algorithm that predicts the service life of concrete contains supplementary cementitious materials, SCMs, and determines time of corrosion initiation. The algorithm drives effective diffusivity from an apparent diffusion model, using experimental binding data performed in the lab, temperature, free ion concentration, and carbonation, and generates free chloride profiles for concrete with and without SCMs by using Fick's law in a finite element model. Adjusting diffusion coefficient at each step of the solution, by addressing the impact of different parameters, simplifies the algorithm and reduces calculation time without jeopardizing the results' quality. Results generated by the model compare well to the performance of concrete blocks constructed in an exposure site on the east coast of Saudi Arabia. The exposure site hosted five different mixes of Portland cement and SCMs, and the concrete blocks were exposed to harsh weather over the period of two years. Linear polarization and chloride profiling assessed the performance of the mixes against corrosion activities. Lab work identified the performance of the mixes through binding capacity and chloride profiling. Statistical analysis evidenced the accuracy of the model through correlation and regression analysis. Furthermore, a new proposed binding model, produced from binding data in different studies, alters the experimental binding data in the algorithm to decouple the solution from experimental values. The algorithm proves its accuracy when compared to the experimental free chloride profile. The proposed transport model proves that using effective diffusion and binding capacity are enough to generate reliable results, and the effective diffusion can be calibrated with environmental conditions such as temperature, age, and carbonation. Finally, the algorithm presents its features in an object-oriented programming using C# and user friendly web interface.

Keywords: concrete service life; binding capacity; exposure site; transport model; diffusion

1. Introduction

Concrete remains the most widely used material around the world with over six billion tons of concrete used to establish the required infrastructures. This is due, in general, to its reliable structural integrity such as high compressive strength, good fire resistance, relatively low cost, good durability, and availability. Durability of reinforced concrete structures is a great challenge because factors affecting concrete durability are numerous, including the concrete permeability, aggregate content, cement type and quantity, water/cement ratio, and environmental factors such as temperature, humidity, and level of concentrations of aggressive species which lead to concrete deterioration over time.

Among all the durability challenges facing reinforced concrete structures, resistance to chloride diffusion and subsequent reinforcement corrosion is still a major durability index that greatly influences the structural integrity of reinforced concrete members. Chloride attacks concrete structures in aggressive environments, such as marine environments, ground water, and de-icing salts. Chloride itself does not directly result in damage to concrete in normal circumstances, but can cause steel corrosion within concrete if not monitored [1]. Upon reaching chloride threshold value at the steel location, the corrosion initiates and increases the volume of concrete which leads to concrete cracking and cover spalling over time. Consequently, there will be serious economic and social effects as more funds will be diverted to repair damaged areas of structural facilities [2–4]. In extreme cases, the facility might even be completely closed during repair if the extent of damage is huge. To mitigate economic distress and to ensure quality of concrete structures at design stage, it is essential to have a good understanding of chloride diffusion mechanism, measurement methods, and relevant service life prediction modeling.

Over the past decades, researchers have devoted efforts to understanding the chloride diffusion mechanism, measurement, and prediction in Portland cement concrete (PCC) as evidenced by volumes of published high quality papers on chloride diffusion in PCC. Some of the works have focused on explaining the diffusion mechanism and experimental measurements while others devoted effort to specifying various prediction models for concrete service life. A detailed experimental study on the mechanism of chloride transport in non-saturated concrete and its interaction with moisture diffusion was investigated by Ababneh and Xi [5] and Ababneh et al. [6]. The influence of the incorporation of supplementary cementitious materials and chemical admixture on chloride penetration in concrete have also been studied, such as the work of Liu et al. [7], Zhang et al. [8], and Lee and Lee [9]. A good number of the published articles on the subject have sought to address the complexities inherent in the chloride transport in different capacities. A pure chloride ion transport model in saturated concrete was studied by Xi and Bažant [10] while Samson et al. [11], Samson and Marchand [12], and Nguyen et al. [13] focused on the development of models that couple chloride ion transport with moisture transport in unsaturated cement-based materials. Saetta et al. [14] analyzed and experimentally compared results of chloride diffusion in partially saturated concrete under the influence of moisture diffusion and temperature gradient. The improvement of the chloride models to include the chemical activity of other ions in concrete pore solution on chloride diffusion has also been considered in the literature [15,16].

The general equation describing the flux of chloride ions through any medium is given by [1]:

$$J_D = -D_i \left(\frac{\partial C_i}{\partial x} + C_i \frac{\partial \ln \gamma_i}{\partial x} \right) + C_i u_i \frac{\partial \phi}{\partial x}, \quad (1)$$

where J_D is the diffusion flux of the ion (mol/m² s), C_i is the concentration of the ion (mol/m³), D_i is the effective diffusion coefficient of the ion (m²/s), γ_i is the chemical activity coefficient of the ion, u_i is the ion mobility, and ϕ is the electric potential between multi-ions in a solution (Volt). Applying the equation specifically for the diffusion of chloride ion, it describes the diffusion of chloride under concentration gradient (first term in the right-hand side of the equation) taking the ionic strength of the solution (advective flux) and electric field induced (electric flux) into consideration (second and third terms, respectively).

It is generally known that because of the formation of what is known as Friedel's salt in PCC hydration product and the huge surface area of hydration gel, a portion of chloride ions entering into concrete from environmental solutions will be captured by the concrete [1]. This phenomenon is referred to as 'chloride binding' [1]. The binding due to huge surface area is termed physical binding while that due to reaction with Friedel's salt is termed chemical binding [1]. Even though, in previous studies, the bound chloride was thought to be harmless as it is not freely available to induce reinforcement corrosion, Luping and Nilsson [17] argued that the bound chloride might be released in the event of carbonation or sulfate ingress thereby increasing the volume of the free chloride. Since the chloride binding effect is a complex phenomenon in cement-based materials, researchers have generally addressed the relation between free and bound chlorides with what is known as binding isotherm. Various binding isotherms have been proposed in the literature from linear isotherms [18] to non-linear isotherms [19–22].

Since Equation (1) is a steady state description of chloride diffusion in a medium that assumes constant diffusion coefficient and absence of binding capacity of concrete, a more general non-steady state equation considering the variable nature of the diffusion coefficient, the binding capacity, ionic strength, and ionic interaction has been proposed and used in different capacities to describe chloride transport. The equation otherwise known as the Nernst–Planck equation is described next:

$$\frac{\partial C_{fi}}{\partial t}\left(1 + \frac{\partial C_b}{\partial C}\right) = \frac{\partial}{\partial x}\left(D_i\left(\frac{\partial C_i}{\partial x} + C_i\frac{\partial \ln \gamma_i}{\partial x}\right) + C_i u_i \frac{\partial \phi}{\partial x}\right), \tag{2}$$

where $\frac{\partial C_b}{\partial c}$ is the binding capacity of concrete.

Equation (2) is a complex non-steady state description of ion transport in a porous medium as a result of its dependence on various factors, including the extent of hydration products formed, the chemical activity, and electric potential between pore solution ions. The hydration products in turn depend on many factors from material parameters, such as water/binder (w/b) ratio, binder content, admixture quantities, and environmental factors, such as temperature, humidity, and chloride concentration. However, some authors have indicated that chemical activity is only relevant for unsaturated conditions and that it has a weak influence in the case of saturated conditions [23,24].

Due to the inherent complexity of Equation (2), an early development of chloride transport models has ignored many terms in the equation leading to the well-known general form of Fick's law of diffusion:

$$J_D = -D_i \frac{\partial C_i}{\partial x}. \tag{3}$$

The assumptions of Fick's law are obviously questionable, because under the assumptions, chloride ions are treated as uncharged particles [1]. The need to incorporate the effect of surrounding ions in a solution on chloride diffusion has made many researchers utilize the Nernst–Planck equation in their models [11], albeit in different capacities.

Riding et al. [25] developed a model for chloride concentration that used the apparent diffusion coefficient and Fick's second law of diffusion using finite difference and validated their model against results from concrete blocks exposed to marine environment for 25 years. They pointed out that diffusion is the main chloride transport mechanism for concrete exposed to water and chloride.

Recently, Fenaux et al. [16] have shown that using solely diffusion and chloride binding generates good results. In addition, they pointed out the difficulty of distinguishing whether the variation of the apparent diffusion coefficient was based on the microstructure or binding capacity of the concrete. They have shown the superiority of using the effective diffusion coefficient over the apparent diffusion coefficient.

Therefore, the present work uses Fick's law of diffusion and the effective diffusion coefficient, which accounts for effects of age, temperature, carbonation, free chloride, and experimental binding capacity, to predict free chloride concentrations along the depth of concrete using 1D finite element modeling. The outcome of the transport model was validated against results from marine exposure site and coastal zone on shore. Then, the paper proposes a new binding model using the oxides content of the cement and supplementary cementitious materials (SCMs) and uses the model in the transport model. The results show good correlation to the experimental results. Finally, the transport model was incorporated into user friendly software application developed using object-oriented based C-Sharp programming language.

2. Materials and Experimental Investigation

Experimental investigation to study the ingress of chloride ion into reinforced concrete was carried out under a laboratory setting and field exposure, and its details and results were published in previous works [26,27]. In the main study for field exposure, five different mixture proportions were developed to study the effect of cement compositions and supplementary cementitious materials on the resistance of reinforced concrete to chloride ingress. Replicate mixtures were also prepared in the laboratory. Table 1 shows the mixture proportions used in this study. ASTM C150 [28] type I (OPC) and V (Sulfate attack resistant) cements were used in order to quantify the impact of the cement aluminate content and consequent chloride binding amount on the chloride ingress rate. Three different supplementary cementitious materials (SCMs) were used in this project: an ASTM C1240 silica fume (SF) [29], an ASTM C618 class F fly ash (FA) [30], and an ASTM C989 slag cement (SC) [31]. FA, SC, and SF were used at replacement percentages of 25%, 70%, and 6% by weight of cement, respectively. These ratios also were chosen based on the optimal percentages found in the literature and according to the common mixes used by local stakeholders. The high percentage of SC used was similar to the percentage used in King Fahd Causeway constructed to Bahrain. A common fixed water-to-cementitious materials ratio (w/cm) of 0.40 was used because it showed better performance of concrete compared to other values [32]. Fine dune sand was used as fine aggregate in addition to limestone coarse aggregates in all the concrete mixtures.

Table 1. Mix proportioning.

Mix	W/C	Cement kg/m^3	Coarse Aggregate kg/m^3	Sand kg/m^3	Water kg/m^3	Silica Fume kg/m^3	Fly Ash kg/m^3	Slag Cement kg/m^3	Notes
I	0.4	340	1070	775	136	-	-	-	Type OP/CEM 1
V		340	1070	775	136	-	-	-	Type V
SF		320	1100	735	136	21	-	-	OP + SF
FA		255	1090	735	136	-	85	-	OP + FA
SC		100	1095	735	136	-	-	240	OP + SC

Reinforced concrete blocks (230 mm × 460 mm × 1200 mm) and 75 mm × 150 mm cylinders were made in a ready mix company. Concrete temperatures ranged from 26 to 29 °C. Concrete blocks were covered during the first day of curing using plastic sheets and shaded to prevent plastic and drying shrinkage. They were cured using wet burlap layers that covered each block after demolding for 28 days. Then the blocks were located on a marine-exposure site (Figures 1 and 2). Figure 1 shows the specimens located in the tidal zone with apparent fluctuation of the tides—the frontal row shows plain concrete samples and the back row shows the reinforced blocks. Each reinforced block contained four black steel rebars positioned at different cover depths (12.7, 25.4, 38.1, and 50.8 mm) and connected at the top to stainless steel bars which were needed for measurements. The specimens rested on solid marine-based plywood to avoid sinking and toppling. Specimens in Figure 2 were located about 15 m away from the shoreline, positioned on a solid base and laterally tied to each other using galvanized steel. Chloride profiling and linear polarization, using Gamry Reference 3000 Potentiostat [33], determined the performance of the different mixtures to corrosion activities.

Figure 1. Concrete blocks setup in the tidal zone (inner blocks: reinforced; outer blocks: plain).

Figure 2. Concrete blocks setup in the coastal zone, with the proper bracing on top and isolation from sand.

Cylinder specimens with dimensions of 75 mm × 150 mm were exposed to 5% NaCl solution after 28 days of lab curing in water saturated with calcium hydroxide to avoid leaching. The NaCl exposure was carried out according to the specifications of ASTM C1556 [34]. A chloride profile procedure was carried out for the laboratory specimens. The results were produced and averaged from three readings. The chloride binding capacity was determined by taking measurements of both the free chloride and total chloride ion in concentration in exposed specimens according to [27].

3. Methodology

This section is split into three divisions: the formulation of the effective diffusion model, the proposition of a new binding model, and the logic of the solution of the transport model.

General Conventions: The general naming conventions used throughout this work are such that for an ion i, its free concentration is expressed as C_{fi} (kg/m³ of concrete), bound concentration expressed as C_{bi} (kg/m³ of concrete), and total concentration as C_{ti} (kg/m³ of concrete). Its free concentration is also expressed as C_i (kg/m³ or g/l of pore solution). The total concentration is the summation of the free concentration and bound concentration.

Therefore, for the model of ingress of chloride ion i into concrete, the total, free, and bound chloride concentrations are C_{ti}, C_{fi}, and C_{bi}, respectively.

The apparent chloride diffusion coefficient is expressed as D_{ai} (m²/s) and the effective chloride diffusion coefficient is expressed as D_i (m²/s).

3.1. Effective Diffusion Model

The simplest approach to modeling the transport of chloride ion in concrete is the so-called apparent diffusion model that considers only the diffusive flux of chloride ion in pore solution of concrete. By ignoring the terms for ionic interaction, electric potential, and binding capacity in Equation (2), the apparent diffusion model can be expressed as

$$\frac{\partial C_{fi}}{\partial t} = \frac{\partial}{\partial x}\left[D_{ai}\left(\frac{\partial C_i}{\partial x}\right)\right], \qquad (4)$$

where D_{ai} is the apparent chloride diffusion coefficient taken as constant in the model. In order to obtain the apparent chloride diffusion coefficient, the solution of the apparent diffusion model would be fitting a numerical solution of Fick's second law into the chloride concentration profiles obtained from experimental procedure at a specific period. The apparent chloride diffusion coefficient results could then be used for a subsequent prediction of chloride concentration profiles at different ages. In the present study, the apparent diffusion model is used to determine the effective diffusion coefficient by accounting for several parameters, as shown in the following section, and the free chloride concentration profiles in the field exposed specimens.

In the effective diffusion model, the variable nature of the chloride diffusion coefficient and the chloride binding capacity is taken into consideration for the development of the chloride transport model written as

$$\theta_i \frac{\partial C_{fi}}{\partial t} = \nabla \cdot [D_i \nabla C_i], \tag{5}$$

where θ_i is the chloride binding capacity and is expressed as $(1 + \frac{\partial C_{bi}}{\partial C_{fi}})$.

The effective chloride diffusion coefficient in this case is dependent on many factors making the solution of Equation (5) to be more complex. Ababneh et al. [6] expressed the chloride diffusion coefficient in concrete in terms of such multi-factors as given in Equation (6).

$$D_i = f_1 f_2(g_i) f_3(H) f_4(T) f_5(C_{fi}), \tag{6}$$

where the dependence of chloride diffusion coefficient is on water/cement ratio f_1, aggregate and cement paste diffusivities $f_2(g_i)$, humidity $f_3(H)$, temperature $f_4(T)$, and free chloride ion concentration $f_5(C_{fi})$ [6].

In the present study, a simplified version of Equation (6) was used to depict the variability of the chloride diffusion coefficient with age, temperature, and carbonation as given by Equation (7).

$$D_i = D_{ti} f_t(T) f_t(C_{fi}) g_c. \tag{7}$$

D_{ti} is the apparent chloride diffusion coefficient (m^2/s) which depends on factors such as water/cement ratio, cement content, admixture type, and content and age as shown in the empirical Equations (8)–(10) derived by Riding et al. [25].

$$D_{ti} = D_{ao}\left(\frac{28}{t}\right)^m + D_{ult}\left(1 - \left(\frac{28}{t}\right)^m\right), \tag{8}$$

$$D_{ao} = 2.17 \times 10^{-12} \, e^{w/cm}/0.279, \tag{9}$$

$$D_{ult} = D_{ao}\left(\frac{28}{36,500}\right)^m, \tag{10}$$

where t is time (days), D_{ao} is the initial apparent chloride diffusion coefficient, m is a decay constant, and w/cm is the water/cement ratio. It should be mentioned that the parameter m embeds the cementitious materials impact and the parameter D_{ao} accommodates the effect of w/cm ratio [25]. This study uses 0.4 for w/cm ratio and using Equation (9) gives similar results to the experimental ones [27].

$f_t(T)$ accounts for the effect of temperature at time t on chloride diffusion coefficient.

$$f_t(T) = Exp\left[\frac{U}{R}\left(\frac{1}{T_o} - \frac{1}{T}\right)\right], \tag{11}$$

where U is the activation energy of diffusion process (kJ/mol.), R is universal gas constant (8.314 J/mol./K), T_o is the reference temperature (usually taken as 296 K [6]), and T is the temperature (K) at time t.

$f_t(C_f)$ accounts for the effect of free chloride concentration on the chloride diffusion coefficient.

$$f_t(C_f) = 1 - k_{ion}(C_{fi})^n, \tag{12}$$

where k_{ion} and n are constants taken as 8.333 and 0.5, respectively [6].

g_c is the carbonation constant. Ngala and Page [35] have shown that the effective diffusion coefficient increases when the porosity increases. This was confirmed by [36] which also have shown that carbonation of cement-based materials densifies the structure due to formation of calcium carbonate, which results in decrease of diffusion coefficient by a reduction factor. In this study, a reduction factor of 0.58 is calibrated from the experimental results to reduce diffusion coefficient of carbonated nodes. The depth of carbonation d_c can be determined by square root of time equation $k\sqrt{t'}$ [36] where k is the carbonation coefficient (mm/year$^{0.5}$) and is determined from [36] as 0.26 cm/year$^{0.5}$; and t' is the exposure time to CO_2 (year). Mangat and Molloy [37] proposed a relationship between m and the water/cement ratio (w/c) of the concrete (that is, $m = 2.5$ w/cm $- 0.6$). However, other researchers have shown the value of m to be mainly influenced by the nature of the cementitious materials, particularly the presence of fly ash or ground-granulated blast-furnace slag (GGBFS) [38–40] with the w/c mainly influencing the initial diffusion coefficient [16]. Bamforth [38] proposed a value of $m = 0.264$ for plain Portland cement concrete and values of 0.699 and 0.621 for concrete containing fly ash and slag, respectively. In the Life-365 model [41], the value of m varies between 0.2 for concrete without fly ash and slag, up to 0.6 for concrete with either 50% fly ash or 70% slag; the value changes linearly for intermediate levels of fly ash and slag. In the present model, a value of m similar to Life-365 model was used as shown in Equation (13).

$$m = const + 0.4\left(\frac{FA}{50} + \frac{SC}{70}\right), \tag{13}$$

where $const$ is 0.26 and 0.3 for tidal and coastal exposure, respectively, FA is the fly ash content (class F fly ash or ultra-fine fly ash) as a percentage of the total cementitious material content by weight and SC is the slag cement content as a percentage of the total cementitious material content by weight. Equation (13) shows the value of m to be constant for plain Portland cement concrete and Portland cement concrete replaced with specific percentage of cementitious materials except for FA and SC that vary linearly as FA increases from 0% to 50% and SC increases from 0% to 70%.

In the present work, the outcome of the laboratory experiment has been utilized, as well as the new binding model proposed in the following section, to express the bound chloride as function of free chloride Freundlich isotherm.

The chloride binding capacity $\left(1 + \frac{\partial C_{bi}}{\partial C_{fi}}\right)$ can then be derived from Equation (14) as

$$\left(1 + \frac{\partial C_{bi}}{\partial C_{fi}}\right) = 1 + \alpha\beta C_{fi}^{\beta-1}. \tag{14}$$

By using the multi-factor approach to express the diffusion coefficient of chloride ion and by incorporating the chloride binding capacity given in Equation (14) into the effective diffusion model of Equation (5), a more accurate expression for chloride transport in concrete can be achieved leading to more realistic prediction of time to corrosion initiation in reinforced concrete. The results produced using the experimental binding values are shown in Section 4.1.

3.2. New Binding Model

The authors collected binding data from different papers [42–56] and related the binding results with different oxide components of cement quantitatively. Figure 3 shows the impact of alumina content Al_2O_3, classified into five groups according to the ratio to the total content, on the bound chloride.

It is clear from Figure 3 that increasing the Al_2O_3 content increases the binding capacity. Performing the same procedure, the other oxides did not confirm the same trends; they presented, however, mixed results with no consistent trend. Consequently, the model uses Al_2O_3 content as the main driving parameter on the binding capacity. Based on Figure 4 which shows regression analysis, the terms α and β relate to Al_2O_3 content ratio (AC) by the following equations:

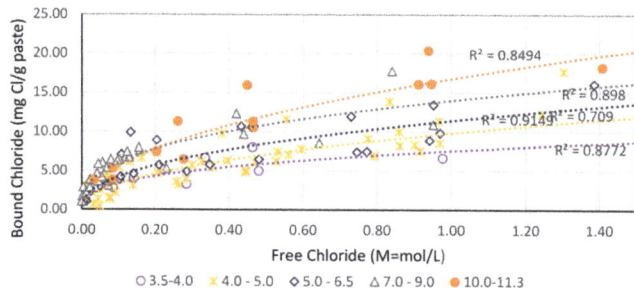

Figure 3. Relationship between bound and free chloride for concrete with different Al$_2$O$_3$ content.

$$\alpha = 1.3\,AC + 3.44, \tag{15}$$

$$\beta = 0.0077\,AC + 0.30. \tag{16}$$

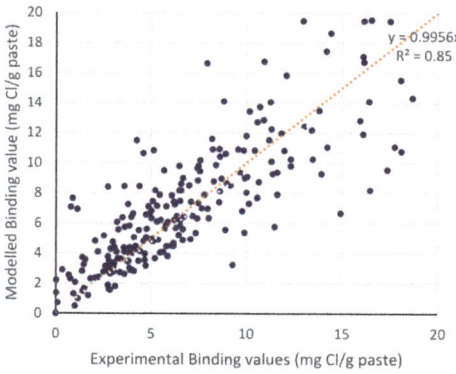

Figure 4. Linear regression analysis of the proposed binding model versus experimental results in the literature.

The proposed binding model solves for α and β in Equations (15) and (16) using the AC as an input, and solves for the bound chloride using the Freundlich isotherm [57]:

$$C_b = \alpha C_f^\beta. \tag{17}$$

Figure 4 shows that the proposed model captures the experimental value very well especially for the realistic moderate values found in water bodies (1.5 M) [58]. The proposed model has a good correlation with $R^2 = 0.85$ to the experimental data given the wide range of the experimental results. Thus, the model is incorporated in the software solution instead of the experimental binding values. The results produced using the binding capacity model are shown in Section 4.2.

3.3. Solution of the Transport Model

Because of the non-linear nature of the effective transport model, it will be difficult to approach its solution analytically. The robust Finite Element Method (FEM) coupled with Finite Difference Method (FDM) was utilized to solve the complex model numerically. For the FEM part, the transformation of the transport model differential equation to FEM compliant equations involved using the Galerkin approach summarized in the steps below:

a. Multiplying the residual (the differential equation) with a weighting function usually taken as the shape function, and integrating over the volume;
b. Rearranging differential terms; and
c. Applying divergence theorem.

Representing Equation (6) with W_1 and multiplying with shape function:

$$W_1 = \int N^T \left\{ \theta_i \frac{\partial C_{fi}}{\partial t} - \nabla[D_i \nabla C_i] \right\} d\Omega. \tag{18}$$

By carrying out the integrals in Equation (18) explicitly over a 1D bar element (with length L and cross-section area A), the equation could be transformed to a linear system shown in Equation (19):

$$M\dot{d} + Kd = F, \tag{19}$$

where for each finite element, d is the nodal unknown concentrations, K is the stiffness matrix, M is the capacitance matrix, and F is the flux vector, such that

$$d = \begin{pmatrix} C_{11} \\ C_{21} \\ C_{i1} \\ \vdots \\ C_{1j} \\ C_{2j} \\ C_{ij} \end{pmatrix},$$

$$\dot{d} = \begin{pmatrix} \dot{C}_{11} \\ \dot{C}_{21} \\ \dot{C}_{i1} \\ \vdots \\ \dot{C}_{1j} \\ \dot{C}_{2j} \\ \dot{C}_{ij} \end{pmatrix},$$

$$K = \begin{pmatrix} \frac{AD_i}{L} & \frac{-AD_i}{L} \\ \frac{-AD_i}{L} & \frac{AD_i}{L} \end{pmatrix},$$

$$M = \begin{pmatrix} \frac{AL\theta_i}{3} & \frac{AL\theta_i}{6} \\ \frac{AL\theta_i}{6} & \frac{AL\theta_i}{3} \end{pmatrix}.$$

The time derivative of the chloride ion concentration ($\dot{d}$) can be evaluated using the FDM. After evaluation, Equation (19) becomes

$$(M + \Delta t\, K) d_{t-\Delta t} = M\, d_t + \Delta t F_{t-\Delta t}, \tag{20}$$

where Δt is the time step chosen for numerical evaluation and t is the current time step.

4. Results and Discussion

4.1. Transport Model Using Experimental Binding Data

Figures 5 and 6 show the experimental readings and the theoretical chloride profile for the concrete blocks in coastal and tidal zones, respectively. The plots are for duration of exposure for 6, 12, and 24 months. In general, the theoretical plots match well the experimental readings for all of the mixes. Mix SC slightly gives conservative results. This might be attributed to the curing duration of the mixes, which was not sufficient to activate the slag completely [26]. However, there was improvement in the two years readings when the slag functioned effectively in the block.

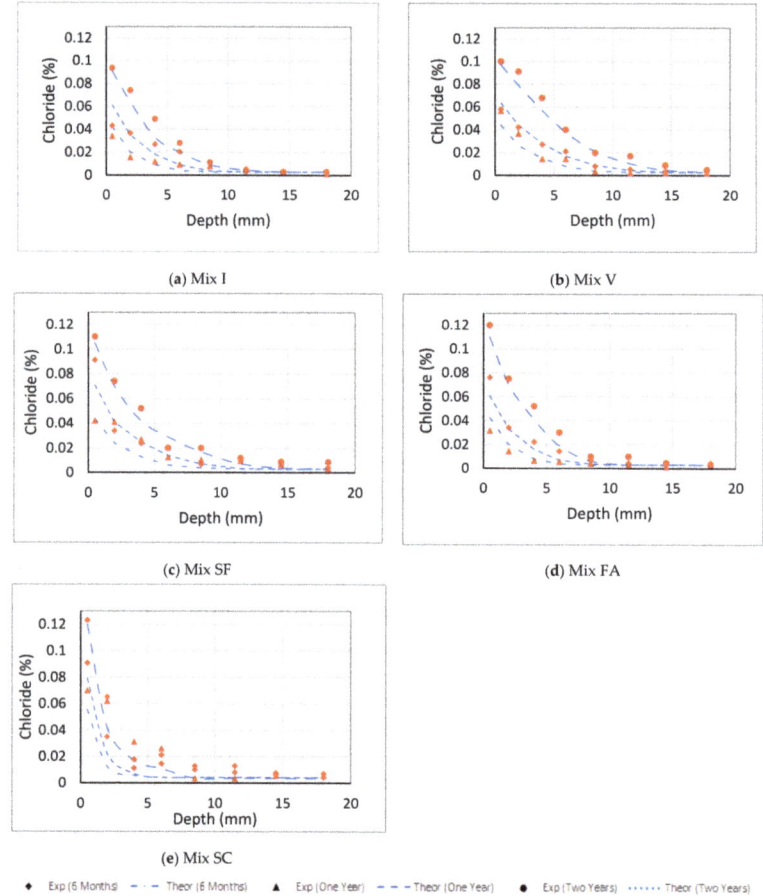

Figure 5. Chloride profiles for mixes in the coastal zone using the experimental binding results. (**a**) concrete with 100% ordinary Portland cement (OPC), (**b**) concrete with 100% type V cement (Sulfate resistance), (**c**) concrete with 6% silica fumes and 94% OPC, (**d**) concrete with 25% fly ash and 75% OPC, (**e**) concrete with 70% slag cement and 30% OPC.

Owing to the complex nature of chloride diffusion models as a result of the dependence of chloride ion diffusion in concrete on many factors, obtaining a perfectly matching result between the numerical models and experimental investigation is not feasible as evident by results obtained in other studies, such as those by Liu et al. [59], Pradelle et al. [24], Riding et al. [25], and Bernal et al. [2]. However, Table 2 shows the statistical analysis conducted to validate the theoretical results against experimental ones. The $β$-coefficient and R^2 (Figure 7) are equal to 1.00 and 0.94 for coastal zone, and 0.89 and 0.89 for tidal zone. This evidenced the accuracy of the model. It can also be seen that the $β$-coefficient increases with exposure increase, which implies more accuracy is obtained overtime.

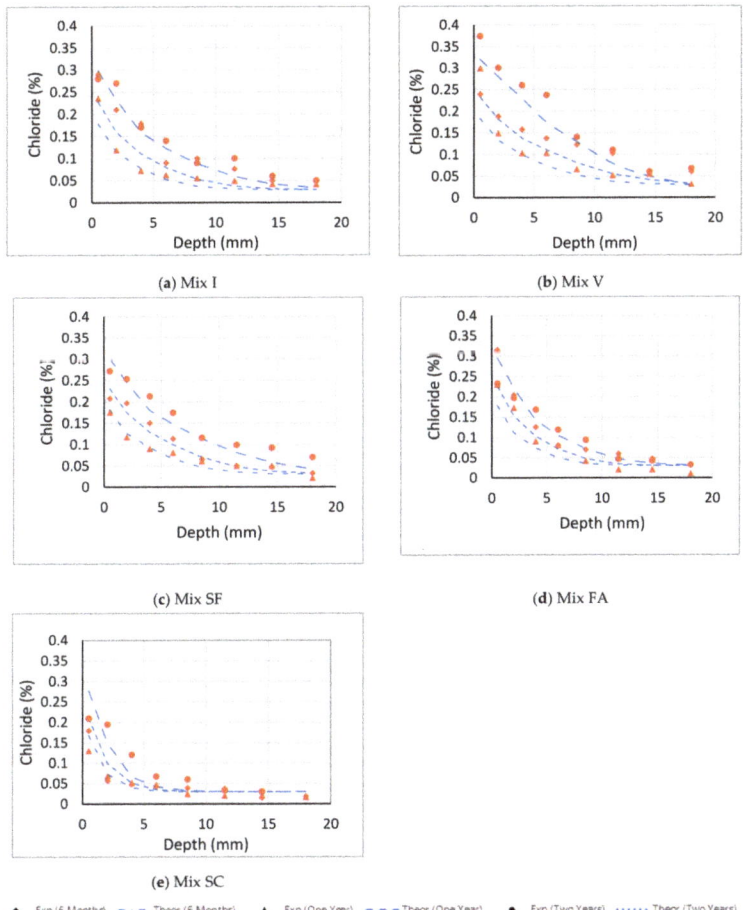

Figure 6. Chloride profiles for mixes in the tidal zone using the experimental binding results. (**a**) concrete with 100% ordinary Portland cement (OPC), (**b**) concrete with 100% type V cement (Sulfate resistance), (**c**) concrete with 6% silica fumes and 94% OPC, (**d**) concrete with 25% fly ash and 75% OPC, (**e**) concrete with 70% slag cement and 30% OPC.

4.2. Transport Model Using the New Binding Model

Figures 8 and 9 show the experimental readings and the theoretical chloride profile for the concrete blocks, using the experimental binding data and the binding model, in coastal and tidal zones, respectively. The chloride profile plots are well correlated to the theoretical plots using the experimental binding data. This shows the accuracy of the proposed binding model in predicting the effective diffusion coefficient.

Table 2. Statistical analysis results.

	Coastal					
	6 Months		12 Months		24 Months	
	β-Coefficient	R^2	β-Coefficient	R^2	β-Coefficient	R^2
Mix 1	0.693	0.918	0.713	0.969	0.928	0.988
Mix 2	0.737	0.990	0.905	0.996	0.937	0.994
Mix 3	0.979	0.914	0.816	0.968	0.968	0.989
Mix 4	0.701	0.992	0.837	0.991	0.941	0.996
Mix 5	0.714	0.850	0.776	0.990	0.962	0.980
	Tidal					
	6 Months		12 Months		24 Months	
	β-Coefficient	R^2	β-Coefficient	R^2	β-Coefficient	R^2
Mix 1	0.789	0.974	0.806	0.983	0.895	0.972
Mix 2	0.612	0.957	0.851	0.985	0.920	0.988
Mix 3	0.864	0.978	0.959	0.978	0.965	0.980
Mix 4	0.650	0.978	0.725	0.994	0.779	0.966
Mix 5	0.758	0.955	0.795	0.965	0.761	0.907

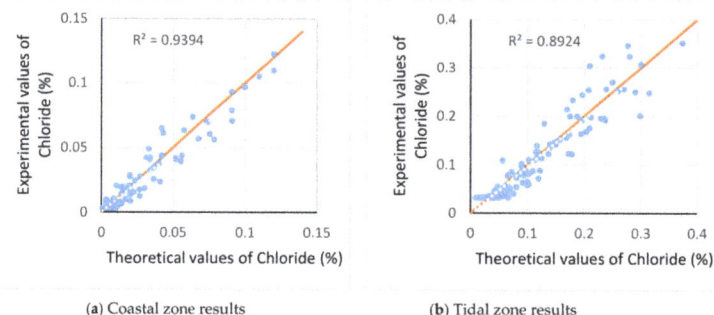

(a) Coastal zone results   (b) Tidal zone results

Figure 7. Regression analysis of the theoretical results of the transport model against experimental results. (**a**) Results from blocks located in coastal zone, (**b**) Results from blocks located in tidal zone.

It is noteworthy that using the proposed binding model produces conservative results. The parameter m serves as a shape function of the chloride profile, and increasing its value yields more concave profiles which indicate the existence of high SCMs in the concrete. The calculations of the initial apparent diffusion and the apparent diffusivity over time in Equations (8) and (9) are good approximations of their counterparts in the experimental testing. Using these and the binding values produces reasonable values of the effective diffusivity [16]. Adopting the effective diffusivity in the transport model can be a direct caliber for prediction of concrete threshold at which steel starts corrosion. On the other hand, using the apparent diffusivity results in determining approximately the total chloride without identifying the state of chlorides. Therefore, free chloride profiling is more realistic in service life modeling and assessing durability of concrete structures. Moreover, the algorithm embeds different environmental parameters, such as temperature and carbonation, in calibrating the effective diffusion coefficient at each step of the solution. This approach simplifies the analysis by decoupling the impacts of the environmental parameters and reduces the time needed for obtaining results. Yet, it produces accurate results. Therefore, the developed algorithm supports using effective diffusion coefficient with calibration along with binding capacity only to predict the free chloride profile in concrete at different depths. Hence, the other terms in the transport laws, such as chemical activities and electrical flux, could be omitted from solution because the expected change in the results might not be noticeable and is not worth the expensive processing.

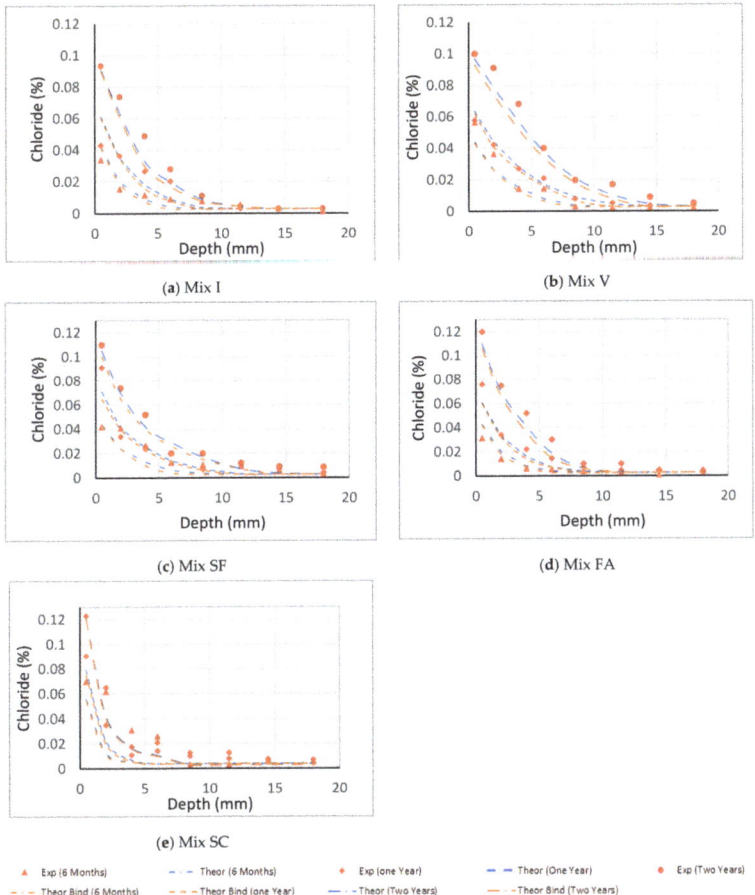

Figure 8. Chloride profiles for mixes in the coastal zone using the new binding model. (**a**) concrete with 100% ordinary Portland cement (OPC), (**b**) concrete with 100% type V cement (Sulfate resistance), (**c**) concrete with 6% silica fumes and 94% OPC, (**d**) concrete with 25% fly ash and 75% OPC, (**e**) concrete with 70% slag cement and 30% OPC.

Figure 10 shows, through statistical analysis, that β-coefficient and R^2 are equal to 0.9 and 0.93 for coastal zone, and 0.87 and 0.83 for tidal zone. This evidenced the accuracy of the model using the new binding model when compared with the results generated using experimental binding data.

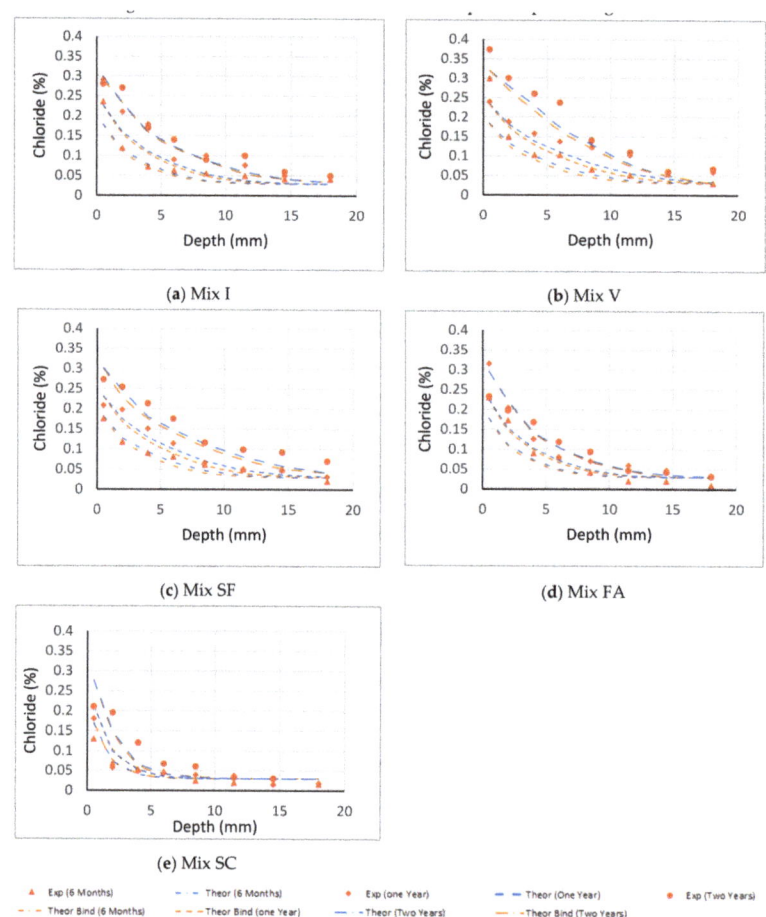

Figure 9. Chloride profiles for mixes in the tidal zone using the new binding model. (**a**) concrete with 100% ordinary Portland cement (OPC), (**b**) concrete with 100% type V cement (Sulfate resistance), (**c**) concrete with 6% silica fumes and 94% OPC, (**d**) concrete with 25% fly ash and 75% OPC, (**e**) concrete with 70% slag cement and 30% OPC.

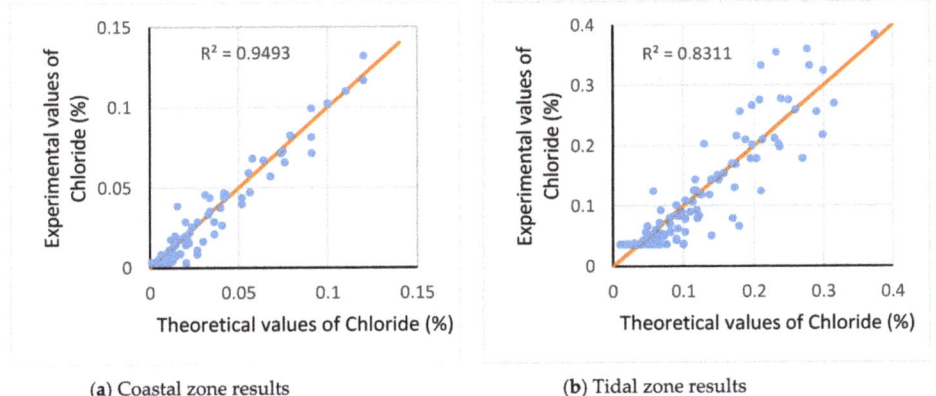

(a) Coastal zone results (b) Tidal zone results

Figure 10. Regression analysis of the theoretical results of the transport model using the new binding model against experimental results. (**a**) Results from blocks located in coastal zone, (**b**) Results from blocks located in tidal zone.

5. Software Implementation

The finite element formulation derived in the previous section was developed into programming code using C-Sharp programming language. The developed software is an object-oriented program that utilizes "classes" to achieve the objective. Figure 11 shows the high-level diagram of the classes utilized in the software. The Exposed Cross Section comprised implementation of various other classes including Concrete, Cement, Binding Capacity, Ion, and Weather Condition. Numerical solution code was implemented using the Assembly Matrices of the Finite Elements of the Exposed Concrete Section by taking the Accuracy Level specified by the user (number of nodes in a Cross-Section) into account.

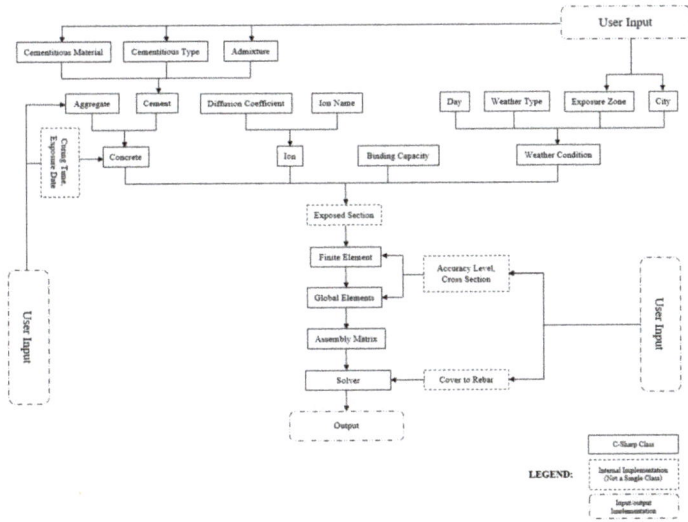

Figure 11. Class diagram/software implementation flowchart.

5.1. Development of GUI

In order to implement the code for practical purpose, various inputs relating to the weather condition, concrete composition, and cross-section must be retrieved from the user. A web-based interface designed for this purpose will be described in this section.

5.1.1. Weather Condition Details

The modeling engine is equipped with weather data for three cities, namely: Dammam, Dhahran, and Jeddah in Saudi Arabia. The data contains temperature and humidity values for the cities for previous years. An algorithm was written to enable extrapolation to future years using the previous years' weather data. A user will choose the city of interest and the exposure zone, whether tidal or splash, using the interface.

5.1.2. Reinforced Concrete Properties and Member Geometric Details

Reinforced concrete properties, including water/cement ratio, cement type, cementitious material type and ratio, admixture type and ratio, aggregate type and content, as well as concrete curing time and exposure date are collected from the user using the interface. The interface is also used to gather data relating to the geometric properties of the exposed reinforced concrete, including the concrete member cross-sectional area and cover to reinforcement.

5.2. Modeling Outputs

The details collected from the user through the interfaces described in the previous section were utilized for the modeling since all the parameters are embedded in the ionic diffusion model described in the previous section. The output of the calculation is the chloride ion concentration at various depths from the member surface to the top of the reinforcement in the concrete member. From Equation (5), it is obvious that the diffusion model is time dependent, and hence, the calculation will continue in time until a specified chloride threshold is attained at which point the calculation will stop. The chloride concentration at this point in time will then be saved together with the total period (in days) before the chloride threshold is reached (known as corrosion initiation time) in the reinforced concrete member. Chloride ion threshold in reinforced concrete member varies depending on the code of practice being adopted. ACI specifies a chloride threshold of 0.5% by weight of concrete member being analyzed. This value was utilized in the current model to produce results similar to Figure 12.

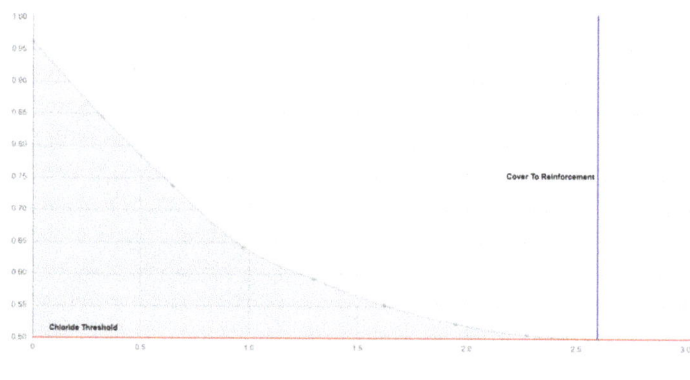

Figure 12. Typical modeling results of chloride profile and duration to corrosion initiation.

6. Conclusions

This paper develops a new model for predicting the effective concrete diffusivity which uses the apparent diffusion along with binding, age, temperature, carbonation, and free chloride effects. Then, the paper develops an algorithm to predict the free chloride concentration at different depths in the concrete. Different OPC and SCMs mixes prepared in the lab determined the experimental binding values which are used in the algorithm. Chloride profiles were measured on two sets of concrete blocks: one exposed daily to seawater, and the other exposed to harsh weather along the coast for two years at the east coast of Saudi Arabia. The proposed transport model proves reliability by comparing its results to the exposure site results. To completely decouple the transport model from any experimental values, the paper introduces a new binding model which relies on the oxides of cements and level of chlorides. The binding model introduces itself in the transport model, and the results compare well to the exposure site results, and the results of the transport model which uses the experimental binding data. The following remarks conclude the work:

- A reliable value of the effective diffusion coefficient is developed from the apparent diffusion coefficient by considering the binding capacity of concrete and other environmental impacts. The effective diffusivity decouples the impact of different environmental impact and embeds their effects in its calibration at each step of solution.
- A reliable FEM transport model is developed to predict chlorides concentrations for concrete in tidal and harsh environments. The model uses solely calibrated diffusion coefficient and binding data in Fick's law, and produces accurate results. Hence, the other terms in the transport laws, such as chemical activities and electrical flux, might not be valuable to the solution because the expected refinement against time of processing is marginal.

The transport model predicts the chloride profile effectively using the new proposed binding model. Statistical analysis evidence the accuracy of the proposed model, using experimental binding data, by recording β-coefficient and R^2 of 1.00 and 0.94 for coastal zone, and 0.89 and 0.89 for tidal zone; while when using the new binding model, statistical analysis show β-coefficient and R^2 of 0.9 and 0.93 for coastal zone, and 0.87 and 0.83 for tidal zone.

Author Contributions: Methodology, Project administration, Supervision; Validation, Resources, Writing—Review & editing, A.A.E.F.; Methodology, Software implementation Writing—original draft, M.O.A. All authors have read and agreed to the published version of the manuscript.

Funding: This research is funded by the Deanship of Scientific Research (DSR) at King Fahd University of Petroleum and Minerals (KFUPM) through project no. IN161050.

Acknowledgments: The authors acknowledge the support provided by the Deanship of Scientific Research (DSR) at King Fahd University of Petroleum and Minerals (KFUPM) for funding this work through project no. IN161050.

Conflicts of Interest: The authors declare no conflict of interest.

References

1. Luping, T.; Nilsson, L.-O.; Basheer, P.A.M. *Resistance of Concrete to Chloride Ingress*, 1st ed.; Spon Press: London, UK, 2012.
2. Bernal, J.; Fenaux, M.; Moragues, A.; Reyes, E.; Gálvez, J.C. Study of chloride penetration in concretes exposed to high-mountain weather conditions with presence of deicing salts. *Constr. Build. Mater.* **2016**, *127*, 971–983. [CrossRef]
3. Kirkpatrick, T.J.; Weyers, R.E.; Anderson-Cook, C.M.; Sprinkel, M.M. Probabilistic model for the chloride-induced corrosion service life of bridge decks. *Cem. Concr. Res.* **2002**, *32*, 1943–1960. [CrossRef]
4. Guzmán, S.; Gálvez, J.C.; Sancho, J.M. Modelling of chloride ingress into concrete through a single-ion approach: Application to an idealized surface crack pattern. *Int. J. Numer. Anal. Methods Geomech.* **2014**, *38*, 1683–1706. [CrossRef]

5. Ababneh, A.; Xi, Y. An experimental study on the effect of chloride penetration on moisture diffusion in concrete. *Mater. Struct.* **2002**, *35*, 659–663. [CrossRef]
6. Ababneh, A.; Benboudjema, F.; Xi, Y. Chloride penetration in Nonsaturated concrete. *J. Mater. Civ. Eng.* **2003**, *15*, 183–191. [CrossRef]
7. Liu, J.; Liu, J.; Huang, Z.; Zhu, J.; Liu, W.; Zhang, W. Effect of Fly Ash as Cement Replacement on Chloride Diffusion, Chloride Binding Capacity, and Micro-Properties of Concrete in a Water Soaking Environment. *Appl. Sci.* **2020**, *10*, 6271. [CrossRef]
8. Lee, T.; Lee, J. Evaluation of chloride resistance of early-strength concrete using blended binder and polycarboxylate-based chemical admixture. *Appl. Sci. Switz.* **2002**, *10*, 2972–2974. [CrossRef]
9. Zhang, J.; Zhou, X.Z.; Zheng, J.J.; Ye, H.L.; Yang, J. Experimental investigation and analytical modeling of chloride diffusivity of fly ash concrete. *Materials* **2020**, *13*, 862. [CrossRef]
10. Xi, Y.; Bažant, Z.P. Modeling Chloride Penetration in Saturated Concrete. *J. Mater. Civ. Eng.* **1999**, *11*, 58–65. [CrossRef]
11. Samson, E.; Marchand, J.; Snyder, K.A.; Beaudoin, J.J. Modeling ion and fluid transport in unsaturated cement systems in isothermal conditions. *Cem. Concr. Res.* **2005**, *35*, 141–153. [CrossRef]
12. Samson, E.; Marchand, J. Modeling the transport of ions in unsaturated cement-based materials. *Comput. Struct.* **2007**, *85*, 1740–1756. [CrossRef]
13. Nguyen, T.Q.; Petković, J.; Dangla, P.; Baroghel-Bouny, V. Modelling of coupled ion and moisture transport in porous building materials. *Constr. Build. Mater.* **2008**, *22*, 2185–2195. [CrossRef]
14. Saetta, A.V.; Scotta, R.V.; Vitaliani, R.V. Analysis of chloride diffusion into partially saturated concrete. *ACI Mater. J.* **1993**, *90*, 441–451.
15. Samson, E.; Marchand, J. Numerical Solution of the Extended Nernst-Planck Model. *J. Colloid Interface Sci.* **1999**, *215*, 1–8. [CrossRef]
16. Fenaux, M.; Reyes, E.; Gálvez, J.C.; Moragues, A. Modelling the transport of chloride and other ions in cement-based materials. *Cem. Concr. Compos.* **2019**, *97*, 33–42. [CrossRef]
17. Luping, T.; Nilsson, L.O. Chloride Binding Capacity, Penetration and Pore Structures of Blended Cement Pastes with Slag and Fly Ash. In Proceedings of the Blended Cement in Construction, International Conference, University of Sheffield, Sheffield, UK, 9–12 September 1991.
18. Tuutti, K. *Corrosion of Steel in Concrete*; Swedish Cement and Concrete Research Institute: Stockholm, Sweden, 1982.
19. Blunk, G.; Gunkel, P.; Smolczyk, H.G. On the distribution of chloride between the hardening cement paste and its por solution. In Proceedings of the 8th International Congress on the Chemistry of Cement, Rio de Janeiro, Brazil, 22–27 September 1986; pp. 85–89.
20. Tritthart, J. Chloride binding in cement I. Investigations to determine the composition of porewater in hardened cement. *Cem. Concr. Res.* **1989**, *19*, 586–594. [CrossRef]
21. Tritthart, J. Chloride binding in cement II. The influence of the hydroxide concentration in the pore solution of hardened cement paste on chloride binding. *Cem. Concr. Res.* **1989**, *19*, 683–691. [CrossRef]
22. Luping, T.; Nilsson, L.-O. Chloride binding capacity and binding isotherms of OPC pastes and mortars. *Cem. Concr. Res.* **1993**, *23*, 247–253. [CrossRef]
23. Samson, E.; Lemaire, G.; Marchand, J.; Beaudoin, J.J. Modeling chemical activity effects in strong ionic solutions. *Comput. Mater. Sci.* **1999**, *15*, 285–294. [CrossRef]
24. Pradelle, S.; Thiéry, M.; Baroghel-Bouny, V. Comparison of existing chloride ingress models within concretes exposed to seawater. *Mater. Struct. Mater. Constr.* **2016**, *49*, 4497–4516. [CrossRef]
25. Riding, K.A.; Thomas, M.D.; Folliard, K.J. Apparent diffusivity model for concrete containing supplementary cementitious materials. *ACI Mater. J.* **2013**, *110*, 705–714.
26. Abd El Fattah, A.; Al-Duais, I.; Riding, K.; Thomas, M.; Al-Dulaijan, S. Field Validation of Concrete Transport Property Measurement Methods. *Materials* **2020**, *13*, 1166. [CrossRef] [PubMed]
27. Abd El Fattah, A.; Al-Duais, I.; Riding, K.; Thomas, M. Field evaluation of corrosion mitigation on reinforced concrete in marine exposure conditions. *Constr. Build. Mater.* **2018**, *165*, 663–674. [CrossRef]
28. ASTM. *Standard C150, 2020, Standard Specification for Portland Cement*; ASTM International: West Conshohocken, PA, USA, 2020; Available online: https://www.astm.org/ (accessed on 10 September 2020).

29. ASTM. *Standard C1240, 2020, Standard Specification for Silica Fume Used in Cementitious Mixtures*; ASTM International: West Conshohocken, PA, USA, 2020; Available online: https://www.astm.org/ (accessed on 10 September 2020).
30. ASTM. *Standard C618, 2019, Standard Specification for Coal Fly Ash and Raw or Calcined Natural Pozzolan for Use in Concrete*; ASTM International: West Conshohocken, PA, USA, 2019; Available online: https://www.astm.org/ (accessed on 7 September 2020).
31. ASTM. *Standard C989, 2018, Standard Specification for Slag Cement for Use in Concrete and Mortars*; ASTM International: West Conshohocken, PA, USA, 2018; Available online: https://www.astm.org/ (accessed on 7 September 2020).
32. Thomas, M.D.; Scott, A.; Bremner, T.; Bilodeau, A.; Day, D. Performance of slag concrete in marine environment. *ACI Mater. J.* **2008**, *105*, 628.
33. GAMRY. Referene 3000 Potentiostat. GAMRY Webpage. Available online: https://www.gamry.com/potentiostats/reference-3000/ (accessed on 12 September 2020).
34. ASTM. *Standard C1556, 2016 Standard Test Method for Determining the Apparent Chloride Diffusion Coefficient of Cementitious Mixtures by Bulk Diffusion*; ASTM International: West Conshohocken, PA, USA, 2016; Available online: https://www.astm.org/ (accessed on 10 September 2020).
35. Ngala, V.T.; Page, C.L. Effects of Carbonation on Pore Structure and Diffusional Properties of *Hydrated Cem. Pastes* **1997**, *27*, 995–1007.
36. Moffatt, E.T.G.; Thomas, M.D. Effect of Carbonation on the Durability and Mechanical Performance of Ettringite-Based Binders. *ACI Mater. J.* **2019**, 116. [CrossRef]
37. Mangat, P.S.; Molloy, B.T. Prediction of long term chloride concentration in concrete. *Mater. Struct.* **1994**, *27*, 338–346. [CrossRef]
38. Bamforth, P.B. CONCRETE DURABILIITY BY DESIGN: Limitations of the current prescriptive approach and alternative methods for durability design. In Proceedings of the NZ Concrete Society Conference, Taupo, New Zealand, 4–6 October 2002.
39. Maage, M.; Helland, S.; Carlsen, J.E. Practical non-steady state chloride transport as a part of a model for predicting the initiation period. In Proceedings of the RILEM International Workshop on Chloride Penetration into Concrete, Saint-Rémy-lès-Chevreuse, France, 15–18 October 1995; pp. 398–406.
40. Thomas, M.D.A.; Bamforth, P.B. Modelling chloride diffusion in concrete effect of fly ash and slag. *Cem. Concr. Res.* **1999**, *29*, 487–495. [CrossRef]
41. Ehlen, M.; Thomas, M.D.; Bentz, E.C. Life-365 Service Life Prediction Model. *Concr. Int.* **2009**, *31*, 41–46.
42. Thomas, M.D.A.; Hooton, R.D.; Scott, A.; Zibara, H. The effect of supplementary cementitious materials on chloride binding in hardened cement paste. *Cem. Concr. Res.* **2012**, *42*, 1–7. [CrossRef]
43. Zibara, H. *Binding of External Chlorides by Cement Pastes*; University of Toronto: Toronto, ON, Canada, 2001.
44. Delagrave, A.; Marchand, J.; Ollivier, J.; Julien, S.; Hazrati, K. Chloride Binding Capacity of Various Hydrated Cement Paste Systems. *Adv. Cem. Based Mater.* **1997**, *6*, 28–35. [CrossRef]
45. Guo, Y.; Zhang, T.; Tian, W.; Wei, J.; Yu, Q. Physically and chemically bound chlorides in hydrated cement pastes: A comparison study of the effects of silica fume and metakaolin. *J. Mater. Sci.* **2019**, *54*, 2152–2169. [CrossRef]
46. Saillio, M.; Baroghel-Bouny, V.; Barberon, F. Chloride binding in sound and carbonated cementitious materials with various types of binder. *Constr. Build. Mater.* **2014**, *68*, 82–91. [CrossRef]
47. Ogirigbo, O.R.; Black, L. Chloride binding and diffusion in slag blends: Influence of slag composition and temperature. *Constr. Build. Mater.* **2017**, *149*, 816–825. [CrossRef]
48. Saillio, M.; Baroghel-Bouny, V.; Pradelle, S. Effect of carbonation and sulphate on chloride ingress in cement pastes and concretes with supplementary cementitious materials. In Proceedings of the 8th International Conference on Concrete under Severe Conditions–Environment and Loading, Lecco, Italy, 12–14 September 2016; Volume 711, pp. 241–248.
49. Song, H.W.; Lee, C.H.; Jung, M.S.; Ann, K.Y. Development of chloride binding capacity in cement pastes and influence of the pH of hydration products. *Can. J. Civ. Eng.* **2008**, *35*, 1427–1434. [CrossRef]
50. Yang, Z.; Gao, Y.; Mu, S.; Chang, H.; Sun, W.; Jiang, J. Improving the chloride binding capacity of cement paste by adding nano-Al_2O_3. *Constr. Build. Mater.* **2019**, *195*, 415–422. [CrossRef]
51. Ann, K.Y.; Hong, S.I. Modeling chloride transport in concrete at pore and chloride binding. *ACI Mater. J.* **2018**, *115*, 595–604. [CrossRef]

52. Jung, M.S.; Kim, K.B.; Lee, S.A.; Ann, K.Y. Risk of chloride-induced corrosion of steel in SF concrete exposed to a chloride-bearing environment. *Constr. Build. Mater.* **2018**, *166*, 413–422. [CrossRef]
53. Ann, K.Y.; Kim, T.S.; Kim, J.H.; Kim, S.H. The resistance of high alumina cement against corrosion of steel in concrete. *Constr. Build. Mater.* **2010**, *24*, 1502–1510. [CrossRef]
54. Ramírez-Ortíz, A.E.; Castellanos, F.; de Cano-Barrita, P.F. Ultrasonic Detection of Chloride Ions and Chloride Binding in Portland Cement Pastes. *Int. J. Concr. Struct. Mater.* **2018**, *12*, 20. [CrossRef]
55. Wang, Y.; Shui, Z.; Gao, X.; Yu, R.; Huang, Y.; Cheng, S. Understanding the chloride binding and diffusion behaviors of marine concrete based on Portland limestone cement-alumina enriched pozzolans. *Constr. Build. Mater.* **2019**, *198*, 207–217. [CrossRef]
56. Qiao, C.; Suraneni, P.; Ying, T.N.W.; Choudhary, A.; Weiss, J. Chloride binding of cement pastes with fly ash exposed to CaCl$_2$ solutions at 5 and 23 °C. *Cem. Concr. Compos.* **2019**, *97*, 43–53. [CrossRef]
57. Freundlich, H. *Kapillarchemie: Eine Darstellung der Chemie der Kolloide und verwandter Gebiete*; Akademische Verlagsgesellschaf: Leipzig, Germanny, 1909.
58. Lee, J.S.; Ray, R.I.; Little, B.J. Comparison of Key West and Persian Gulf Seawaters. In Proceedings of the NACE–International Corrosion Conference Series, Nashville, TN, USA, 11–15 March 2007.
59. Liu, Q.F.; Hu, Z.; Y, X.L.; Yang, J.; Azim, I.; Sun, W. Prediction of chloride distribution for offshore concrete based on statistical analysis. *Materials* **2020**, *13*, 174. [CrossRef] [PubMed]

Publisher's Note: MDPI stays neutral with regard to jurisdictional claims in published maps and institutional affiliations.

© 2020 by the authors. Licensee MDPI, Basel, Switzerland. This article is an open access article distributed under the terms and conditions of the Creative Commons Attribution (CC BY) license (http://creativecommons.org/licenses/by/4.0/).

Article

Hardening Parameter Homogenization for J2 Flow with Isotropic Hardening of Steel Fiber-Reinforced Concrete Composites

Petr V. Sivtsev [1],* and Piotr Smarzewski [2]

[1] Department of Computational Technologies, Institute of Mathematics and Information Science, Ammosov North-Eastern Federal University, 58 Belinskogo, 6677013 Yakutsk, Russia
[2] Department of Structural Engineering, Faculty of Civil Engineering and Architecture, Lublin University of Technology, Nadbystrzycka 40, 20-618 Lublin, Poland; p.smarzewski@pollub.pl
* Correspondence: sivkapetr@mail.ru; Tel.: +7-924-169-17-47

Citation: Sivtsev, P.V.; Smarzewski, P. Hardening Parameter Homogenization for J2 Flow with Isotropic Hardening of Steel Fiber-Reinforced Concrete Composites. *Crystals* **2021**, *11*, 776. https://doi.org/10.3390/cryst11070776

Academic Editor: Tomasz Sadowski

Received: 8 June 2021
Accepted: 30 June 2021
Published: 2 July 2021

Publisher's Note: MDPI stays neutral with regard to jurisdictional claims in published maps and institutional affiliations.

Copyright: © 2021 by the authors. Licensee MDPI, Basel, Switzerland. This article is an open access article distributed under the terms and conditions of the Creative Commons Attribution (CC BY) license (https://creativecommons.org/licenses/by/4.0/).

Abstract: Numerical modeling of the stress–strain state of composite materials such as fiber-reinforced concrete is a considerable computational challenge. Even if a computational grid with the resolution of all inclusions is built, it will take a great amount of time for the most powerful clusters to calculate the deformations of one concrete block with ideal parallelization. To solve this problem, the method of numerical homogenization is actively used. However, when plastic deformations are taken into account, the numerical homogenization becomes much more complicated due to nonlinearity. In this work, the description of the anisotropic nature of the hardening of the composite material and the numerical homogenization for the J2 flow with isotropic hardening is proposed. Here, the deformation of a composite material with a periodic arrangement of inclusions in the form of fibers is considered as a model problem. In this case, the assumption is made that inclusions have pure elastic properties. Numerical homogenization of the elasticity and plasticity parameters is performed on the representative element. The novelty of the work is related to the attempt at hardening parameter homogenization. The calculated effective parameters are used to solve the problem on a coarse mesh. The accuracy of using the computational algorithm is checked on model problems in comparison with the hardening parameters of the base material. The finite element implementation is built using the FEniCS computing platform and the fenics-solid library.

Keywords: elastoplastic; hardening; homogenization; J2 flow; composites; fiber reinforcement; finite element

1. Introduction

Composite materials are used in various industries. Over time, their structures change, and their production is optimized in terms of operational properties. Composite materials have the advantage of a high strength-to-weight ratio, improved thermal conductivity, or isolation. Most composites are materials with elastoplastic properties, for example, reinforced concrete [1–3]. The behavior of traditional materials has been studied well enough, and there are mathematical models verified over time that describe the deformation properties of a particular natural material, whereas the description of composite materials is not so simple. In particular, the description of the deformations of a composite material outside the theory of linear elasticity can be handled by a second-order radial return algorithm [4].

In our study, we consider metal inclusions. For the mathematical description of the deformation of concrete, we use the theory of plastic flow with isotropic hardening and a description of the plastic flow region using the Mises yield criterion. Numerical modeling of a composite material with a mesh resolution of all inclusions in the form of reinforcement and distributed fibers is a considerable computational problem [5,6]. Modern computer equipment has impressive power due to a large number of cores with a high clock speed. However, the complexity of the problem increases nonlinearly with an increase in the structure size under consideration.

Previous work [7] proposed the use of the effective tensor of elasticity, obtained from numerical homogenization, the parameters of plasticity, such as the yield stress, and the hardening parameter as in concrete. At small values of plastic deformations, this approximation demonstrates good accuracy and can be used to describe plastic deformations until the body begins to collapse. In other words, our model considers concrete stress–strain dependence as piecewise linear, as shown in Figure 1. This model is supposed to be used for stresses below material strength [8].

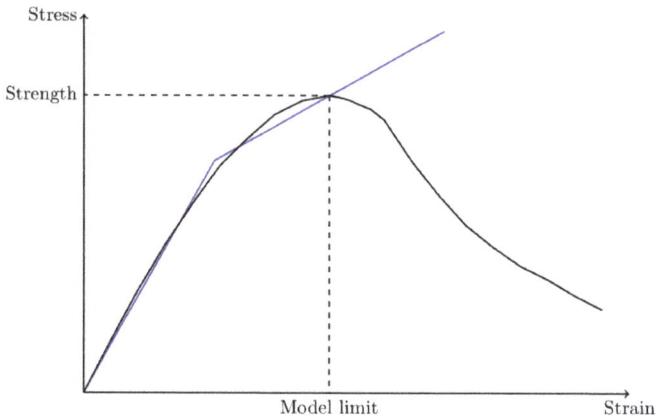

Figure 1. Plastic flow with isotropic hardening representation (blue).

In this work, a description of the anisotropic nature of the hardening of the composite material and numerically homogenizing the hardening parameter for the J2 flow with isotropic hardening is proposed. Deformation of a composite material with a periodic arrangement of inclusions in the form of fibers is considered as a model problem.

On the representative element, the elastic parameters and the hardening parameter are numerically homogenized. The calculated effective parameters are used to solve the problem on a coarse mesh. The accuracy of the computational algorithm is checked on model problems in comparison with the use of the hardening parameter of the base material, i.e., concrete. The finite element implementation is built using the FEniCS [9] computing platform and the fenics-solid library [10].

2. Materials and Methods

2.1. Problem Statement

A two-dimensional mathematical model describing the stress–strain state of a concrete composite with steel fibers is considered, described by the equilibrium equation

$$\text{div}\,\sigma = 0, \quad x \in \Omega = \Omega_1 \cup \Omega_2, \tag{1}$$

where Ω_1 is a concrete subdomain, Ω_2 is a subdomain of fiber inclusions, σ is a stress tensor. For the plasticity model [11,12], the elastic tensor consists of elastic and plastic parts:

$$\varepsilon = \begin{pmatrix} \frac{\partial u_1}{\partial x_1} \\ \frac{\partial u_2}{\partial x_2} \\ \frac{\partial u_2}{\partial x_1} + \frac{\partial u_1}{\partial x_2} \end{pmatrix} = \varepsilon^e + \varepsilon^p. \tag{2}$$

Next, the connection between the stress tensor and the elastic strain tensor using the generalized Hooke's law is added. For convenience, Voight notation is used. In a two-dimensional case, Hooke's law can be presented as:

$$\sigma = \begin{pmatrix} \sigma_{11} \\ \sigma_{22} \\ \sigma_{12} \end{pmatrix} = C\varepsilon^e, \quad C = \begin{pmatrix} C_{1111} & C_{1122} & C_{1112} \\ C_{2211} & C_{2222} & C_{2212} \\ C_{1211} & C_{1222} & C_{1212} \end{pmatrix}, \quad (3)$$

where C is an elasticity tensor, which is as follows for isotropic materials:

$$C = \begin{pmatrix} \lambda + 2\mu & \lambda & 0 \\ \lambda & \lambda + 2\mu & 0 \\ 0 & 0 & \mu \end{pmatrix}, \quad (4)$$

where λ and μ are the Lamé coefficients determined through Young's modulus E and Poisson's ratio ν. Model values of these parameters for concrete and steel are presented in Table 1 [13,14].

Table 1. Model elastic parameters of concrete matrix, basalt, and steel fibers.

Material	Young Modulus E, GPa	Poisson Ratio, ν
Concrete	40	0.15
Steel	200	0.3

Yield criterion of plastic flow with isotropic hardening is following:

$$f(\sigma, \varepsilon^p, \kappa) = \phi(\sigma) - q_{iso}(\kappa) - \sigma_y \leq 0, \quad (5)$$

where $\phi(\sigma)$ is a scalar effective stress measure, $q_{iso}(\kappa)$ is a scalar stress-like internal variable used to model isotropic hardening, κ is an internal variable, and σ_y is initial yield stress. In this work, a von Mises model (also known as J_2 flow) with linear isotropic hardening is used. For this model, ϕ and q_{iso} are written as:

$$\phi(\sigma) = \sqrt{\frac{3}{2} s_{ij} s_{ij}}, \quad (6)$$

$$q_{iso}(\kappa) = H\kappa, \quad (7)$$

where $s_{ij} = \sigma_{ij} - \sigma_{kk}\delta_{ij}/3$ is the deviatoric stress and the constant H is a hardening parameter. For associative von Mises plasticity with isotropic hardening, there are the following rates:

$$\dot{\varepsilon}^p = \dot{\lambda}\frac{\partial f}{\partial \sigma}, \quad (8)$$

$$\kappa = \sqrt{\frac{2}{3}\dot{\varepsilon}_{ij}^p \dot{\varepsilon}_{ij}^p} \quad (9)$$

Concrete hardening parameters H and yield stress σ_y are taken as equal to 21.5 GPa and 30.0 MPa, respectively. These values are related to compression and used for all kinds of deformation in the modeling purposes of this study.

2.2. Hardening Parameter Homogenization Algorithm

2.2.1. Effective Hardening Parameter

The main assumption of this work is that the effective hardening parameter depends on the tensor of plastic deformations and takes a certain numerical value for each of its components. In this case, for the classical model of isotropic hardening, the description of compression and tension is described in the same way. The simplest form of the effective hardening factor satisfying these criteria will be as follows:

$$h_{eff} = \frac{|\varepsilon^p_{11}|h_1 + |\varepsilon^p_{22}|h_2 + |\varepsilon^p_{12}|h_3}{|\varepsilon^p_{11}| + |\varepsilon^p_{22}| + |\varepsilon^p_{12}|} \text{ for } |\varepsilon^p_{11}| + |\varepsilon^p_{22}| + |\varepsilon^p_{12}| \neq 0,$$
$$h_{eff} = h_c \text{ for } |\varepsilon^p_{11}| + |\varepsilon^p_{22}| + |\varepsilon^p_{12}| = 0,$$
(10)

the following is true for it:

$$h_{eff} = h_1 \text{ for } \varepsilon^p_{22} = \varepsilon^p_{12} = 0 \text{ and } \varepsilon^p_{11} \neq 0,$$
$$h_{eff} = h_2 \text{ for } \varepsilon^p_{11} = \varepsilon^p_{12} = 0 \text{ and } \varepsilon^p_{22} \neq 0,$$
$$h_{eff} = h_3 \text{ for } \varepsilon^p_{11} = \varepsilon^p_{22} = 0 \text{ and } \varepsilon^p_{12} \neq 0,$$
(11)

Let us take the effective value of the yield point equal to the concrete yield point. Next, consider the algorithm with which one can calculate the values of h_1, h_2, h_3.

2.2.2. Elastic Parameter Homogenization

First, the representative elements are selected. For the unidirectional location of fiber, a unit cell is used. For the bidirectional location of fibers, a 2 × 2 cell is used. In the case with a random distribution of fibers, the accuracy of the elasticity problem highly depends on representative volume size. Therefore, an optimal representative volume with 16 fibers is chosen. The effective elastic tensor of the composite material C_{eff} is calculated using representative volume, as in previous work [15].

2.2.3. Hardening Parameter Homogenization

Second, three problems of elastic plasticity are solved using very large values of the boundary condition, for example, when the deformations on whole boundary are equal:

$$u_D = (10^6 \cdot x_1, \ 0),$$
$$u_D = (0, \ 10^6 \cdot x_2),$$
$$u_D = (10^6 \cdot x_2/2, \ 10^6 \cdot x_1/2).$$
(12)

Additionally, for each case, we obtain the averaged stress tensors σ^1, σ^2, σ^3, which must correspond to solutions with effective parameters. Given the large value of deformation and, accordingly, the large value of stresses, it can be assumed that purely elastic deformations are negligible, and the ratio of stress and strain is described by the elastic-plasticity tensor. Thus, by solving the same problems on a homogeneous medium using the effective parameters, we must obtain the same stress values. However, since complete correspondence to the stress tensor by varying the value of the hardening parameter alone is hardly obtained, the agreement that it is necessary to obtain correspondence with the most essential components of the tensors is taken. Namely, σ^1_{11}, σ^2_{22}, σ^3_{12}.

Further, by solving the problem on a coarse mesh with a uniform distribution of the effective parameters C_{eff}, h_{eff} with boundary conditions (12), correspondence with σ^1_{11}, σ^2_{22}, σ^3_{12} must be obtained. When solving the problem with boundary conditions (12), smallness of the elastic deformations occurs and, accordingly, the following becomes true:

$$\sigma \approx C^{ep}\varepsilon$$
(13)

At the same time, from the theory of elastic flow with isotropic hardening, the elasto-plastic tangent is equal to

$$C^{ep} = \Xi - \frac{\Xi : \partial_\sigma g \otimes \partial_\sigma f : \Xi}{\partial_\sigma f : \Xi : \partial_\sigma g + h},$$
(14)

where $\Xi = C/(I + \Delta\lambda C : \partial^2_{\sigma\sigma}g)$.

As long as the values of the strain tensor for each point are the same, a hardening parameter that will correspond to the selected direction of deformation can be calculated at any point. To calculate the parameters, the following relations can be used:

$$h_1 = \frac{R^1[1,1]}{\Xi^1[1,1] - \frac{\sigma_{11}^1}{\varepsilon_{11}^1}} - \alpha^1, \quad h_2 = \frac{R^2[2,2]}{\Xi^2[2,2] - \frac{\sigma_{22}^2}{\varepsilon_{22}^2}} - \alpha^2, \quad h_3 = \frac{R^3[3,3]}{\Xi^3[3,3] - \frac{\sigma_{12}^3}{\varepsilon_{12}^3}} - \alpha^3, \quad (15)$$

where $R = \Xi : \partial_\sigma g \otimes \partial_\sigma f : \Xi$, and $\alpha = \partial_\sigma f : \Xi : \partial_\sigma g$ (top index corresponds to each problem). Thus, by solving three problems with different boundary conditions (12), the sought-for hardening parameter values can be obtained.

2.3. Research Object

A two-dimensional concrete structure with inclusions in the form of fibers is considered for different cases of fibers:

1. Evenly distributed and unidirectional;
2. Bidirectionally distributed;
3. Randomly distributed.

The aspect ratio of fibers is 1 to 10 and the concentration is equal to 2.5%. The locations of the fibers for each case are shown in Figure 2. The number of fibers is 256.

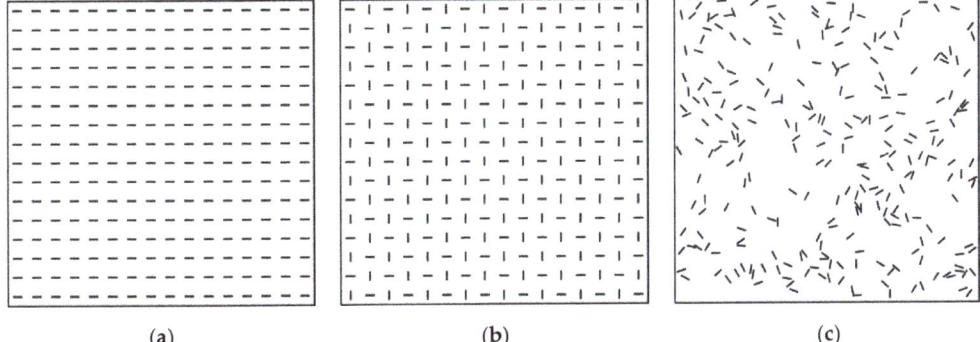

Figure 2. Research object geometries: (**a**) unidirectional location of fibers; (**b**) bidirectional location of fibers; (**c**) random location of fibers.

3. Results

For the object of study, the values of the elastic tensor coefficients and the components of the anisotropic hardening coefficient in the case of unidirectional, bidirectional, and random steel fiber distribution were calculated.

To check the effective parameters, the following tests were performed:

1. Compression along the horizontal axis (Figure 3a);
2. Compression along the vertical axis (Figure 3b);
3. Tangential stress (Figure 4a)
4. All-round compression (Figure 4b).

The purpose of the tests was to identify the scope of the proposed model.

In the first test, the left border is completely fixed, i.e., a Dirichlet boundary condition equal to zero vector is applied, and normal stress is applied to the right border, which increases from 0 to 200 MPa over 100 time steps. The second test is similar to the first, but anchoring is at the bottom, and tensile stress is on the top. The third test is related to tangential stress at the right side with anchoring on the left side.

In contrast to other tests, the fourth test features compression along two axes. A quarter of the body with 1024 fibers is considered. For this problem, the random distribution is modeled as symmetric for vertical and horizontal axes. Thus, the simulated geometry is the same. Normal stress is set above and on the right, and it also increases from 0 to 200 MPa over 100 time steps. The corresponding symmetry boundary conditions are set on the left and below.

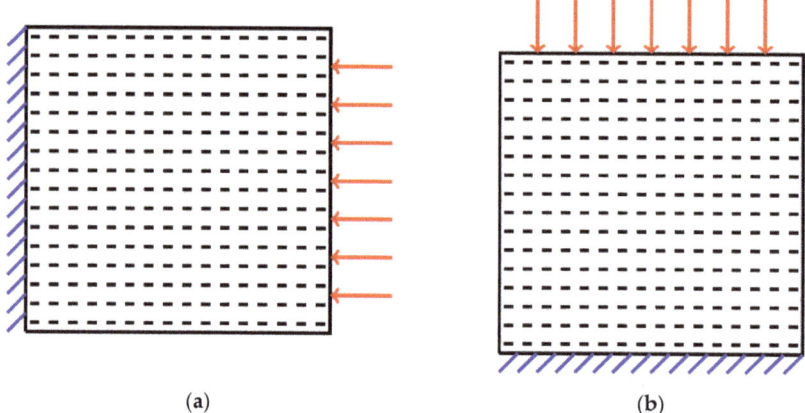

Figure 3. Schematic description of the tests: (**a**) Schematic description of test 1. Full anchoring of the left side and normal compression from the right. (**b**) Schematic description of test 2. Full anchoring from the bottom edge and normal compression from the top.

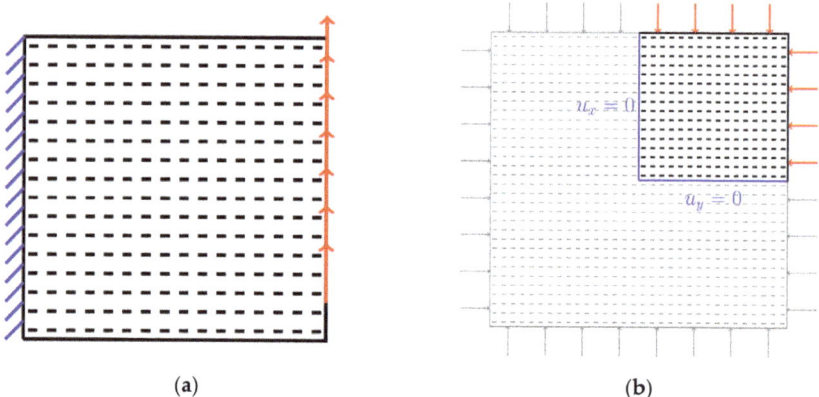

Figure 4. Schematic description of the tests: (**a**) Schematic description of test 3. Full anchoring of the left side and tangential compression from the right. (**b**) Schematic description of test 4. Symmetry conditions on the left and bottom faces and normal tension on the right and top faces.

In addition to material variation, the solution of the elastic plasticity problem with isotropic hardening is performed on an inhomogeneous region on a computational grid with full geometry (Figure 5a–c). This solution is considered as an exact solution. The errors of the compared models are calculated referring to it. A coarse grid (Figure 5d) requires a solution made using the effective elastic tensor. The models using the effective hardening factor and the concrete hardening parameter are compared. When using return mapping, the value of the plastic strain tensor for calculating the effective coefficient is taken from the previous iteration.

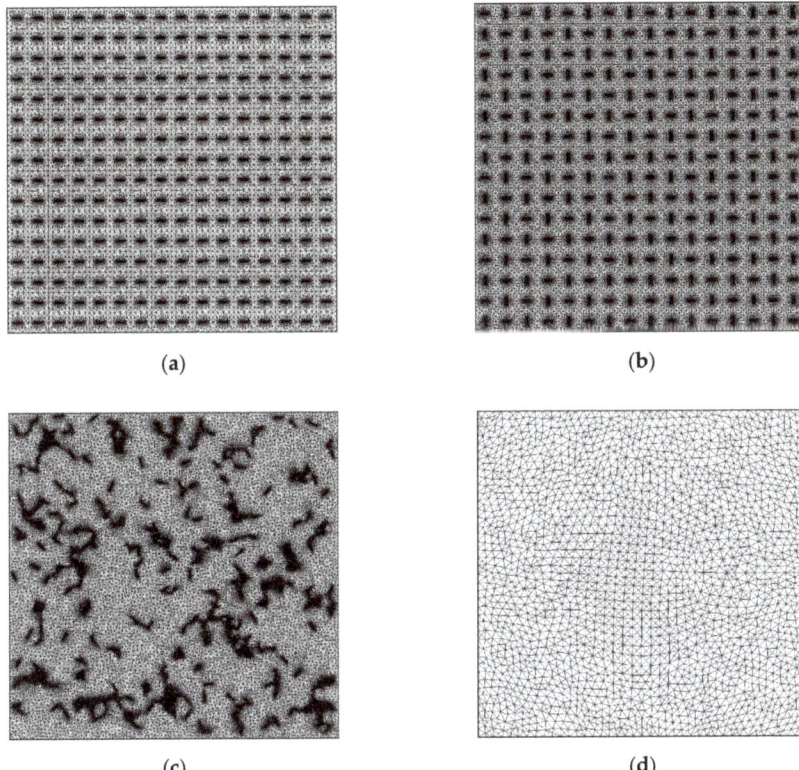

Figure 5. Computational grids: (**a**) Computational grid for case with unidirectional fiber distribution. (**b**) Computational grid for case with bidirectional fiber distribution. (**c**) Computational grid for case with random fiber distribution. (**d**) A coarse computational grid on which the numerical solution was performed using the effective parameters.

The computational mesh of the complete geometry for unidirectional, bidirectional, and random fiber distribution and the coarse computational mesh, on which the solution is performed using the effective parameters, have 33,963 vertices, 78,361 cells, 44,814 vertices, 100,063 cells, 47,822 vertices, 106,910 cells, and 2319 vertices, 4640 cells, respectively.

3.1. Calculation of Effective Parameters

To calculate the effective parameters, for the unidirectional location of fibers, a unit cell is used. For the bidirectional location of fibers, we used a 2 × 2 cell. In the case with a random distribution of fibers, the accuracy of the elasticity problem highly depends on the representative volume size. It is thought that a 16-fiber case would be optimal. The computational meshes for calculating the elastic parameters are shown in Figure 6a–f and the computational mesh for calculating the hardening parameters is shown in Figure 6c.

The computational mesh for calculating the elastic parameters for unidirectional, bidirectional, and random distribution and the coarse computational mesh for calculating the hardening parameters have 528 vertices, 1126 cells, 2020 vertices 4319 cells, 2627 vertices, 5960 cells and 177 vertices, 312 cells, respectively.

After applying the algorithm, the effective parameters presented in Table 2 were obtained.

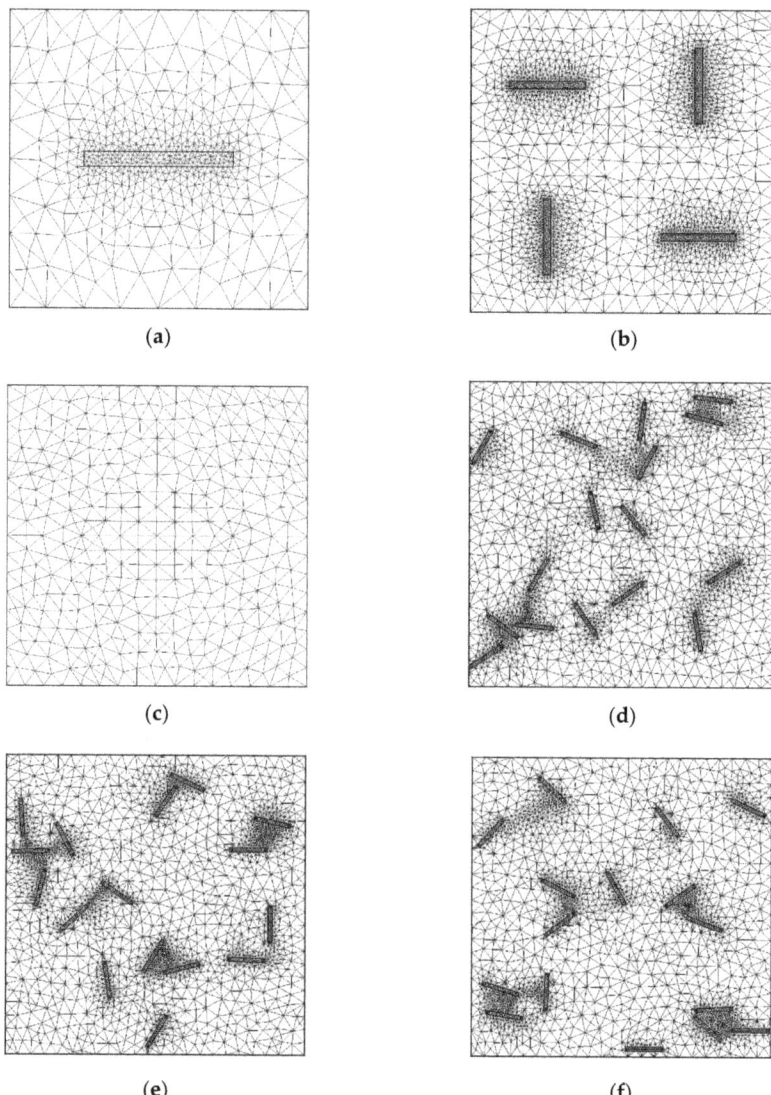

Figure 6. Computational grids used to calculate the effective parameters: (**a**) Computational mesh of a representative element for unidirectional case. (**b**) Computational mesh of a representative element for bidirectional case. (**c**) Coarse computational mesh used to calculate the effective hardening parameters. (**d**) Computational mesh of a representative element with random distribution 1. (**e**) Computational mesh of a representative element with random distribution 2. (**f**) Computational mesh of a representative element with random distribution 3.

Table 2. Calculated effective elasticity and plasticity parameters for all cases: unidirectional, bidirectional, and random distribution of steel fibers.

Coefficient	Unidirectional	Bidirectional	Random 1	Random 2	Random 3
C_{1111}	45.23 GPa	44.20 GPa	44.04 GPa	44.11 GPa	44.43 GPa
C_{2222}	43.23 GPa	44.20 GPa	44.05 GPa	44.09 GPa	43.69 GPa
C_{1212}	17.77 GPa	17.77 GPa	18.01 GPa	17.95 GPa	18.01 GPa
$C_{1122}\ C_{2211}$	7.87 GPa	7.86 GPa	8.09 GPa	8.03 GPa	8.09 GPa
$C_{1112}\ C_{1211}$	147 KPa	−112 KPa	112 KPa	4.12 MPa	−140 MPa
$C_{2212}\ C_{1222}$	−62.3 KPa	−75.8 KPa	559 KPa	74.5 MPa	−86.4 MPa
h_1	25.69 GPa	22.91 GPa	21.96 GPa	22.28 GPa	23.00 GPa
h_2	20.98 GPa	22.91 GPa	21.91 GPa	22.22 GPa	21.21 GPa
h_3	22.17 GPa	22.17 GPa	23.44 GPa	23.11 GPa	23.38 GPa

3.2. Compression along the Horizontal Axis

Based on the results of test 1, the dependences of the maximum value of the displacement modulus in all cases were obtained. They are presented in Figures 7–9.

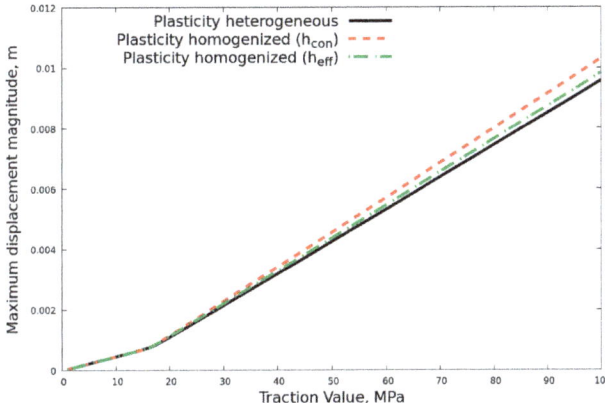

Figure 7. A graph of the dependence of the maximum value of the deformation modulus in Table 1 for unidirectional fiber location case.

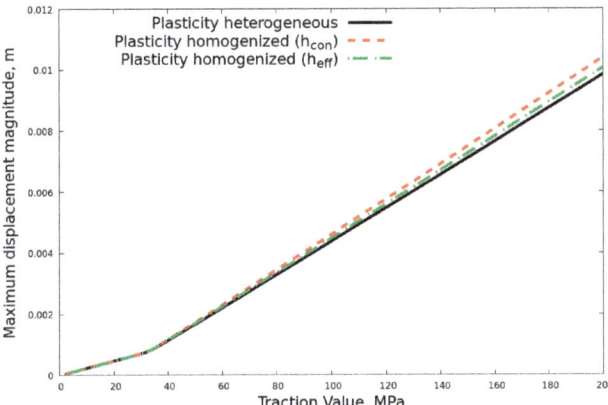

Figure 8. A graph of the dependence of the maximum value of the deformation modulus on the traction value for solving the problem of plasticity, taking into account inclusions, using effective parameters, effective parameters of elasticity, and plastic flow parameters of concrete for test 1 for bidirectional fiber location case.

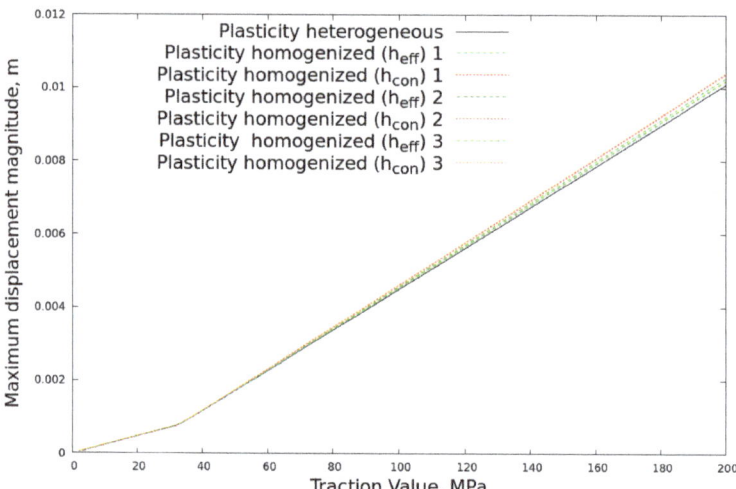

Figure 9. A graph of the dependence of the maximum value of the deformation modulus on the traction value for solving the problem of plasticity, taking into account inclusions, using effective parameters, effective parameters of elasticity, and plastic flow parameters of concrete for test 1 for random fiber distribution case.

The distributions of the deformation error modulus $|u_{err}|$, which can be expressed by fine mesh solution u_{fine} taken as exact and homogenized solution u_{hom} as $|u_{err}| = \sqrt{\left(u_{fine}^X - u_{hom}^X\right)^2 + \left(u_{fine}^Y - u_{hom}^Y\right)^2}$, for test 1 for unidirectional, bidirectional, and random distribution of fibers, are shown in Figures 10–14, respectively.

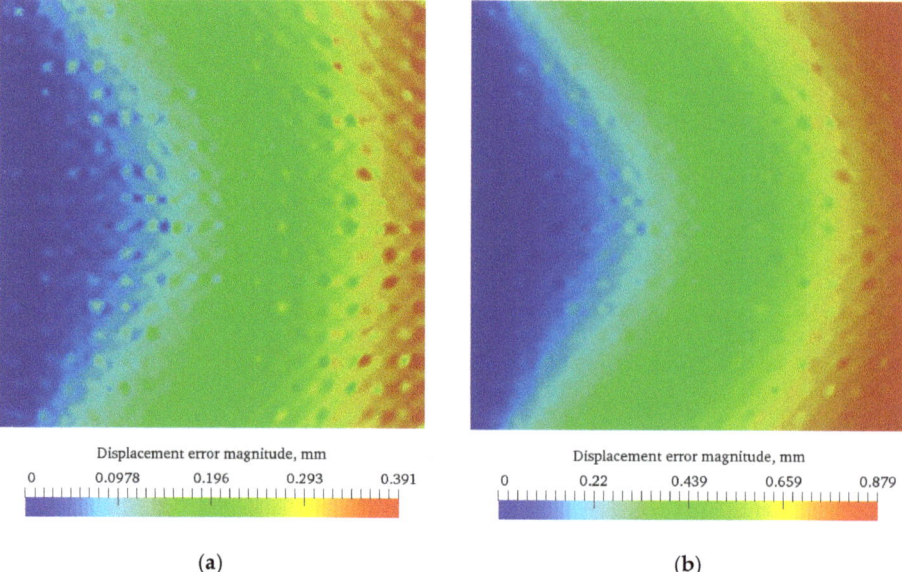

Figure 10. Distribution of the deformation error modulus for test 1 for unidirectional case: (**a**) When using effective hardening parameters; (**b**) when using the concrete hardening parameter.

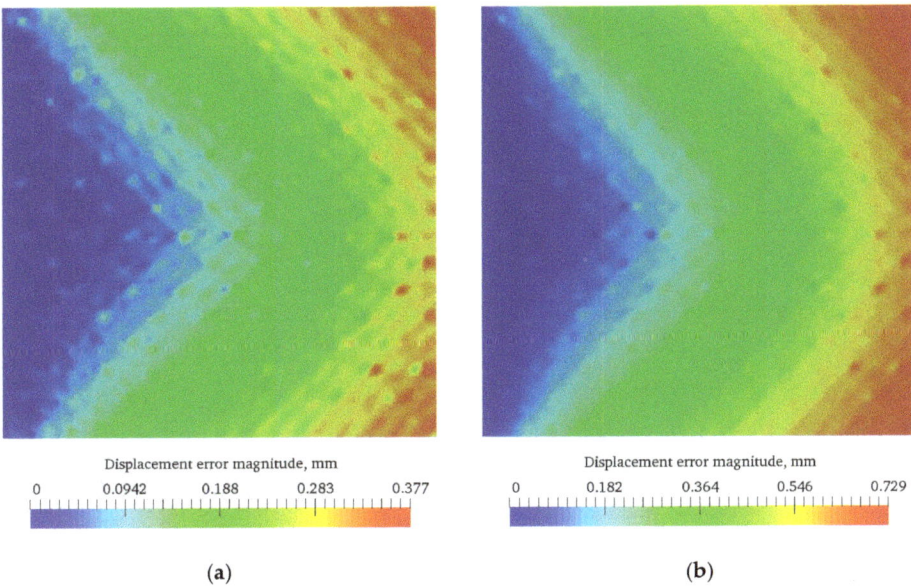

Figure 11. Distribution of the deformation error modulus for test 1 for bidirectional case: (**a**) When using effective hardening parameters; (**b**) when using the concrete hardening parameter.

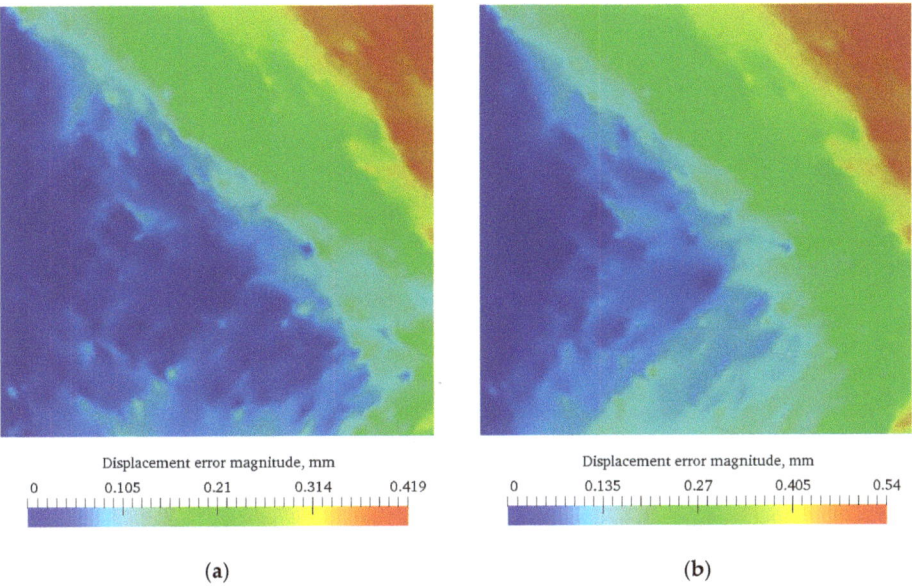

Figure 12. Distribution of the deformation error modulus for test 1 for random distribution case 1: (**a**) When using effective hardening parameters; (**b**) when using the concrete hardening parameter.

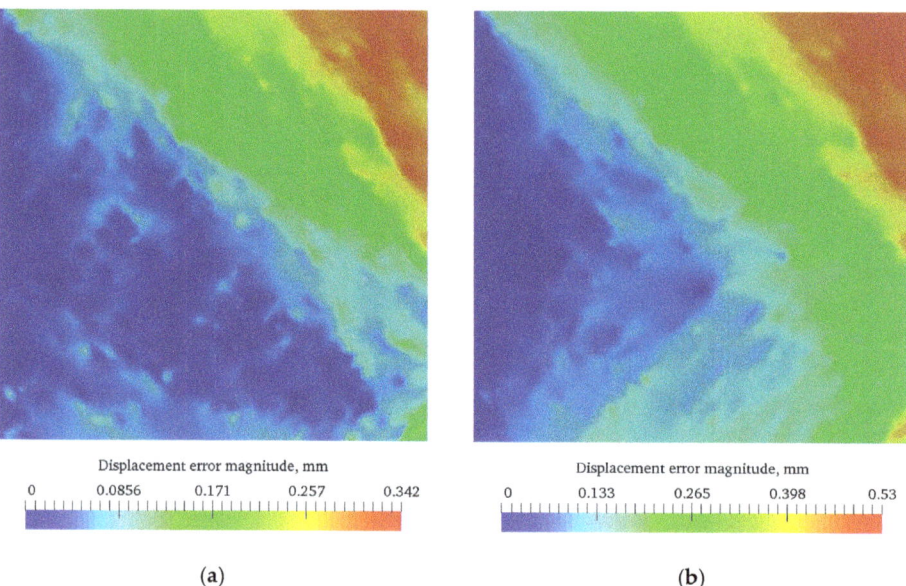

Figure 13. Distribution of the deformation error modulus for test 1 for random distribution case 2: (**a**) When using effective hardening parameters; (**b**) when using the concrete hardening parameter.

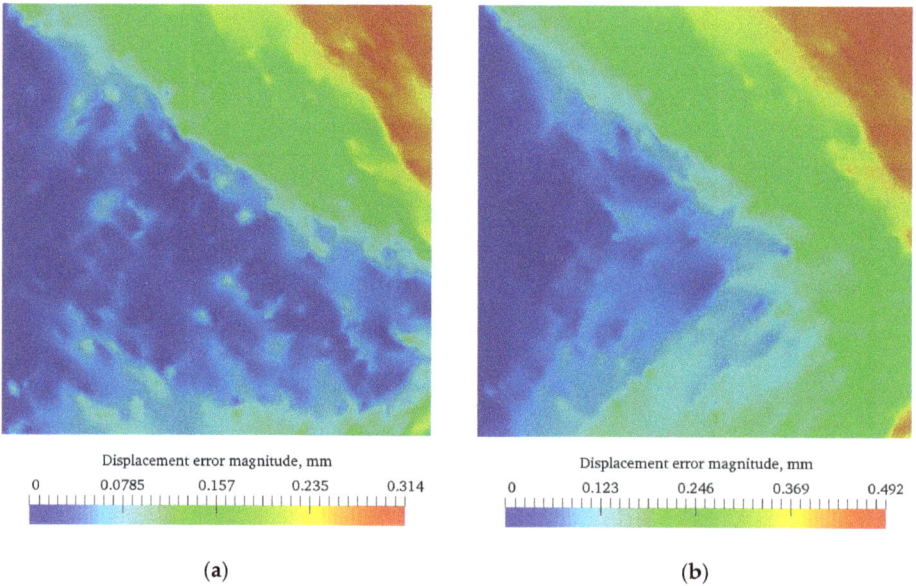

Figure 14. Distribution of the deformation error modulus for test 1 for random distribution case 3: (**a**) When using effective hardening parameters; (**b**) when using the concrete hardening parameter.

In all cases, it can be seen that when using the effective hardening parameter, the solution has a significantly lower value of the deformation error.

3.3. Compression along the Vertical Axis

Based on the results of test 2, the dependences of the maximum value of the displacement modulus in all cases were obtained. They are presented in Figures 15–17.

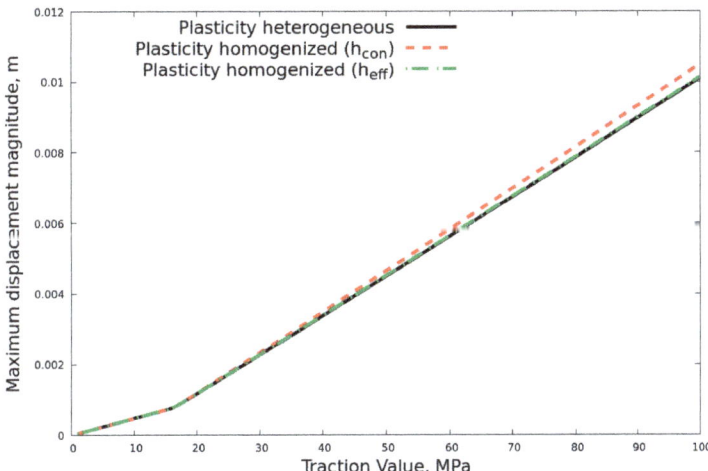

Figure 15. A graph of the dependence of the maximum value of the deformation modulus on the traction value for solving the problem of plasticity, taking into account inclusions, using effective parameters, effective parameters of elasticity, and plastic flow parameters of concrete for test 2 for unidirectional fiber location case.

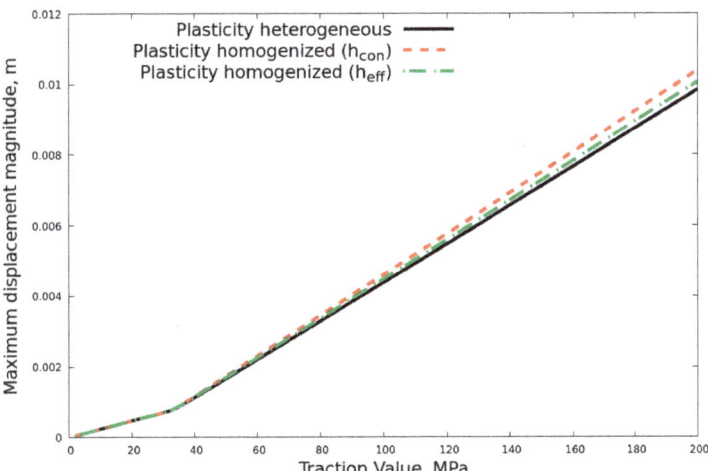

Figure 16. A graph of the dependence of the maximum value of the deformation modulus on the traction value for solving the problem of plasticity, taking into account inclusions, using effective parameters, effective parameters of elasticity, and plastic flow parameters of concrete for test 2 for bidirectional fiber location case.

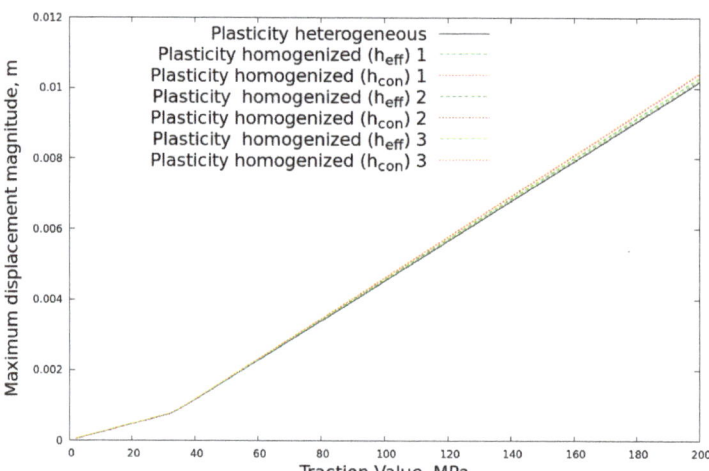

Figure 17. A graph of the dependence of the maximum value of the deformation modulus on the traction value for solving the problem of plasticity, taking into account inclusions, using effective parameters, effective parameters of elasticity, and plastic flow parameters of concrete for test 2 for random fiber distribution case.

The distributions of the deformation modulus error for test 2 for unidirectional, bidirectional, and random distribution of fibers are shown in Figures 18–22, respectively.

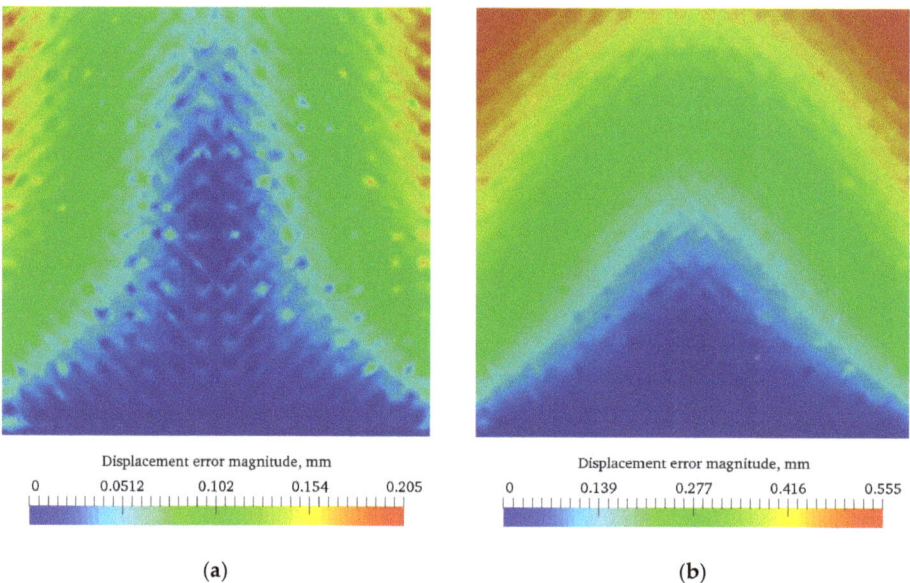

Figure 18. Distribution of the deformation error modulus for test 2 for unidirectional case: (**a**) When using effective hardening parameters; (**b**) when using the concrete hardening parameter.

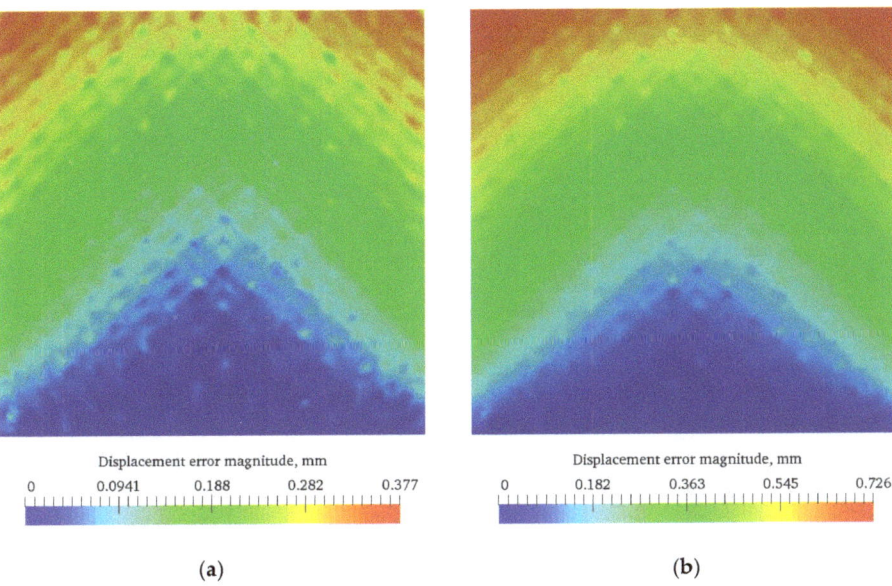

Figure 19. Distribution of the deformation error modulus for test 2 for bidirectional case: (**a**) When using effective hardening parameters; (**b**) when using the concrete hardening parameter.

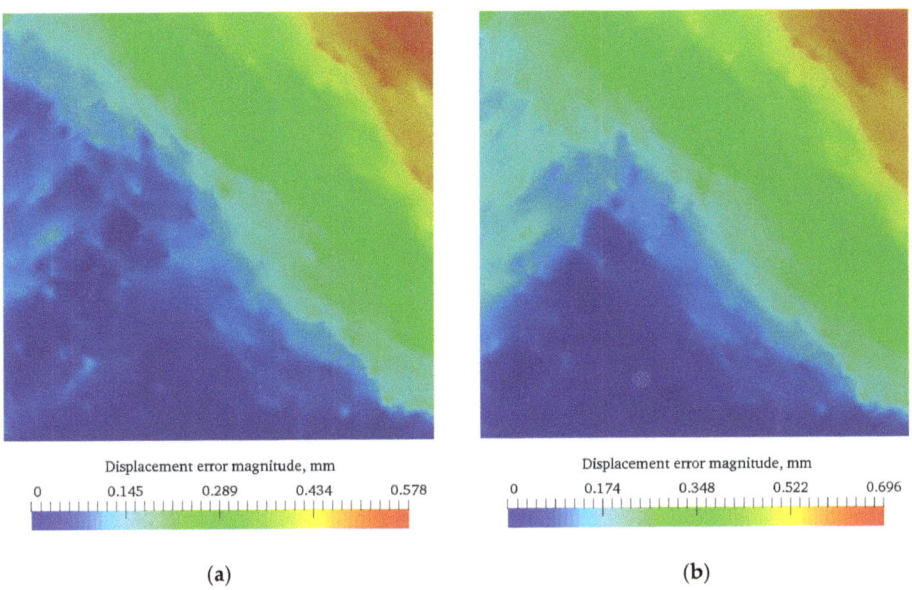

Figure 20. Distribution of the deformation error modulus for test 2 for random distribution case 1: (**a**) When using effective hardening parameters; (**b**) when using the concrete hardening parameter.

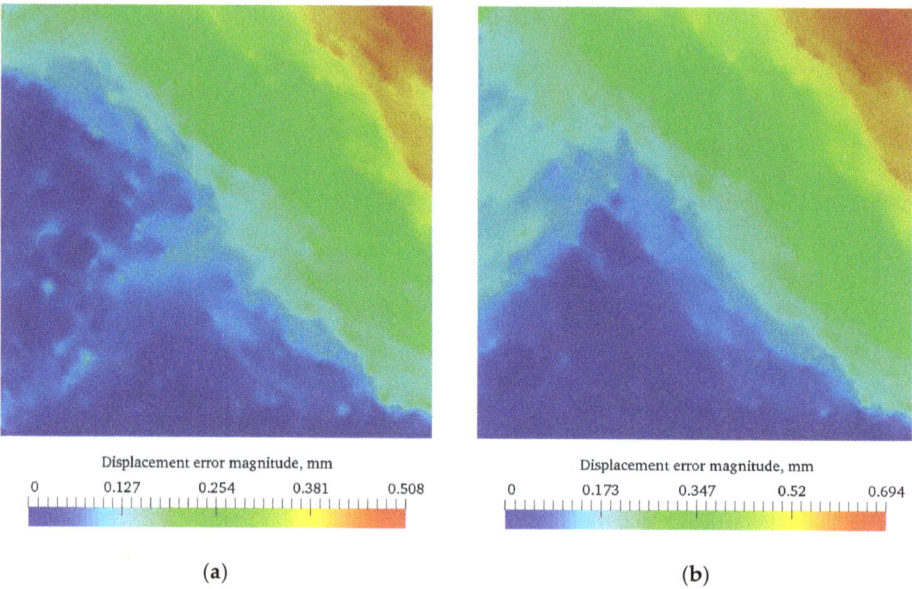

Figure 21. Distribution of the deformation error modulus for test 2 for random distribution case 2: (**a**) When using effective hardening parameters; (**b**) when using the concrete hardening parameter.

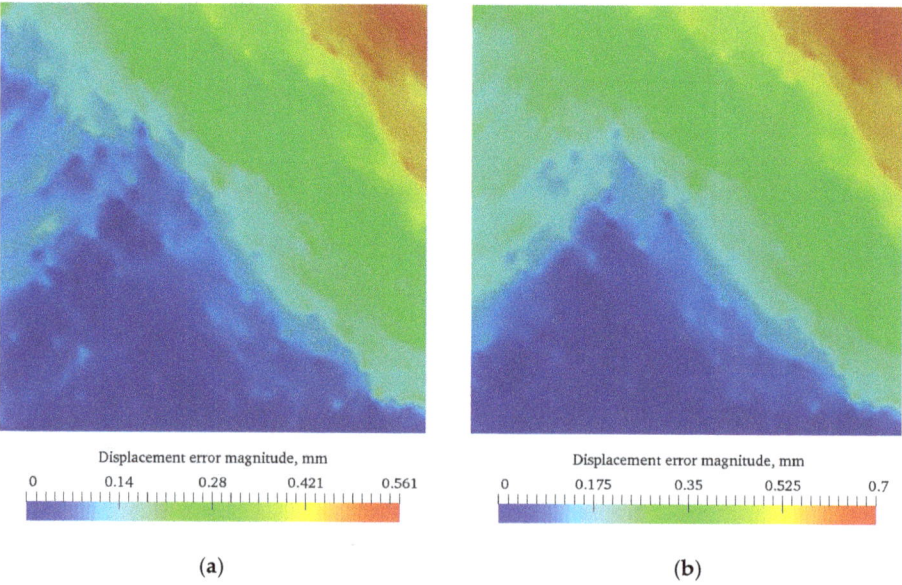

Figure 22. Distribution of the deformation error modulus for test 2 for random distribution case 3: (**a**) When using effective hardening parameters; (**b**) when using the concrete hardening parameter.

In both unidirectional and bidirectional cases, as in test 1, it can be seen that the method proposed in the article gives a solution with a significantly smaller error value. However, for the case with random distribution error values when using effective and base hardening, parameters are similar. This could be due to a bad representative volume.

3.4. Tangential Stress

Based on the results of test 3, the dependences of the maximum value of the displacement modulus in all cases were obtained. They are presented in Figures 23–25.

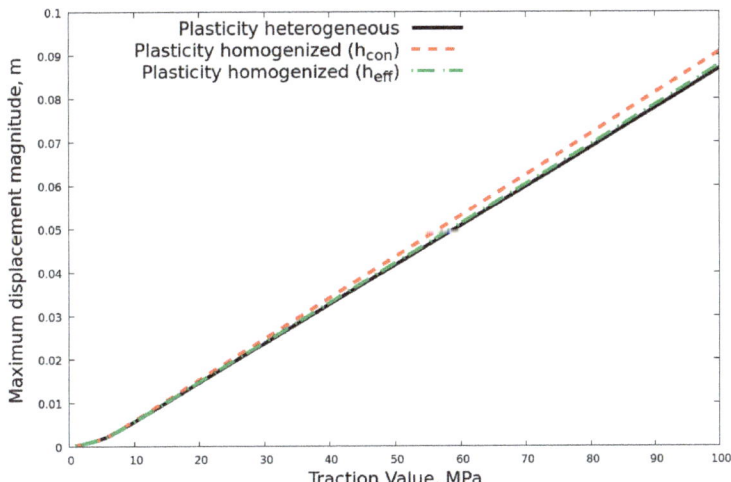

Figure 23. A graph of the dependence of the maximum value of the deformation modulus on the traction value for solving the problem of plasticity, taking into account inclusions, using effective parameters, effective parameters of elasticity, and plastic flow parameters of concrete for test 3 for unidirectional fiber location case.

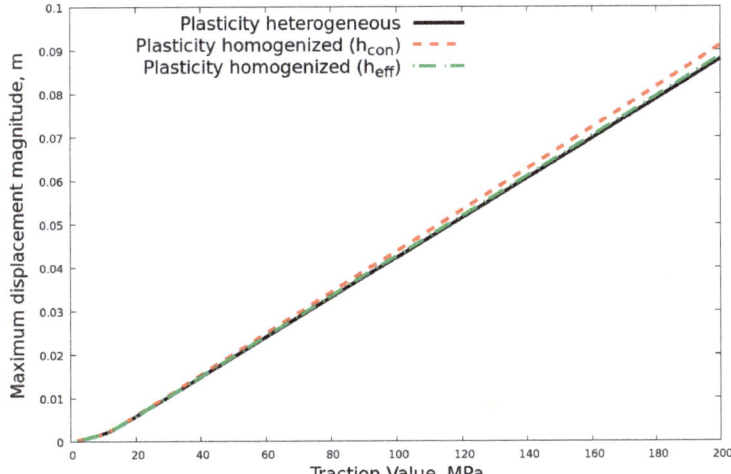

Figure 24. A graph of the dependence of the maximum value of the deformation modulus on the traction value for solving the problem of plasticity, taking into account inclusions, using effective parameters, effective parameters of elasticity, and plastic flow parameters of concrete for test 3 for bidirectional fiber location case.

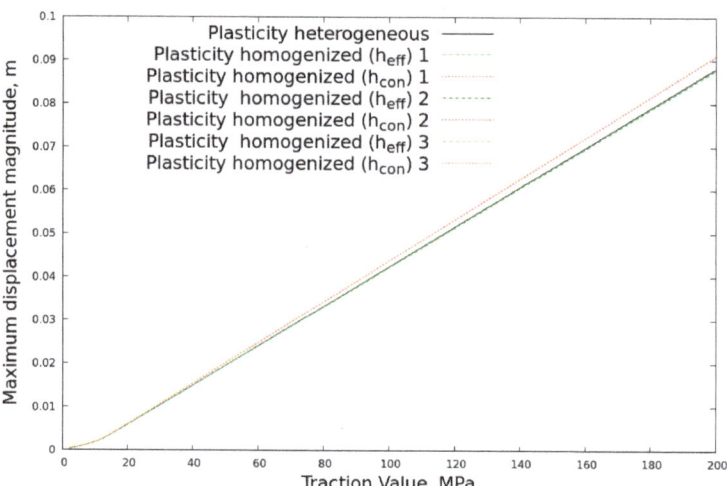

Figure 25. A graph of the dependence of the maximum value of the deformation modulus on the traction value for solving the problem of plasticity, taking into account inclusions, using effective parameters, effective parameters of elasticity, and plastic flow parameters of concrete for test 3 for random fiber distribution case.

The distributions of the deformation modulus error for test 3 for unidirectional, bidirectional, and random distribution of fibers are shown in Figures 26–30, respectively.

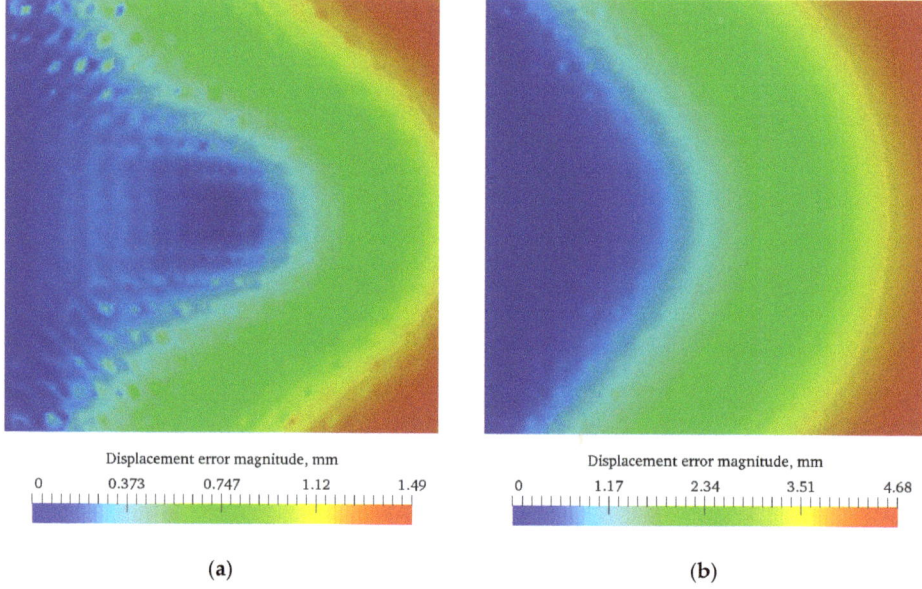

Figure 26. Distribution of the deformation error modulus for test 3 for unidirectional case: (**a**) When using effective hardening parameters; (**b**) when using the concrete hardening parameter.

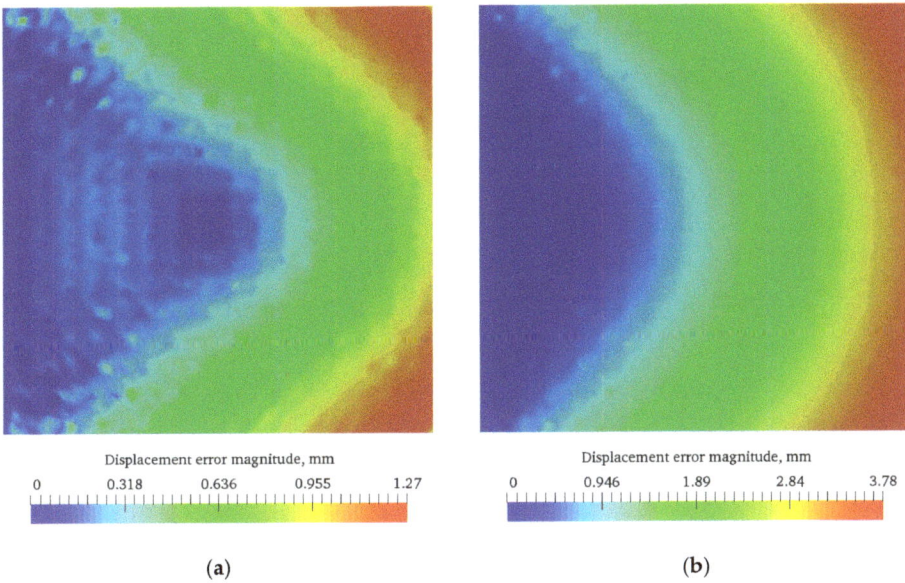

Figure 27. Distribution of the deformation error modulus for test 3 for bidirectional case: (**a**) When using effective hardening parameters; (**b**) when using the concrete hardening parameter.

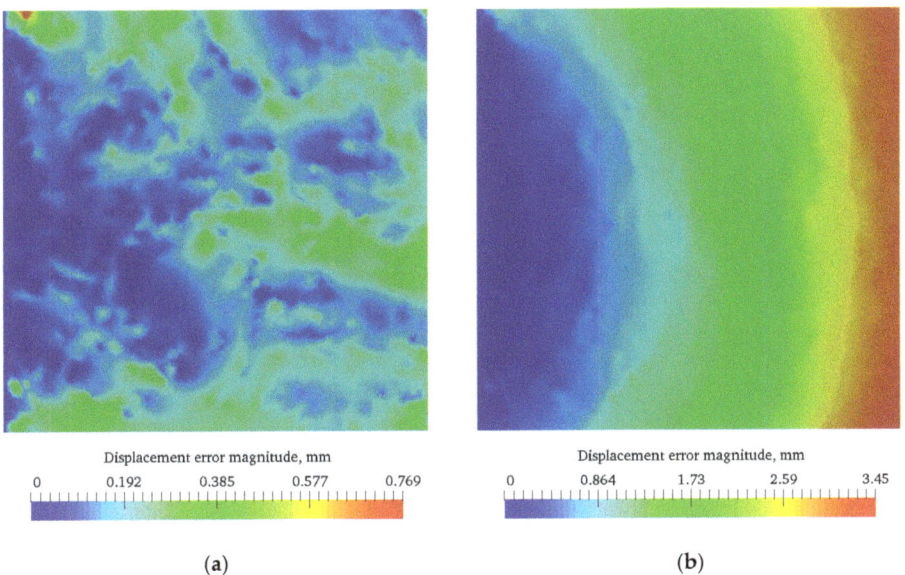

Figure 28. Distribution of the deformation error modulus for test 3 for random distribution case 1: (**a**) When using effective hardening parameters; (**b**) when using the concrete hardening parameter.

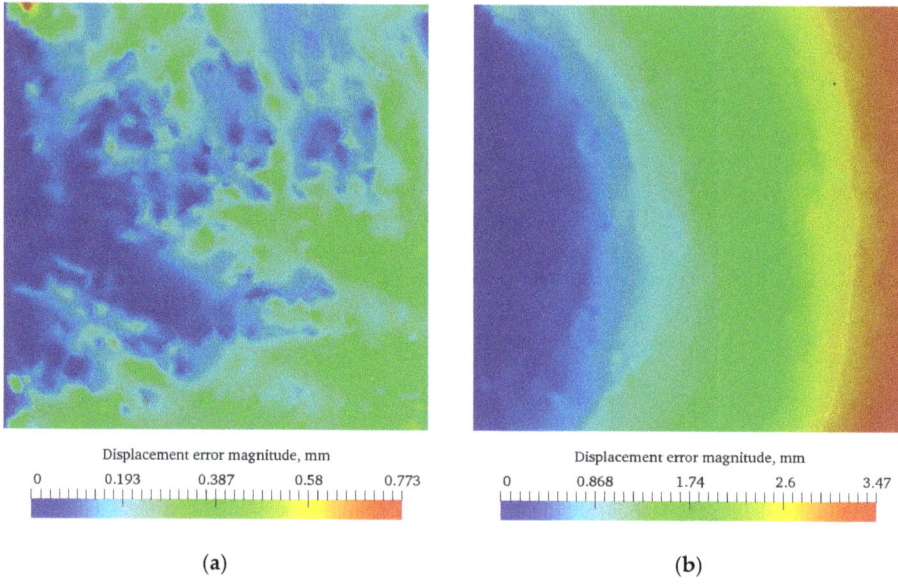

Figure 29. Distribution of the deformation error modulus for test 3 for random distribution case 2: (**a**) When using effective hardening parameters; (**b**) when using the concrete hardening parameter.

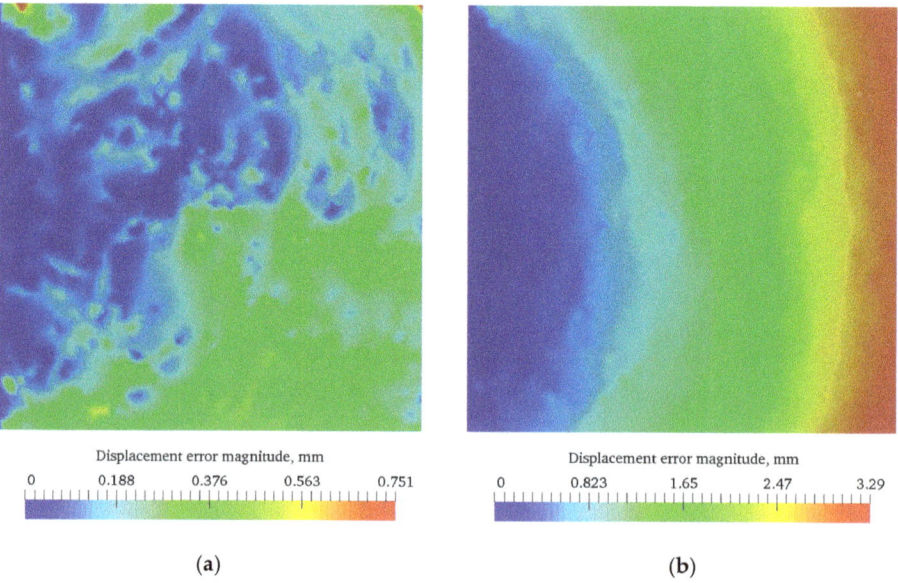

Figure 30. Distribution of the deformation error modulus for test 3 for random distribution case 3: (**a**) When using effective hardening parameters; (**b**) when using the concrete hardening parameter.

In all cases, as in previous tests, we see that the method proposed in the article gives a solution with better accuracy. In particular, this can be seen for random distribution. However, accuracy could be worse for other representative volumes.

3.5. All-Round Compression

Based on the results of test 4, the dependences of the maximum value of the displacement modulus in all cases were obtained. They are presented in Figures 31–33.

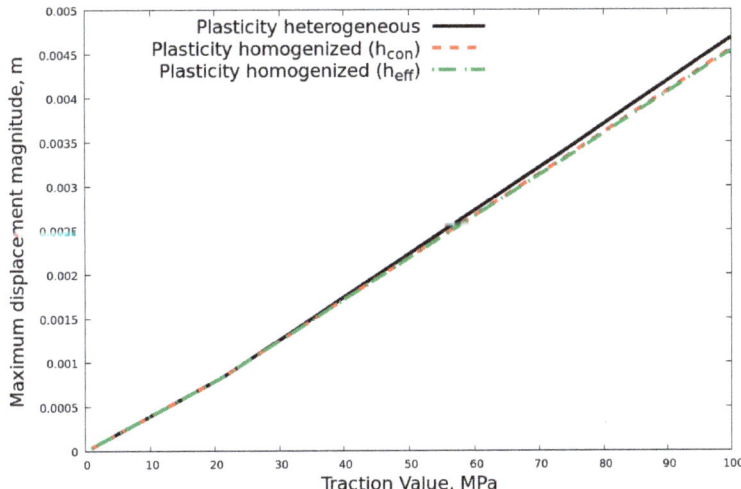

Figure 31. A graph of the dependence of the maximum value of the deformation modulus on the traction value for solving the problem of plasticity, taking into account inclusions, using effective parameters, effective parameters of elasticity, and plastic flow parameters of concrete for test 4 for unidirectional fiber location case.

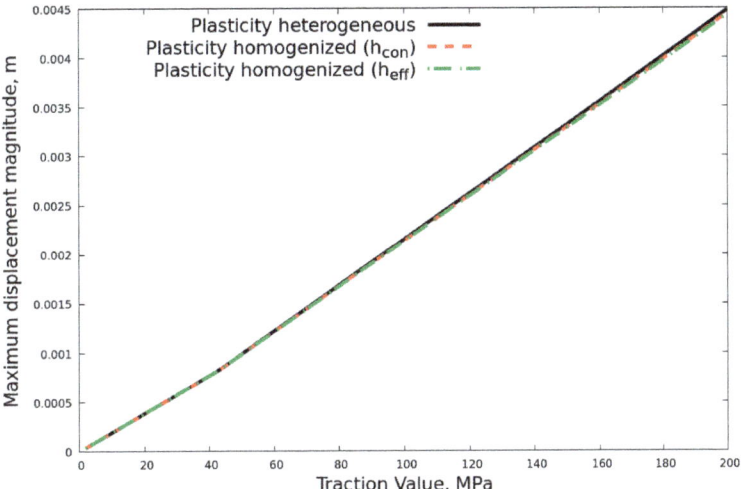

Figure 32. A graph of the dependence of the maximum value of the deformation modulus on the traction value for solving the problem of plasticity, taking into account inclusions, using effective parameters, effective parameters of elasticity, and plastic flow parameters of concrete for test 4 for bidirectional fiber location case.

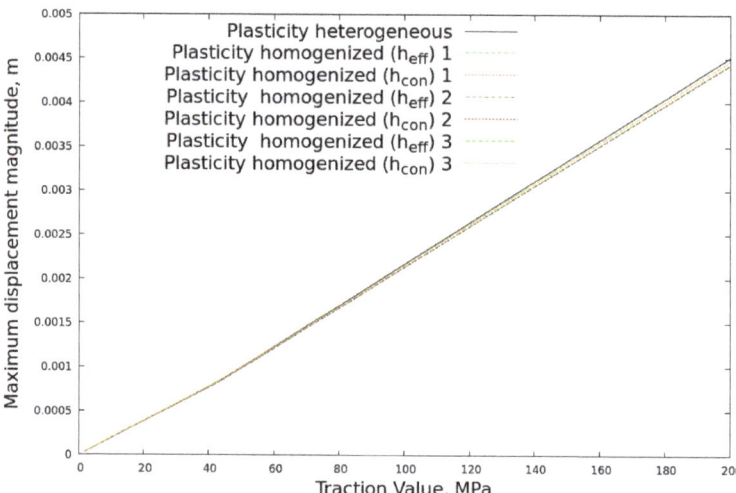

Figure 33. A graph of the dependence of the maximum value of the deformation modulus on the traction value for solving the problem of plasticity, taking into account inclusions, using effective parameters, effective parameters of elasticity, and plastic flow parameters of concrete for test 4 for random fiber distribution case.

The distributions of the deformation modulus error for test 4 for unidirectional, bidirectional, and random distribution of fibers are shown in Figures 34–38, respectively.

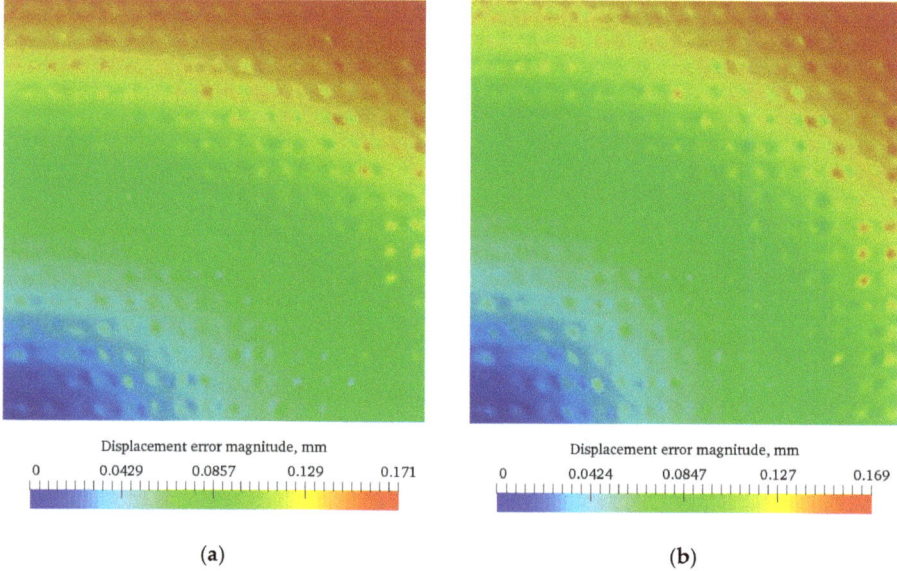

Figure 34. Distribution of the deformation error modulus for test 4 for unidirectional case: (**a**) When using effective hardening parameters; (**b**) when using the concrete hardening parameter.

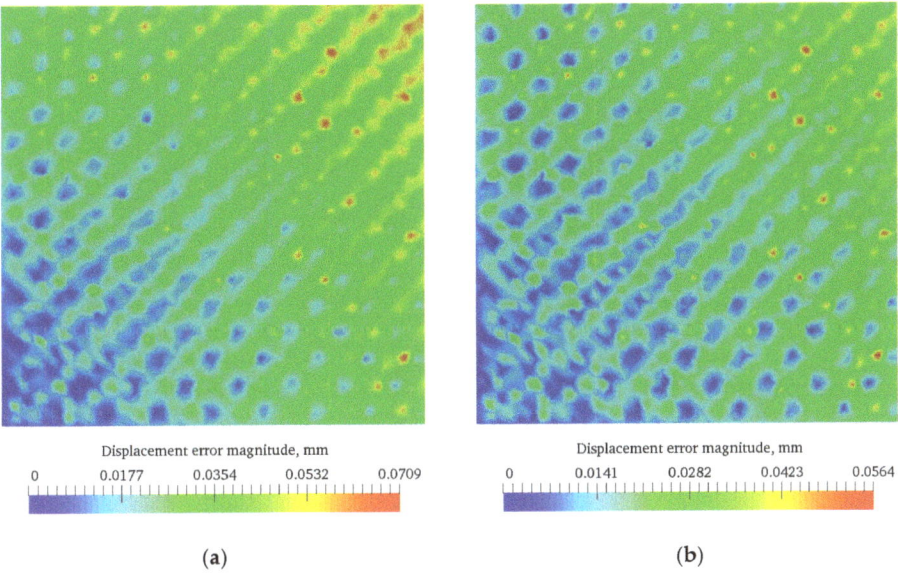

Figure 35. Distribution of the deformation error modulus for test 4 for bidirectional case: (**a**) When using effective hardening parameters; (**b**) when using the concrete hardening parameter.

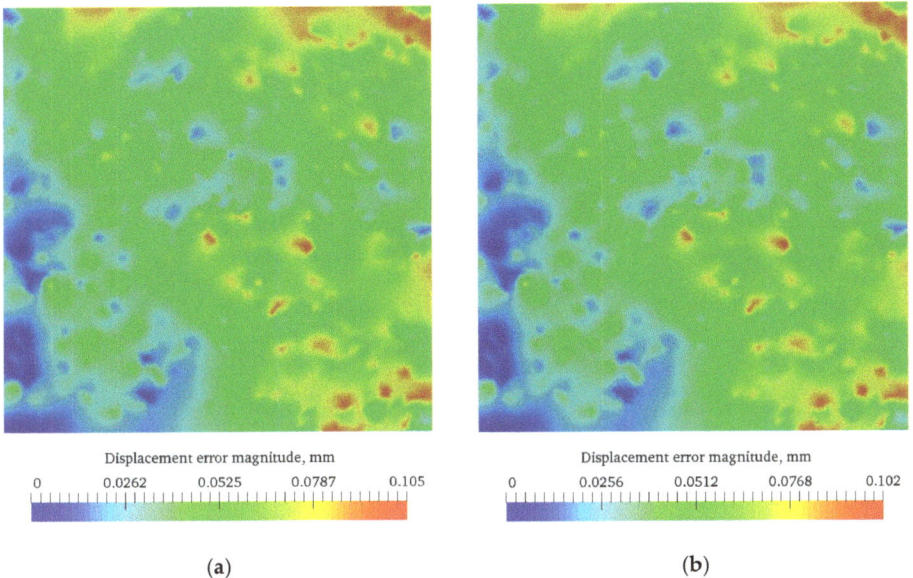

Figure 36. Distribution of the deformation error modulus for test 4 for random distribution case 1: (**a**) When using effective hardening parameters; (**b**) when using the concrete hardening parameter.

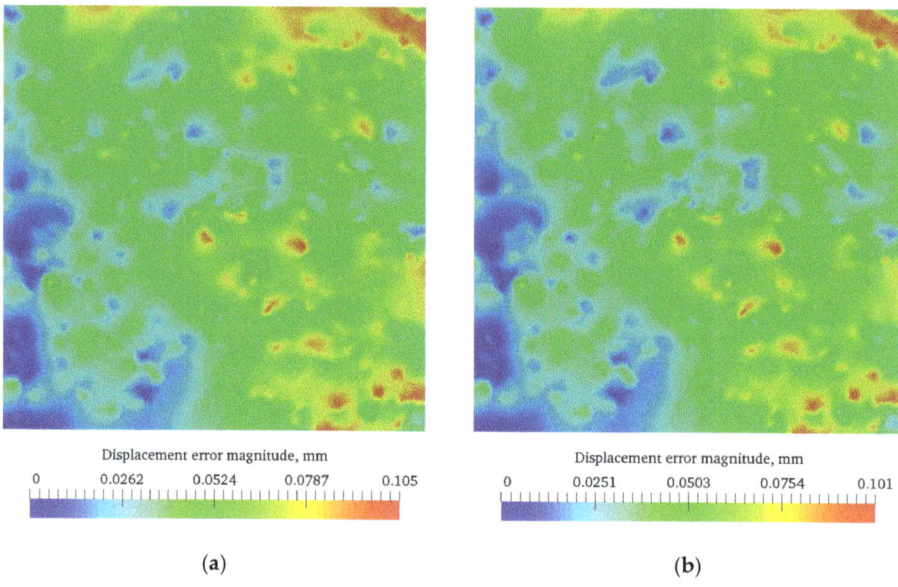

Figure 37. Distribution of the deformation error modulus for test 4 for random distribution case 2: (**a**) When using effective hardening parameters; (**b**) when using the concrete hardening parameter.

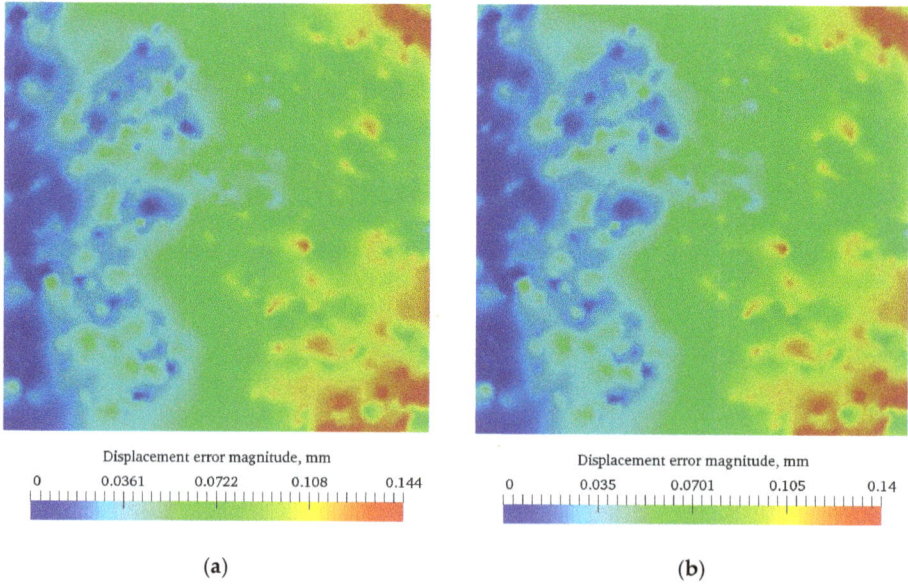

Figure 38. Distribution of the deformation error modulus for test 4 for random distribution case 3: (**a**) When using effective hardening parameters; (**b**) when using the concrete hardening parameter.

In contrast to all other tests, the results of the fourth test show the weak side of the proposed model, namely, when comparable values of the strain tensor components occur. In this case, it is impossible to take into account the whole anisotropic nature of the elastic–plastic flow of the composite material.

4. Discussion

J2 flow with isotropic hardening can be used to model plastic behavior of a great variety of materials, for example, concrete. Fiber-reinforced concrete is currently of great interest. Additionally, to model building blocks made of fiber-reinforced concrete, proper homogenization or multiscale methods are required. The main focus of such methods is to homogenize the elastoplastic tangent tensor. The algorithm presented here of hardening coefficient homogenization is simple to understand in comparison to other plasticity problem homogenization algorithms.

In a previous study, the algorithm of the homogenization of fiber-reinforced concrete with elasticity tensor homogenization using a hardening parameter and the yield stress of concrete without reinforcement was investigated. A satisfactory result with good accuracy was obtained. In this study, we tried to add hardening parameter homogenization in order to improve accuracy. A comparison with the previous algorithm was performed using three different fiber locations and four tests.

5. Conclusions

According to the results obtained, the proposed model of the hardening coefficient homogenization demonstrates satisfactory results when one of the components of the strain tensor prevails. Under this condition, the results obtained have an error less than when using the concrete hardening coefficient, as shown by tests 1, 2, and 3.

However, when the strain tensor components have comparable values, error values become similar to the values of error of the simple model from previous work. This is due to the anisotropy of the plastic flow that cannot be fully taken into account in a simple numerical change in the hardening coefficient. In this case, we note that both approximations work quite accurately at small plastic deformations.

In further work, it will be necessary to obtain a method of numerical homogenization, which will have more degrees of freedom and describe the anisotropy of plastic deformations better. Different values of plastic parameters of concrete, for compression and tension, should also be addressed in future work.

Author Contributions: Conceptualization, P.V.S. and P.S.; Data curation, P.V.S.; Formal analysis, P.V.S. and P.S.; Funding acquisition, P.S.; Investigation, P.V.S. and P.S.; Methodology, P.V.S. and P.S.; Software, P.V.S.; Supervision, P.S.; Validation, P.V.S. and P.S.; Visualization, P.V.S.; Writing—original draft, P.V.S.; Writing—review and editing, P.S. All authors have read and agreed to the published version of the manuscript.

Funding: This research was funded by a grant of the President of the Russian Federation, grant number MK-2249.2021.1.1 and "Subvention for Science" (MEiN), project no. FN-4/2021.

Acknowledgments: The authors would like to thank every person/department who helped thorough out the research work. The careful review and constructive suggestions by the anonymous reviewers were gratefully acknowledged.

Conflicts of Interest: The authors declare no conflict of interest.

References

1. Sivtsev, P.V. *Chislennoe Modelirovanie Zadachi Uprugosti Zhelezobetonnyh Plit*; Vestnik Severo-Vostochnogo Federal'nogo Universiteta im. MK Ammosova: Yakutsk, Russia, 2015.
2. Smarzewski, P. Influence of basalt-polypropylene fibres on fracture properties of high performance concrete. *Compos. Struct.* **2019**, *209*, 23–33. [CrossRef]
3. Kolesov, A.E.; Sivtsev, P.V.; Smarzewski, P.; Vabishchevich, P.N. Numerical Analysis of Reinforced Concrete Deep Beams. In *Proceedings of the International Conference on Numerical Analysis and Its Applications*; Springer: Cham, Switzerland, 2016; pp. 414–421. [CrossRef]
4. Sivtseva, A.V.; Sivtsev, P.V. Numerical Simulation of Deformations of Basalt Roving. In *Proceedings of the International Conference on Finite Difference Methods*; Springer: Cham, Switzerland, 2018; pp. 501–508. [CrossRef]
5. Sivtsev, P.V.; Kolesov, A.E.; Sirditov, I.K.; Stepanov, S.P. The Numerical Solution of Thermoporoelastoplasticity Problems. In *Proceedings of the AIP Conference Proceedings*; AIP Publishing LLC: Albena, Bulgaria, 2016; Volume 1773, p. 110010. [CrossRef]

6. Zhang, L.; Yu, W. Variational asymptotic homogenization of elastoplastic composites. *Compos. Struct.* **2015**, *133*, 947–958. [CrossRef]
7. Sivtsev, P.V.; Kolesov, A.E.; Zakharov, P.E.; Yang, Y. Numerical homogenization of elastoplastic deformations of composite material with small proportion of inclusions. In *Proceedings of the Journal of Physics: Conference Series*; IOP Publishing: Moscow, Russia, 2019; Volume 1392, p. 012074.
8. Popovics, S. A numerical approach to the complete stress-strain curve of concrete. *Cem. Concr. Res.* **1973**, *3*, 583–599. [CrossRef]
9. Logg, A.; Mardal, K.-A.; Wells, G. *Automated Solution of Differential Equations by the Finite Element Method: The FEniCS Book*; Springer: Berlin/Heidelberg, Germany, 2012; Volume 84, ISBN 3-642-23099-7.
10. Ølgaard, K. *Automated Computational Modelling for Complicated Partial Differential Equations*; Delft University of Technology, Faculty of Civil Engineering and Geosciences: Delft, The Netherlands, 2013.
11. Simo, J.C.; Hughes, T.J.R. *Computational Inelasticity*; Springer Science and Business Media: New York, NY, USA, 1998; Volume 7.
12. De Souza Neto, E.A.; Peric, D.; Owen, D.R. *Computational Methods for Plasticity: Theory and Applications*; John Wiley & Sons: Chichester, UK, 2011; ISBN 1-119-96454-7.
13. Valentino, P.; Furgiuele, F.; Romano, M.; Ehrlich, I.; Gebbeken, N. Mechanical Characterization of Basalt Fibre Reinforced Plastic with Different Fabric Reinforcements-Tensile Tests and FE-Calculations with Representative Volume Elements (RVEs). Gruppo Italiano Frattura. Convegno Nazionale IGF. *Acta Fract.* **2013**, *22*, 231.
14. Smarzewski, P. Processes of Cracking and Crushing in Hybrid Fibre Reinforced High-Performance Concrete Slabs. *Processes* **2019**, *7*, 49. [CrossRef]
15. Zakharov, P.E.; Sivtsev, P.V. Numerical Calculation of the Effective Coefficient in the Problem of Linear Elasticity of a Com-posite Material. *Math. Notes NEFU* **2017**, *24*, 75–84. [CrossRef]

Article

Bond Behavior of Cleaned Corroded Lap Spliced Beams Repaired with Carbon Fiber Reinforced Polymer Sheets and Partial Depth Repairs

Hisham Alabduljabbar [1,*], Rayed Alyousef [1,*], Hossein Mohammadhosseini [2,*] and Tim Topper [3]

1. Department of Civil Engineering, College of Engineering, Prince Sattam bin Abdulaziz University, Alkharj 16273, Saudi Arabia
2. Institute for Smart Infrastructure and Innovative Construction (ISIIC), School of Civil Engineering, Faculty of Engineering, Universiti Teknologi Malaysia (UTM) Skudai, Johor 81310, Malaysia
3. Department of Civil and Environmental Engineering, University of Waterloo, 200 University Avenue West, Waterloo, ON N2L 3G1, Canada; topper@uwaterloo.ca
* Correspondence: h.alabduljabbar@psau.edu.sa (H.A.); r.alyousef@psau.edu.sa (R.A.); hofa2018@yahoo.com (H.M.)

Received: 11 October 2020; Accepted: 6 November 2020; Published: 9 November 2020

Abstract: The present research investigated the bond behavior of a cleaned corroded reinforcing bar repaired with a partial depth concrete repair and a partial depth concrete repair followed by the application of fiber-reinforced polymer (FRP) sheets. Twelve lap splice beams were cast and tested under static loading. The test variables considered were a partial depth repair with prepackaged self-consolidating concrete (SCC) for six lap splice beams and additional confinement with carbon fiber reinforced polymer (CFRP) sheets for another six beams. The test results for the repaired lap splice beams were compared with those for a monolithic lap splice beam. This research found that the average bond strength increased as the bar mass loss increased for all bonded lengths. The lap splice beams repaired with partial depth were able to repair concrete with similar properties to those of the monolithic concrete. However, they had higher concrete strength than the monolithic beams which showed a higher average bond strength than the monolithic lap splice beams. The beams confined with FRP sheets showed a rise in the bond strength and the equivalent slip by 34–49%, and 56–260% as compared to the unconfined beams, respectively.

Keywords: bond; corrosion; lap splice beam; carbon fiber-reinforced polymer; concrete; partial depth repair; stirrups

1. Introduction

The lap splice is an effective and simple method to connect reinforcing bars in order to transfer forces between them. However, lap spliced reinforced concrete (RC) members are commonly subject to bond problems. This is because the lap splice region of the bars is in a bond critical region, which may cause an RC member to exhibit a brittle bond failure. Bond failure can occur by splitting of the concrete cover between a spliced bar and the side face of the lap splice beam. The bond between the reinforcing bar and the surrounding concrete might fail through shearing off of the concrete keys between reinforcing bar ribs, resulting in bar slip which is called pull out bond failure [1–3]. Splitting bond failure is the most common bond failure, especially in the absence of confinement. Many parameters influence the bond strength between the reinforcing bar and the surrounding concrete, including bar diameter, anchorage/splice length, concrete strength, concrete cover and the presence of confinement such as transverse reinforcement of fiber-reinforced polymer (FRP) sheets [4–6]. Azizinamini et al. [7] found that providing a bond critical region with confinement is an important way to increase bond strength and change failure mode, resulting in ductile instead of brittle failure. Many experimental studies indicated that providing transverse reinforcement or wrapping an RC member with FRP sheets are the most effective methods to confine the critical bond region [8].

FRP composite materials are heterogeneous and anisotropic materials that do not exhibit plastic deformation. FRP composites have been used in a wide range of modern applications mainly in space and aviation, automotive and maritime. Carbon fiber reinforced polymer (CFRP) and glass fiber reinforced polymer (GFRP) composites, among other fiber reinforced materials, have been gradually substituting conventional materials with their high strength and low specific weight properties. Their manufacturability in varying combinations with customized strength properties—as well as their high fatigue, toughness, high temperature wear and oxidation resistance capabilities—render these materials an excellent choice in engineering applications [9,10].

A reinforcing bar provides strength and ductility to an RC member through good bonding and adequate anchorage to the concrete. The efficiency of the bond between the reinforcing bar and the concrete can be decreased due to corrosion that might deteriorate concrete, the reinforcing bar or both. Corrosion of the reinforcement in the RC structure is a durability issue that may affect the structural capacity and ductile behavior of the structure. Many researchers have investigated the influence of corrosion on the flexural strength of RC members. Once the reinforcing bar is subjected to corrosion, the area of the bar decreases, leading to a reduction in the yield strength and ultimate capacity of the RC member [11]. However, if the corrosion occurred in the critical bond region of the RC member, the mode of failure might change from a flexural failure to a bond splitting failure [12–16].

Corrosion affects the performance of the RC member in three ways: it decreases the efficacy of concrete cover through cracks caused by corrosion; it reduces the cross-sectional area of the reinforcing bar, leading to decreased mechanical performance of the bar; it may lead to losses in the interface bond between the concrete and the reinforcing bar [17–19]. The corrosion operation decreases the cross-sectional area of the reinforcing bar, which results in a reduction in the load-carrying capacity of the RC member. The volume of the rust that is produced by the corrosion operation is also higher than the volume of the steel that formed the corroded material. As the corrosion products expand at the interface between the rebar and the concrete, the concrete is subject to tensile forces, resulting in cracking at the level of the reinforcing bar that leads to reduced bonding amongst the surrounding concrete and the reinforcing bars. The crack width and the number of cracks increase with increasing corrosion levels and the bar mass loss also increases [20].

Once the RC member is subject to a given level of corrosion, the fatigue lifetime of that member can be maintained through an effective repair operation such as partial depth repair or/and FRP sheets. The FRP sheet method is a common method used for repairing corroded RC members or strengthening an existing RC structure due to an altered use of that structure which leads to an increase in the live load. The FRP materials are popular for rehabilitation and strengthening purposes owing to their high strength to weight ratio, excellent durability performance, easy application, and the fact that they do not need heavy equipment to make repairs and are noncorrosive materials that eliminate corrosion problems [21].

Garcia et al. [22] found that wrapping a lap splice beam with an FRP sheet decreased the width of splitting cracks and delayed the propagation of these cracks compared to the results for an unwrapped control beam. The average bond strength, peak load and deflection at the failure of a lap splice beam wrapped with a GFRP sheet increased compared to the results for an unwrapped control beam [23]. The CFRP wrap enhanced the bond capacity of the corroded lap splice beam by up to 33% compared to the corroded unwrapped lap splice beam [24].

Epoxy composites reinforced with carbon or glass fibers are generally used in applications with high structural demand, such as marine structures directly exposed to seawater [25]. Polymer composites in marine structures under seawater environment may be extremely affected, decreasing durability performance. Therefore, in such an environment, the curing process is time and energy consuming, raising the cost of the product. Consequently, approaches to decrease the curing time for composite systems are of great interest. In this regard, Antunes et al. [26] investigated the effect of seawater exposure at 80 °C for up to 28 days on filament-wound glass fiber-reinforced polymer (GFRP) cylinders partially cured by passing saturated steam through them just after winding seeking a faster curing route. The results of their study revealed that using GFRP as filament wound, significantly enhanced the mechanical and durability performance of composite components exposed to seawater.

Malumbela and Alexander [27] concluded that repairing a corroded RC member with an FRP sheet must be done after partial depth repair. FRP wrapping without the partial depth repair did not prevent further corrosion. However, applying only the partial depth repair protected the reinforcing rebar from further corrosion and restored the serviceability state of the beams. The partial depth repair method is widely used for rehabilitating corroded RC members. Before applying the partial depth repair, certain steps must be completed which include removing the damaged concrete, removing the rust and cleaning the reinforcing bar by sandblasting or water pressure and roughening the surface of the original concrete [28,29].

The current paper studied the bond behavior of lap splice beams with a corroded reinforcing bar that was cleaned and repaired with partial depth repair and/or FRP sheets. The partial depth repair used in this study was a commercial prepackaged self-compacting concrete (SCC) extended with 50% of coarse aggregate by mass with a 13 mm maximum aggregate size. This proportion of coarse aggregate was chosen to simulate the proportion of coarse aggregate to the monolithic beams that were cast with ready mix normal concrete (M). In this study, the FRP sheets were used to confine the lap splice zone, which was 300 mm in length. Two corrosion levels (7.5% and 15% mass loss) were studied and compared to non-corroded bars to assess the effect of corrosion severity on the bond behavior.

Many reinforced concrete structures with lap splice beams are vulnerable to corrosion, which is a serious issue that may deteriorate the bond amongst the surrounding concrete and reinforcing bars. However, a few researchers have studied the effect of rehabilitating a corroded RC member with FRP sheets on bond or flexural capacity. Hence, to the author's best knowledge, there is a lack of studies on the bond behavior of cleaned corroded rebar and repaired partial depth repair. Furthermore, repairing the corroded RC members to enhance and restore the bond efficiency with a combination of FRP sheets and partial depth repair has not been studied. To cover this gap, this research studies the effect of repairing the cleaned corroded rebar with partial depth repair and/or FRP sheets on the bond behavior of the lap splice beam.

2. Experimental Program

2.1. Test Specimen

In this study, twelve lap splice beams were made and tested statically under four-point loading with 1800 mm clear span length, 600 mm constant moment region and 600 mm shear spans each. All the lap splice beams had the same dimensions, with a rectangular cross-section of 250 mm × 350 mm with a total length of 2200 mm. The lap splice length was designed to be 300 mm to lead to bar slippage to ensure that all beams would fail in a bond splitting mode before the reinforcing bar reached the yield strength (flexural failure). In addition, all of the beams were reinforced with two 20 M flexural reinforcing bars spliced at the mid-span region. All beams were also reinforced with two continuous 10 M steel bars at the top (compression zone) of the beam. To avoid a brittle shear failure, each beam was provided with 10 M transverse reinforcement in the shear spans spaced at 100 mm center-to-center (Figure 1a).

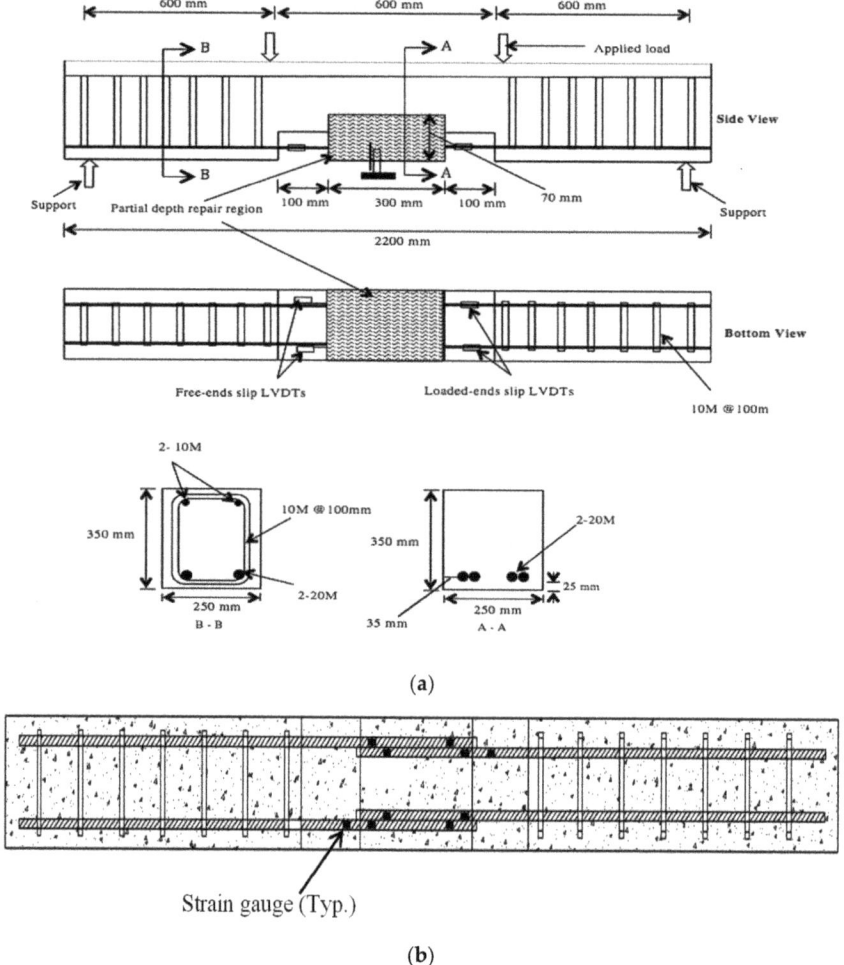

Figure 1. (a) The cross section and reinforcement details of the lap splice beam; (b) Strain gauge layout.

Six lap splice beams were unconfined and six were wrapped with carbon fiber reinforced polymer (CFRP) sheets. For the wrapped condition, the CFRP sheet was U-wrapped in a single layer of 950 mm length and 300 mm width to confine the lap splice zone. For each confinement case, three beams were monolithic at different mass loss levels (control, 7.5% and 15% mass loss) and three beams were repaired with a partial depth repair concrete at different mass loss levels (control, 7.5% and 15% mass loss).

For the beams that were repaired with a partial depth repair, a pocket was designed for the repair purposes, with dimensions of 300 mm in length, 250 mm in width and 70 mm in depth. The concrete cover for all beams was 25 mm. A 25 mm clearance was kept under the spliced bars to meet the minimum clearance reinforcement of 19 mm or 6 mm larger than maximum coarse aggregate size in the partial depth repair material in accordance with the specification of ACI 364-2014.

All the beams were fabricated with two notches with dimensions of 100 mm × 50 mm at the bottom of the beam that exposed the flexural reinforcing bars. The two notches were placed at the two ends of the lap splice region to evaluate the bond stress-slip behavior. Five linear variable differential transformers (LVDTs) were used, two to measure the loaded end slip, two to measure the free end slip and one to measure the deflection of the beam, as revealed in Figure 1b.

A total of 10 electrical resistance strain gauges were used on the spliced bars for each beam to measure the strain distribution along the spliced length. Two strain gauges were fixed on each bar at the lap splice zone at distances of 50 mm and 225 mm from the beginning of the splice zone. Moreover, one strain gauge was placed on each side of the splice zone on the exposed reinforcing bar at the notches to measure the bar stress in this region where the tensile force was taken only by the bar in the beams. The increase in stress at other strain gauge locations toward the stress measured at this location as a test was used to describe the progress of the bar debonding.

2.2. Test Procedure

All of the lap splice beams were supported with a 1800 mm span length and tested statically under four-point loading until failure. The load was applied to the beam through a steel spreader beam connected to the actuator. The steel beam then transferred the load to two locations on the top of the beam to produce a constant moment in the central region of the lap splice beams. The test frame was servo-controlled with a capacity of 500 kN, and was run in displacement control at the rate of 0.3 mm/min. The displacement at the center of the lap splice beam was recorded as the load was applied until the failure occurred. Figure 2 shows the lap splice beam installed in the test frame.

Figure 2. Lap splice beam installed in the test frame.

2.3. Specimen Fabrication

The lap spliced bars that needed to be corroded were cast first into small concrete prisms with dimensions of 100 mm in depth, 100 mm in width and 300 mm in length, which was the lap splice length. To accelerate the corrosion rate, the concrete mix used to cast those prisms contained 3.8% salt of the cement mass, which was equivalent to 2.3% chloride. The prisms were kept in a special chamber subject to a continuous combination of oxygen and moisture, which is essential to facilitate the corrosion operation. Power supplies were used to accelerate the corrosion through an impressed current. The prisms were divided into two groups: 7.5% and 15% mass loss, and each group that was to be corroded to the same level was connected in series to ensure a constant current. Every prism was cast with one reinforcing bar (anode) and one stainless steel bar (cathode). To reach the desired corrosion level (7.5% or 15% mass loss), the period of accelerated corrosion was estimated using Faraday's law [20]:

$$m_l = \frac{MIT}{zF} \qquad (1)$$

where m_l is the mass loss (g); M is the atomic weight of the metal = 56 g for Fe; I is the corrosion current (A) = $i \times s_a$; i is the corrosion current density (μA/cm^2); s_a is the surface area of the corroded steel (cm^2); T is the time (s); z is the valence of the corroding metal (2 for iron); F is Faraday's constant = 96,500 A·s.

From Faraday's law, the estimated time to achieve 7.5% and 15% mass loss was 75 and 150 days, respectively. Once the desired level of corrosion was achieved, the prisms were cut, and the reinforcing bars were removed and cleaned by sandblasting. Subsequently, the bars were weighed, and the mass loss of each bar was calculated as follows:

$$m\% = \frac{mass\ of\ original\ steel - mass\ of\ corroded\ steel}{mass\ of\ original\ steel} \qquad (2)$$

Consequently, the lap splice beams were fabricated and divided into two groups: monolithic lap splice beams and repaired lap splice beams. For the monolithic lap splice beams, the reinforcing bars were placed and cast with ready mix concrete. However, for the lap splice beams to be repaired with partial depth repair, the splice zone was isolated with high-density foam to form the pocket of the partial depth repair region. After curing, the high-density foam was removed, and the partial depth repair concrete was installed.

2.4. Material Properties

The partial depth repair material of SCC was used and extended with 50% of 13 mm coarse aggregate. Table 1 shows the proportions for the monolithic concrete. Table 2 shows the hardened properties such as compressive strength, splitting strength and fracture energy of the monolithic concrete (M) and self-consolidating concrete (SCC).

Table 1. Mixture properties design for the monolithic concrete.

Concrete Type	CA * (kg/m^3)	FA * (kg/m^3)	Cement (kg/m^3)	WR * (%C)	AEA * (%C)	W * (kg/m^3)	W/C	CA%
M	1110	865	280	0.29	0.003	155	0.55	46

* CA: coarse aggregate (19 mm); FA: fine aggregate; WR: water reducer; AEA: air-entrained admixture, W: water, W/C: water/cement ratio.

Table 2. Hardened properties of monolithic concrete (M) and self-consolidating concrete (SCC).

Concrete Mechanical Properties	M	SCC
Compressive strength, f'$_c$ (MPa)	42	48
Splitting strength, f$_t$ (MPa)	4.1	4.1
Fracture energy, G$_f$ (N/m)	135	137

The 20 M deformed reinforcing bar had a 479 MPa yield strength and 612 MPa ultimate tensile strength, and the supplier provided this data. The repair material had a rapid strength gain, a fast turnover of repair area and a flowable consistency. The CFRP wrapping sheet was SikaWrap Hex 103C with a weight of 610 g/m^2. Two types of epoxies were used for the CFRP sheets installation: Sikadur 330 and Sikadur 300. Prior to placing the CFRP, the surface of the concrete was sealed with Sikadur 330, and the CFRP sheets were then saturated manually with Sikadur 300. The properties of CFRP sheets and the two epoxies are shown in Tables 3 and 4, respectively.

Table 3. Properties of carbon fiber reinforced polymer (CFRP) sheets.

Property	Typical Properties of SikaWrap Hex 103C	Cured Laminated Properties of SikaWrap Hex 103C
Tensile strength (MPa)	3.7	1.055
Tensile modulus (MPa)	234,500	64,828
Elongation (%)	1.5	0.89
Thickness (mm)	0.34	1.016

Table 4. Properties of epoxies.

Property	Sikadur 330	Sikadur 300
Tensile strength (MPa)	30	55
Tensile modulus (MPa)	4500	1724
Flexural modulus (MPa)	3800	3450
Elongation (%)	0.9	3

3. Results and Discussion

3.1. Mode of Failure and Cracking Pattern

All of the lap splice beams failed by splitting bond failure (as intended). The first flexural crack in the beams occurred at the upper corners of the notches at the end of the lap spliced region, still within the constant moment region. As the applied load increased, more flexural cracks developed in the spliced zone together with longitudinal cracks along the lap spliced bars. It can be seen from Figure 3a that the longitudinal splitting cracks were developed on the bottom face of the beams from both ends of the splice regions. The unconfined beams had a sudden brittle failure associated with a loud sound resulting from the splitting of the final length of the concrete cover. The bottom face cracks in the unconfined beams formed a V shape (Figure 3b). Sizable chunks of concrete were formed and in some cases spalled off from the concrete cover at failure because of the absence of confinement.

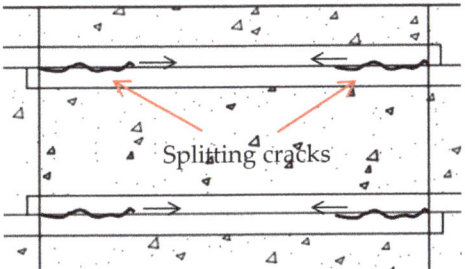
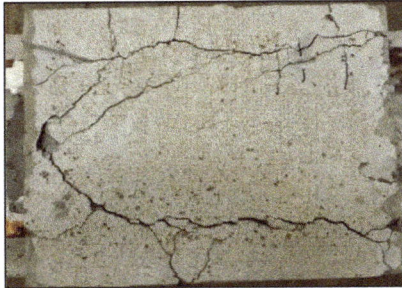

Figure 3. (**a**) Initiation of splitting cracks on the bottom face of the beam. (**b**) Bottom face bond splitting cracks for an unconfined beam.

The mode of failure was different for the unconfined beams than for the beams confined with FRP sheets. The confined beams had a more ductile mode of failure related to the unconfined beams in which the confinement delayed the splitting of the concrete cover. Unlike the unconfined beams, the concrete cover did not spall off in the beams confined with the FRP sheet. The FRP sheets arrested the concrete cover in place after splitting occurred. The confined beams experienced an increase in their ultimate load, corresponding deflection at failure and flexural crack width compared to the unconfined beams. For the beams confined with the FRP sheet, the crack pattern could not be monitored during testing since the cracks did not penetrate the FRP sheet. At failure, the unconfined beam formed horizontal splitting cracks on the side of the spliced zone along the lap spliced bars. After the test, when the FRP sheets were removed for inspection, it was found that the failure produced smaller pieces of concrete compared to unconfined beams, as illustrated in Figure 4.

(a) Unconfined beam

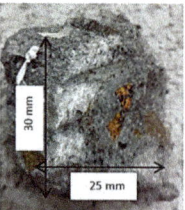

(b) Beam repaired with FRP sheet

Figure 4. Pieces of concrete of (**a**) unconfined beam, (**b**) beam confined with fiber-reinforced polymer (FRP) sheet.

3.2. Lap Splice Beam

Table 5 shows a summary of all lap splice beam test results, including the load at failure, the average bond strength at failure, the loaded end slip and the failure mode. The lap splice beams were labeled as follows: the first part (LS) refers to a lap splice and the second part refers to the type of concrete for monolithic concrete (M) and SCC for commercial prepackaged self-consolidating concrete extended with 50% of 13 mm coarse aggregate. The third part refers to the confinement condition (UN for the unconfined beams and F for the beams confined with FRP sheets). The last part represents the corrosion level (C for the control specimens (non-corroded), 7.5 for the specimens with a 7.5% mass loss level and 15 for the specimens with a 15% mass loss level). The splice length and the concrete cover for all beams were fixed as 300 mm and 25 mm, respectively.

Table 5. Summary of lap splice beams test results.

Specimen	Corrosion Level (%)	Load on Beam at Failure (kN)	f_s (MPa)	Bond Strength τ_b (MPa)	Bar Slip at Failure (mm)	Failure Mode *
LS-M-UN-C	0	172.3	269	4.42	1.30	S
LS-M-F-C	-	231.0	361	5.91	2.04	S
LS-SCC-UN-C	-	186.4	291	4.78	1.44	S
LS-SCC-F-C	-	250.2	391	6.41	2.52	S
LS-M-UN-7.5	7.5	173.9	294	4.63	1.41	S
LS-M-F-7.5	-	245.0	414	6.51	2.42	S
LS-SCC-UN-7.5	-	192.5	325	5.13	1.59	S
LS-SCC-F-7.5	-	261.7	442	6.97	2.81	S
LS-M-UN-15	15	176.8	324	4.91	1.52	S
LS-M-F-15	-	258.1	473	7.16	2.93	S
LS-SCC-UN-15	-	191.8	352	5.32	1.71	S
LS-SCC-F-15	-	285.7	524	7.93	6.25	Y + S

* S: Splitting failure. Y: Bar yield.

Based on the splice length, the concrete cover, the mass loss level and the confinement condition, all lap splice beams were designed to fail by a splitting failure. The average bond strength was calculated using Equation (3).

$$\tau_b = \frac{d_b \, f_s}{4 L_s} \qquad (3)$$

where τ_b is the average bond strength (MPa); d_b is the bar diameter account for actual mass loss (mm); f_s is the steel stress at failure (MPa); and L_s is the splice length (mm) = 300 mm.

3.3. Effect of Partial Depth Repair with and without FRP Wrapping

The load-deflection curves for the lap splice beams (monolithic specimens (M) and partial depth repaired specimens (SCC) with and without FRP wrapping (F)) are revealed in Figure 5a–c for the control beam sample and the 7.5% rebar mass loss and 15% rebar mass loss beams, respectively. The beams confined with FRP sheets experienced an increase in the ultimate load and the corresponding deflection at failure compared to the unconfined beams. Moreover, the maximum load and corresponding deflection were recorded for the beams wrapped with FRP sheets by 55% and 191%, respectively.

The load-deflection curves increased linearly after initial flexural cracking until the maximum load was achieved in coincidence with bond splitting failure. After the splitting failure, the load-deflection behavior of the unconfined beams was different from the beams confined with FRP sheets. The splitting failure of unconfined beams was accompanied by a sudden drop in the applied load and rise in deflection. Nevertheless, the beams confined with FRP sheets revealed a ductile post-failure behavior, in which the load dropped gradually as the deflection increased. This could be attributed to the FRP sheets limiting the bond crack widths and preventing spalling of the concrete cover. Besides, the partial depth repair was able to restore and increase the capacity of the repaired beams, which had a higher maximum load at failure than that of the monolithic beams.

The LS-SCC-15-F beam failed by bar yielding followed by a splitting failure. It was observed that the stress on the bar exceeded the manufactured reported yield strength of 479 MPa before the splitting failure occurred. After the bar yielded, the deflection of the beam increased with a slight rise in the applied load before failure occurred. Except for this beam, all the other beams failed by splitting bond failure. The re-bond state (constant or slight increase in bond after failure) in the unconfined beams could be from the residual friction between the reinforcing bar and the surrounding concrete where the concrete keys did not shear off, unlike the confined beams. It was also observed that beams confined with FRC sheets sustained a higher load than unconfined beams. The higher yield strength of confined beams could be attributed to the existence of FRP sheets, which prevented the sudden failure of the beams and therefore, sustained higher loads.

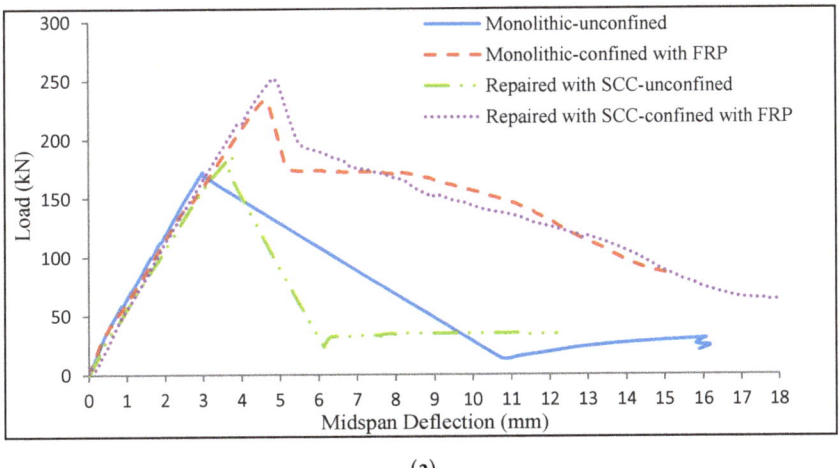

(a)

Figure 5. *Cont.*

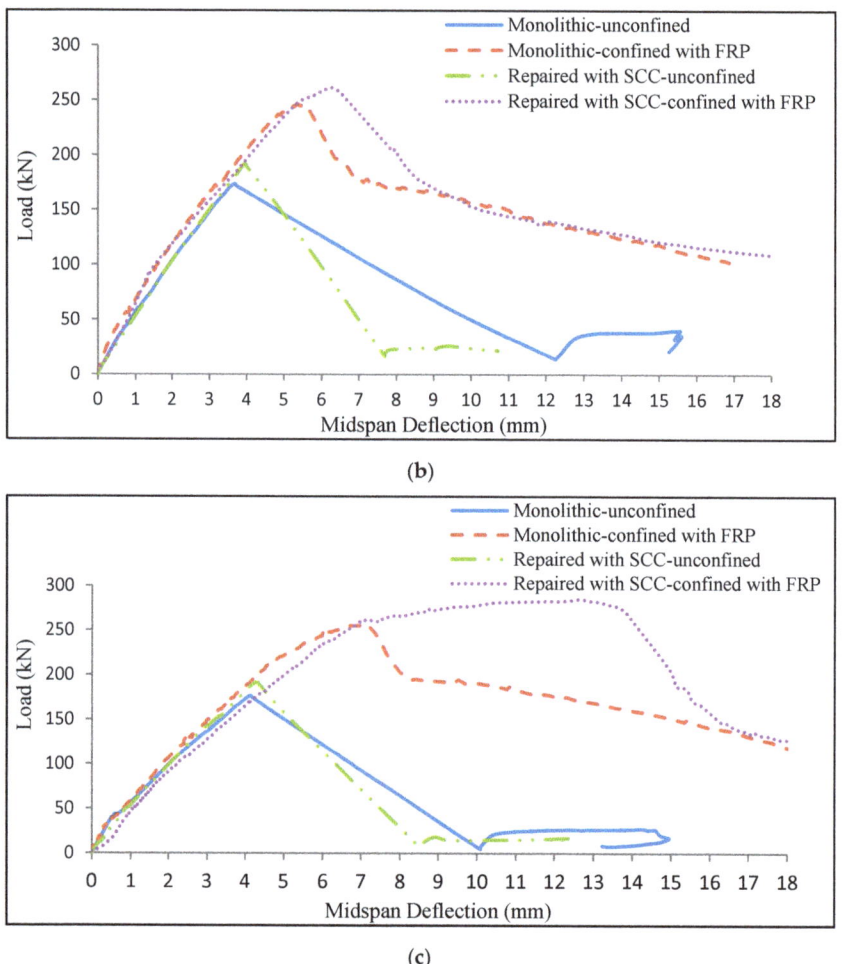

Figure 5. Load deflection curves of monolithic beams and beams repaired with partial depth repair self-consolidating concrete (SCC) with and without FRP confinement. (**a**) Control beams; (**b**) beams with 7.5% mass loss; (**c**) beams with 15% mass loss.

3.4. Effect of Corrosion on Load-Deflection Response

Figure 6 shows typical load-deflection curves of beams with different levels of mass loss. The beams in the same mass loss category had almost the same stiffness. However, the stiffness decreased slightly with increased reinforcing bar mass loss. As the bar mass loss increased, the total cross-sectional area of the bars decreased, reducing the effective stiffness of the beam, thereby increasing the mid-span deflection. Despite the minor differences in load at failure, the average bond strength increased as the mass loss increased, as shown in Figure 7 due to improved bond strength between the concrete and reinforced bars. This occurred because as corrosion increased, the rebar surface became rougher, which improved both the mechanical bonds and friction properties.

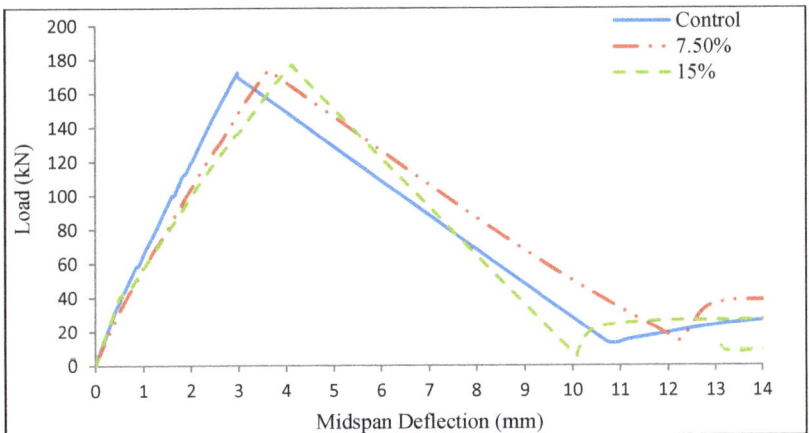

Figure 6. Typical load deflection curve of monolithic unconfined lap splice beams with different levels of mass loss.

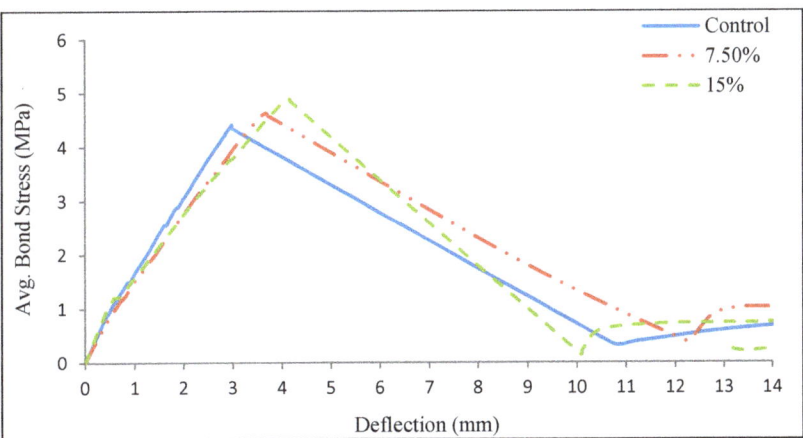

Figure 7. Typical average bond stress versus deflection curve of monolithic unconfined lap splice beams with different levels of mass loss.

3.5. Bond Stress versus Slip Response

The typical bond stress versus slip response for the control monolithic beams (M) and partial depth repaired specimens (SCC) with and without FRP wrapping (F) is shown in Figure 8a. The beams confined with FRP sheets had higher values of bond strength and slipped at maximum load than the unconfined beams. The confined beams experienced a later start and slower growth of splitting cracks. The beams confined with FRP sheets revealed a rise in the bond strength values and the equivalent slip values by about 34–49%, and 56–260% compared to the unconfined beams, respectively. Regardless of the condition of confinement, the partial depth repair with the SCC had a higher bond strength than the monolithic specimens.

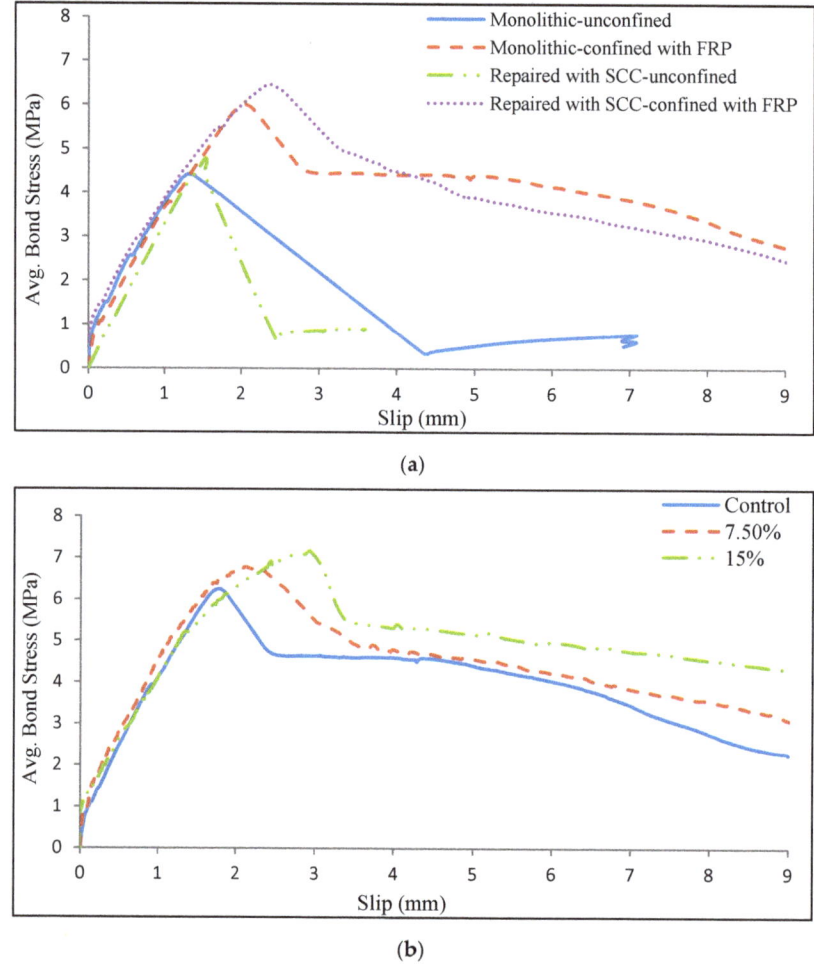

Figure 8. (a) Average bond stress versus slip for control unconfined beams and beams wrapped with FRP. (b) Average bond stress versus slip for beams wrapped with FRP with different rebar mass loss.

The typical bond stress-slip curve of beams confined with FRP sheets for the control beam and the 7.5% and 15% mass loss beams are shown in Figure 8b. The bond strength and the corresponding slip increased with increasing rebar mass loss. This is because the cross sectional area of the bars decreased with increasing corrosion levels; thus, the force on the bar at bond failure decreased more slowly than the area of the bar [30,31].

4. Conclusions

The following conclusions are drawn based on the findings of this study:

1. All of the unconfined beams and the beams confined with FRP sheets failed because of splitting bond failure. However, the beams confined with FRP sheets had a more ductile failure than the unconfined beams. At failure, the lap splice beams confined with FRP sheets produced smaller chunks of concrete than those of the unconfined beams.
2. The average bond strength increased with increasing bar mass loss level due to the decrease in the cross-sectional area of the rebar. Therefore, as the bar diameter decreases, the bond force at

failure decreases more slowly than the decreasing bar area. Also, as the corrosion level increased, the cleaned rebar surface roughness increased, which probably increased the mechanical bond contribution. Repairing the beams with partial depth repair SCC concrete enhanced the average bond strength compared to the monolithic beams.

3. The beams confined with FRP sheets had a delayed bond failure related to that of the unconfined beams. The beams wrapped with FRP sheets had a higher maximum load and corresponding deflection than the unwrapped beams by 49% and 191%, respectively.
4. The beams confined with FRP sheets showed a rise in the bond strength and the equivalent slip by 34–49%, and 56–260% compared to the unconfined beams, respectively.

Author Contributions: Conceptualization, H.A., T.T.; data curation, H.A., and R.A.; formal analysis, H.A., R.A., H.M.; investigation, R.A. and H.M.; methodology, H.A. and R.A.; project administration, H.A.; resources, H.A. and R.A.; supervision, T.T.; validation, H.A. and R.A., visualization, H.A. and K.A.; writing, original draft preparation, H.A. and H.M.; writing, review and editing, H.A., R.A. and H.M. All authors have read and agreed to the published version of the manuscript.

Funding: This research project was supported by the Deanship of Scientific Research at Prince Sattam bin Abdulaziz University (Saudi Arabia) under the research project No. 2020/01/16810.

Acknowledgments: The authors gratefully acknowledge the technical support received from the University of Waterloo and Universiti Teknologi Malaysia (UTM).

Conflicts of Interest: The authors declare no conflict of interest.

References

1. Amaireh, L.; Al-Rousan, R.Z.; Ababneh, A.N.; Alhassan, M. Integration of CFRP strips as an internal shear reinforcement in reinforced concrete beams. *Structures* **2020**, *23*, 13–19. [CrossRef]
2. Junaid, M.T.; Elbana, A.; Altoubat, S. Flexural response of geopolymer and fiber reinforced geopolymer concrete beams reinforced with GFRP bars and strengthened using CFRP sheets. *Structures* **2020**, *24*, 666–677. [CrossRef]
3. Al-Majidi, M.H.; Lampropoulos, A.P.; Cundy, A.B.; Tsioulou, O.T.; Alrekabi, S. Flexural performance of reinforced concrete beams strengthened with fibre reinforced geopolymer concrete under accelerated corrosion. *Structures* **2019**, *19*, 394–410. [CrossRef]
4. Barris, C.; Sala, P.; Gómez, J.; Torres, L. Flexural behaviour of FRP reinforced concrete beams strengthened with NSM CFRP strips. *Compos. Struct.* **2020**, *241*, 112059. [CrossRef]
5. Hadhood, A.; Agamy, M.H.; Abdelsalam, M.M.; Mohamed, H.M.; El-Sayed, T.A. Shear strengthening of hybrid externally-bonded mechanically-fastened concrete beams using short CFRP strips: Experiments and theoretical evaluation. *Eng. Struct.* **2019**, *201*, 109795. [CrossRef]
6. Zuo, J.; Darwin, D. Splice Strength of Conventional and High Relative Rib Area Bars in Normal and High-strength Concrete. *Struct. J.* **2000**, *97*, 630–641.
7. Azizinamini, A.; Pavel, R.; Hatfield, E.; Ghosh, S. Behavior of lap-spliced reinforcing bars embedded in high strength concrete. *Struct. J.* **1999**, *96*, 826–835.
8. Hamad, B.S.; Najjar, S.S. Evaluation of the Role of Transverse Reinforcement in Confining Tension Lap Splices in High Strength Concrete. *Mater. Struct.* **2002**, *35*, 219–228. [CrossRef]
9. Karataş, M.A.; Gökkaya, H. A review on machinability of carbon fiber reinforced polymer (CFRP) and glass fiber reinforced polymer (GFRP) composite materials. *Def. Technol.* **2018**, *14*, 318–326. [CrossRef]
10. Wang, Z.; Almeida Jr, J.H.S.; St-Pierre, L.; Wang, Z.; Castro, S.G. Reliability-based buckling optimization with an accelerated Kriging metamodel for filament-wound variable angle tow composite cylinders. *Compos. Struct.* **2020**, *254*, 112821. [CrossRef]
11. Masoud, S.; Soudki, K.; Topper, T. Postrepair Fatigue Performance of FRP-Repaired Corroded RC Beams: Experimental and Analytical Investigation. *ASCE J. Compos. Constr.* **2005**, *9*, 441–449. [CrossRef]
12. Mohammadhosseini, H.; Yatim, J.M.; Sam, A.R.M.; Awal, A.A. Durability performance of green concrete composites containing waste carpet fibers and palm oil fuel ash. *J. Clean. Prod.* **2017**, *144*, 448–458. [CrossRef]
13. Almusallam, A.; Al-Gahtani, A.; Aziz Rasheeduzzafar, A. Effect of Reinforcement Corrosion on Bond Strength. *Constr. Build. Mater.* **1996**, *10*, 123–129. [CrossRef]

14. Alrshoudi, F.; Mohammadhosseini, H.; Alyousef, R.; Alabduljabbar, H.; Mustafa Mohamed, A. The Impact Resistance and Deformation Performance of Novel Pre-Packed Aggregate Concrete Reinforced with Waste Polypropylene Fibres. *Crystals* **2020**, *10*, 788. [CrossRef]
15. Alyousef, R.; Mohammadhosseini, H.; Alrshoudi, F.; Alabduljabbar, H.; Mohamed, A.M. Enhanced Performance of Concrete Composites Comprising Waste Metalised Polypropylene Fibres Exposed to Aggressive Environments. *Crystals* **2020**, *10*, 696. [CrossRef]
16. El Maadawi, T.; Soudki, K. Carbon-Fibre-Reinforced Polymer Repair to Extend Service Life of Corroded Reinforced Concrete Beams. *J. Compos. Constr. ASCE* **2005**, *9*, 187–194. [CrossRef]
17. Okada, K.; Kobayashi, I.K.; Miyagawa, T. Influence of Longitudinal Cracking Due to Reinforcement Corrosion on Characteristics of Reinforced Concrete Members. *Struct. J.* **1988**, *85*, 134–140.
18. Tahir, M.M.; Mohammadhosseini, H.; Ngian, S.P.; Effendi, M.K. I-beam to square hollow column blind bolted moment connection: Experimental and numerical study. *J. Constr. St. Rese.* **2018**, *148*, 383–398. [CrossRef]
19. Cabrera, J.G. Deterioration of Concrete Due to Reinforcement Steel Corrosion. *Cem. Concr. Compos.* **1996**, *18*, 47–59. [CrossRef]
20. ACI (American Concrete Institute). *Protection of Metals in Concrete against Corrosion*; ACI 222-01: Farmington Hills, MI, USA, 2001.
21. ACI (American Concrete Institute). *Guide for the Design and Construction of Externally Bonded FRP Systems for Strengthening Concrete Structures*; ACI 440.2R-08: Farmington Hills, MI, USA, 2008.
22. Garcia, R.; Helal, Y.; Pilakoutas, K.; Guadagnini, M. Bond Strength of Short Lap Splices in RC Beams Confined with Steel Stirrups or External CFRP. *Mater. Struct.* **2015**, *48*, 277–293. [CrossRef]
23. Alyousef, R.; Topper, T.; Al-Mayah, A. Effect of FRP Wrapping on Fatigue Bond Behavior of Spliced Concrete Beams. *J. Compos. Constr.* **2015**, *20*, 04015030. [CrossRef]
24. 24. Shihata, Ayman. 2011CFRP Strengthening of RC Beams with Corroded Lap Spliced Steel Bars. Master's Thesis, University of Waterloo, Waterloo, ON, Canada, 2011.
25. Almeida, J.H.S.; Souza, S.D.; Botelho, E.C.; Amico, S.C. Carbon fiber-reinforced epoxy filament-wound composite laminates exposed to hygrothermal conditioning. *J. Mater. Sci.* **2016**, *51*, 4697–4708. [CrossRef]
26. Antunes, M.B.; Almeida, J.H.S., Jr.; Amico, S.C. Curing and seawater aging effects on mechanical, thermal, and physical properties of glass/epoxy filament wound composite cylinders. *Compos. Commun.* **2020**, *22*, 100517. [CrossRef]
27. Malumbela, G.; Alexander, M.; Moyo, P. Serviceability of Corrosion-Affected RC Beams after Patch Repairs and FRPs under Load. *Mater. Struct.* **2011**, *44*, 331–349. [CrossRef]
28. Emberson, N.K.; Mays, G.C. Significance of Property Mismatch in the Patch Repair of Structural Concrete. Part 3: Reinforced concrete members in flexure. *Mag. Concr. Res.* **1996**, *48*, 45–57. [CrossRef]
29. Fang, C.; Lundgren, K.; Chen, L.; Zhu, C. Corrosion Influence on Bond in Reinforced Concrete. *Cem. Concr. Res.* **2004**, *34*, 2159–2167. [CrossRef]
30. ACI (American Concrete Institute). *Bond and Development of Straight Reinforcing Bars in Tension*; ACI 408R-03: Farmington Hills, MI, USA, 2003.
31. Alrshoudi, F.; Mohammadhosseini, H.; Tahir, M.M.; Alyousef, R.; Alghamdi, H.; Alharbi, Y.R.; Alsaif, A. Sustainable use of waste polypropylene fibers and palm oil fuel ash in the production of novel prepacked aggregate fiber-reinforced concrete. *Sustainability* **2020**, *12*, 4871. [CrossRef]

Publisher's Note: MDPI stays neutral with regard to jurisdictional claims in published maps and institutional affiliations.

© 2020 by the authors. Licensee MDPI, Basel, Switzerland. This article is an open access article distributed under the terms and conditions of the Creative Commons Attribution (CC BY) license (http://creativecommons.org/licenses/by/4.0/).

Article

Comparative Study of High-Performance Concrete Characteristics and Loading Test of Pretensioned Experimental Beams

Pavlina Mateckova [1],*, Vlastimil Bilek [2] and Oldrich Sucharda [2,3]

1. Department of Structures, Faculty of Civil Engineering, VSB–Technical University of Ostrava, Ludvíka Podéště 1875/17, 708 33 Ostrava-Poruba, Czech Republic
2. Department of Building Materials and Diagnostics of Structures, Faculty of Civil Engineering, VSB–Technical University of Ostrava, Ludvíka Podéště 1875/17, 708 33 Ostrava-Poruba, Czech Republic; vlastimil.bilek@vsb.cz (V.B.); oldrich.sucharda@vsb.cz (O.S.)
3. Centre of Building Experiments, Faculty of Civil Engineering, VSB–Technical University of Ostrava, Ludvíka Podéště 1875/17, 708 33 Ostrava-Poruba, Czech Republic
* Correspondence: pavlina.mateckova@vsb.cz; Tel.: +42-776-781-340

Abstract: High-performance concrete (HPC) is subjected to wide attention in current research. Many research tasks are focused on laboratory testing of concrete mechanical properties with specific raw materials, where a mixture is prepared in a relatively small amount in ideal conditions. The wider utilization of HPC is connected, among other things, with its utilization in the construction industry. The paper presents two variants of HPC which were developed by modification of ordinary concrete used by a precast company for pretensioned bridge beams. The presented variants were produced in industrial conditions using common raw materials. Testing and comparison of basic mechanical properties are complemented with specialized tests of the resistance to chloride penetration. Tentative expenses for normal strength concrete (NSC) and HPC are compared. The research program was accomplished with a loading test of model experimental pretensioned beams with a length of 7 m made of ordinarily used concrete and one variant of HPC. The aim of the loading test was to determine the load–deformation diagrams and verify the design code load capacity calculation method. Overall, the article summarizes the possible benefits of using HPC compared to conventional concrete.

Keywords: high-performance concrete; mechanical properties; loading test; pretensioned beam

1. Introduction

High-performance concrete (HPC) is not a new player in the field of concrete construction. The beginning of the HPC era is connected with the development of superplasticizers together with the use of mineral admixtures finer than cement in the 1970s and 1980s. Though HPC is no longer a novelty [1], it is still an object of interest of many research activities in a few fields.

Scientific work has been focused primarily on the design of suitable concrete mixtures. For HPC, a low water to binder ratio is required. The concrete mixture has to contain a well-compatible superplasticizer and an optimum composition of the binding system [2].

Replacement of part of the cement with mineral admixtures finer than cement helps, among others, to reach significantly better consistency. As a mineral admixture, silica fume is widely used, and another possibility is the utilization of metakaolin. The advantage of metakaolin is the possibility for it to be one of the compounds of a ternary binder, as it contains alumina ions, and together with calcium carbonate (ground limestone), it can create a good base for a ternary binder. In addition, ternary binders with metakaolin are considered as a very effective arrangement for the reduction in expansion caused by the alkali–silica reaction [3].

Considering the fact that HPCs are usually based on a high portion of cement, it is also necessary to deal with impact on the environment (e-CO_2 and e-Energy) [4,5] and compare it with other materials from the point of view of sustainability [6].

The topics of life cycle analysis and environmental impact are closely connected with the material durability and structural service life [7,8]. Concrete durability could be assessed in a few aspects, including testing of carbonation, frost resistance or testing of resistance to chloride penetration. HPC possesses increased durability thanks to its compact microstructure [9,10]. High frost resistance [11], carbonation resistance [12,13] and chloride resistance [14,15] were proved in a few research tasks. Research, which is referred to in [16], indicates that the frost resistance of concretes with metakaolin is more favorable than that of concretes with silica fume.

Parallel with the progress in technology, it is also important to amend design codes and rules for construction [17,18]. Model code 2010 [19] reflects the utilization of HPC with increasing validity for concrete up to the compressive strength of 140 MPa. From this point of view, experimental tests of structural elements made of new types of concrete are very valuable as a verification of load capacity calculation models given in design codes and as a background for numerical modeling. Testing of load-bearing elements is demanding with respect to production, manipulation, transport and the requirement of commonly unavailable testing equipment, and that is why it is less frequentative.

Scientific papers in the field of HPC are usually focused narrowly on mechanical properties [20], compressive strength and microstructure [21], influence of high temperature [22] or, on the other hand, on testing of structural elements [23–25].

The presented paper reflects the mentioned experiences in complex research which starts with the development of an HPC mixture, continues with the description of the microstructure and material mechanical properties and load capacity testing of the experimental structural element, proves the increased resistance to environmental effects and brings forth a possible application in load-bearing structures and the primary quantification of expenses. A complex approach with emphasis on application in the construction industry is the preference of the presented research.

Based on extensive and long-term parametrical laboratory testing, two variants of an HPC mixture were designed using common raw materials. The subobjective was to prove the mechanical properties of concrete developed in the laboratory in industrial conditions with respect to the amount of mixture, fresh concrete transport and casting conditions.

The research program was complemented with the testing of the load capacity of a pretensioned structural element. The aim of the loading test was the verification of the calculation model for bending moment load capacity used in the current code for concrete with compressive strength on the border or outreaching the limits of validity. A comparison of the reserve in load capacities with the calculated value of the structural element made of common concrete and HPC is presented.

Utilization of HPC is also connected with an expected improved durability. The developed variants of HPC are intended to be used preferably in bridge beams which are exposed to chloride aerosols due to traffic on or under the bridge, and that is why testing of chloride resistance was given priority. New variants of HPC show, on the basis of the performed test, higher resistance to chloride penetration, which is promising from the perspective of durability.

Tentative expenses on used materials are compared not only for testing beams but also on an example of a model bridge structure. Though this comparison cannot consider all of the problematic life cycle cost quantification, it has a meaning for the basic notion of the economy aspect of using HPC in the traditional cross-section of pretensioned beams.

2. Concrete Mixture and Material Properties of Concrete

The industrial partner of the presented research is a traditional producer of precast pretensioned and post-tensioned bridge beams. The concrete composition has to be adapted to the volume of the mixture device, which is 1 m³, and fulfill the requirements of workability and compaction of fresh concrete, together with a target of high early strength.

The current concrete mixture used for pretensioned elements is stated in Table 1 and, in this paper, is identified as normal strength concrete (NSC) and listed as the reference concrete. The concrete composition is based on ordinary Portland cement. High-quality granite aggregates 4/8 and 8/16 from the quarry in Litice nad Orlici and sand 0/4 from Lipa nad Orlici belong to local natural resources. The amount of plasticizer is 4.5 l per m³. The water cement ratio of NSC is 0.4.

Table 1. Concrete mixture of normal strength concrete (NSC) and high-performance concrete (HPC).

Raw Material [kg]	NSC Max. Agg. 16 mm	HPC1 Max. Agg. 16 mm	HPC2 Max. Agg. 8 mm
Cement 42.5	450	575	650
Slag	0	40	60
Limestone	0	30	15
Metakaolin	0	80	75
Plasticizer	4.5	-	-
Superplasticizer	-	20	17
Water	180	165	165
Aggregates 0/4 (Lipa nad Orlici)	690	590	830
Aggregates 4/8 (Litice nad Orlici)	215	185	520
Aggregates 8/16 (Litice nad Orlici)	845	725	-
w/c	0.40	0.29	0.25

Based on long-term and extensive laboratory work on composition optimization [26,27], the NSC concrete mixture was modified to develop an HPC mixture. In Table 1, the composition of two variants of the HPC mixture is shown, identified as HPC1 and HPC2. Concrete mixture HPC1 preserves the utilization of the 8/16 aggregate, while in concrete mixture HPC2, aggregate 8/16 is replaced with 0/4 and 4/8 aggregates. Utilization of HPC2 without the 8/16 aggregate is profitable, especially for subtle elements with dense reinforcement. Another advantage is only needing minimum vibration due to the fine-grained composition.

Both variants of HPC were designed with ternary binders using slag and limestone. The ordinarily used plasticizer was replaced with a superplasticizer, based on polycarboxal ethers, due to the low water–cement ratio, which is, respectively, 0.29 and 0.25. Low water–cement and water–binder ratios cause deterioration of workability, and the mixture shows a sticky consistency if only Portland cement is used. Replacing part of the cement with mineral admixtures finer than cement helps to achieve a better consistency. In the presented HPC mixture, metakaolin was used as a mineral admixture. The consistency of all analyzed concretes was classified as superfluid on the basis of the cone slump test, where the slump was over 210 mm. The result of the optimization is a mixture with high early strength as an effect of the ternary binder. Based on previous research, it is supposed that the designed HPC possesses high frost resistance with a minimum risk of the alkali–silica reaction [3,16].

The main advantage of HPC is its dense microstructure, which is a premise for the high strength and high durability. The dense microstructure is a consequence of the very low water to binder ratio. The microstructures of common concrete and HPC are compared in Figure 1.

In Table 2, the test results of the basic mechanical properties of NSC and HPC variants are shown. Laboratory testing was carried out according to valid codes [28–30]. The split tensile strength of HPC was tested on cube specimens both in directions parallel (six specimens) and perpendicular to the filling (three specimens). The mixed mode fracture resistance together with the tensile strength of the analyzed types of concrete is discussed in [31,32].

Figure 1. Concrete microstructure: common concrete with water–cement ratio w/c = 0.55 (**a**), and high-performance concrete, variant 1 with w/c = 0.29 (**b**).

In Figure 2, comparisons of the cylinder compressive strength of NSC and variants of HPC are presented. The increment in cylinder compressive strength is, respectively, 45% and 52%. The value of the cylinder/cube compressive strength ratio is 0.76 for NSC and 0.93 for both variants of HPC. Values of the cylinder/cube compressive strength ratio correspond to the expected values, which are, for NSC, 0.8–0.85 and, for HPC, 0.9–1.0. The increment in the static modulus of elasticity of HPC is 16%, and 18% for the second variant. Values of the modulus of elasticity are slightly lower but roughly correspond to the values expected according to the design code [17] both for NSC and HPC. Values of the split tensile strength are compared in Figure 3. The cube compressive strength of HPC variants in 24 h is compared in Figure 3, and it reaches 50% and 60% of the cube compressive strength in 28 days.

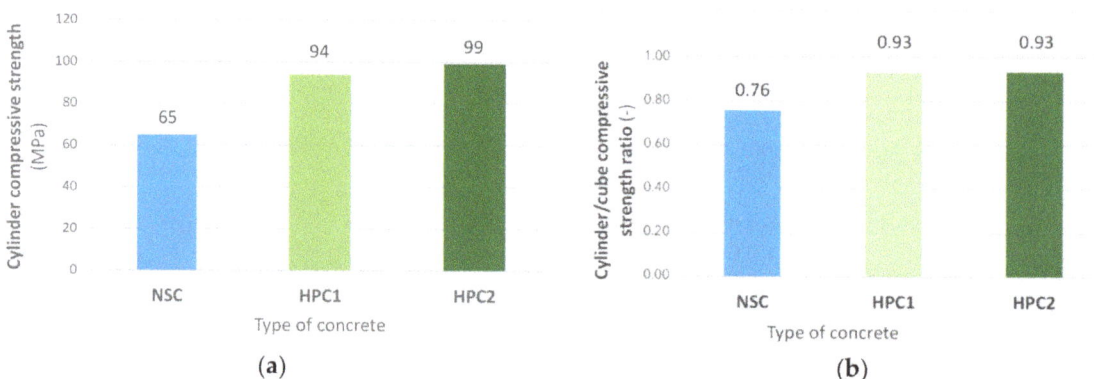

Figure 2. Comparison of cylinder compressive strength (**a**) and cylinder/cube compressive strength ratio (**b**).

Table 2. Tested material characteristics.

	NSC			HPC1		HPC2	
	Number of Specimens	Mean Value	Standard Deviation	Mean Value	Standard Deviation	Mean Value	Standard Deviation
Cube Compressive Strength [MPa]	3	85.8	3.4	101.0	3.3	106.2	3.3
Cylinder Compressive Strength [MPa]	6	65.5	5.7	94.1	5.9	99.1	5.9
Static Modulus of Elasticity [GPa]	3	35.0	1.0	42.1	3.8	41.2	4.3
Split Tensile Strength [MPa]	6+3	5.9	0.3	5.9	0.4	6.4	0.3

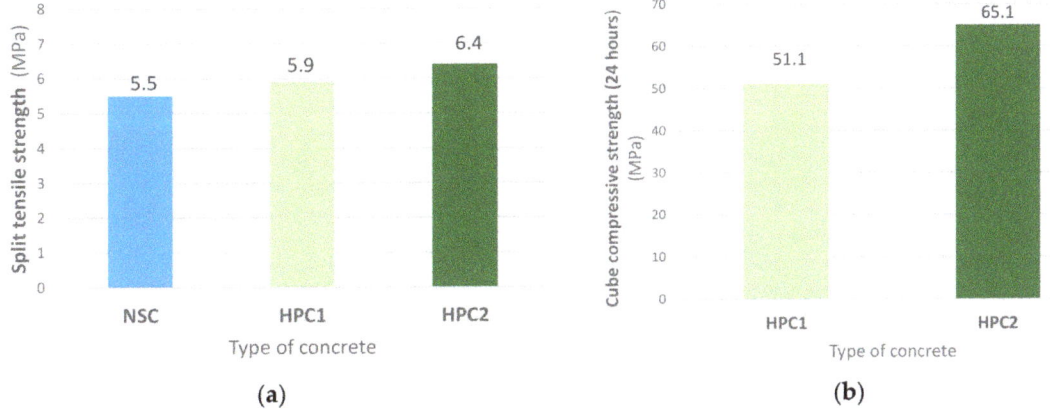

Figure 3. Comparison of spit tensile strength (a) and cube compressive strength in 24 h (b).

3. Durability of Concrete

Resistance to chloride penetration is characterized with a diffusion coefficient D_c and an aging factor m. AASHTO TP-95 tests [33] were carried out (Figure 4), where a surface electrical resistance was measured for six time points for the determination of the diffusion coefficient, which changes with the maturing of concrete, as a function of time. The reference aging factor m was derived based on those values. A higher aging factor represents an increasing resistance to chloride penetration over time, which may be related to a higher persistence of the concrete structure to aggressive substances. The reference diffusion coefficient based on the NT-BUILD 443 method [34] was also evaluated. Results of the tests are stated in Tables 3–5.

Table 3. Diffusion coefficient and aging factor—NSC.

Time	[Days]	7	14	28	56	91	161
Diffusion Coefficient	[m²/s]	1.3×10^{-11}	9.6×10^{-12}	8.79×10^{-12}	6.01×10^{-12}	5.93×10^{-12}	5.89×10^{-12}
Aging Factor m (AASHTO TP-95)			[-]		0.2884		
Diffusion Coefficient (Nord 443)			[m²/s]		10.1×10^{-12}		

Table 4. Diffusion coefficient and aging factor—HPC1.

Time	[Days]	7	14	28	56	91	161
Diffusion Coefficient	[m2/s]	4.66×10^{-12}	3.88×10^{-12}	2.59×10^{-12}	1.97×10^{-12}	1.62×10^{-12}	1.24×10^{-12}
	Aging Factor m (AASHTO TP-95)			[–]		0.4197	
	Diffusion Coefficient (Nord 443)			[m^2/s]		3.46×10^{-12}	

Table 5. Diffusion coefficient and aging factor—HPC2.

Time	[Days]	7	14	28	56	91	161
Diffusion coefficient	[m^2/s]	3.62×10^{-12}	2.93×10^{-12}	2.54×10^{-12}	2.04×10^{-12}	1.73×10^{-12}	1.51×10^{-12}
	Aging Factor m (AASHTO TP-95)			[–]		0.2777	
	Diffusion Coefficient (Nord 443)			[m^2/s]		3.72×10^{-12}	

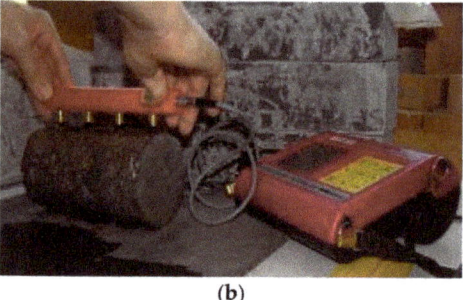

(a) (b)

Figure 4. Specimens in chloride solution prepared for NT BUILD 443 test (a) and diffusion coefficient testing based on AASHTO TP-95 (b).

In Figure 5, the diffusion coefficient and aging factor of the analyzed concretes are compared. The diffusion coefficient is about 3 times lower both for HPC1 and HPC2 than for NSC according to Nord 443. The aging factors of NSC and HPC2 are almost the same, and the aging factor of HPC1 is considerably more favorable than the NSC value. A more detailed description of the testing and discussion of results can be found in [35–37].

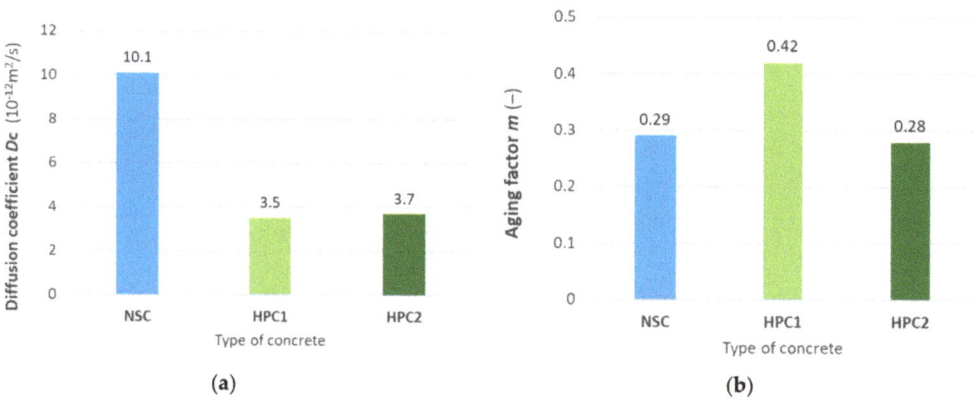

(a) (b)

Figure 5. Comparison of aging factor (a) and diffusion coefficient (b)—NSC and HPC.

4. Experiments of Pretensioned Concrete Beams
4.1. Description and Production of Experimental Beams

The background for the experimental beam design was the portfolio of bridge beams of the industrial partner, which is composed mainly of slab beams and T-shaped beams. It was also necessary to take into consideration the disposition of Centre of Building Experiments at VSB-TUO (CBE), especially the load and space capacity of the testing device, where the limit specimen width is 900 mm and the limit span is 10 m. It was decided to use a modified slab beam, which roughly corresponds to the real structural element, rather than a T-shaped beam, whose cross-section would have to be modified due to the space and loading capacity of the CBE device. Variant HPC2 was selected for experimental beam production.

The criterion for the design of cross-sectional dimensions was the same flexural stiffness of the experimental beam's cross-section both for NSC and HPC2. Slab beams were designed with the width of 900 mm and the height of 560 mm for the beam made of NSC, and 520 mm for the beam made of HPC2, Figure 6. The height of experimental beams corresponds to the expected height of a real slab bridge beam. The length of the beams was 7.0 m, and the span was 6.5 m. The length was limited by the crane load capacity in CBE.

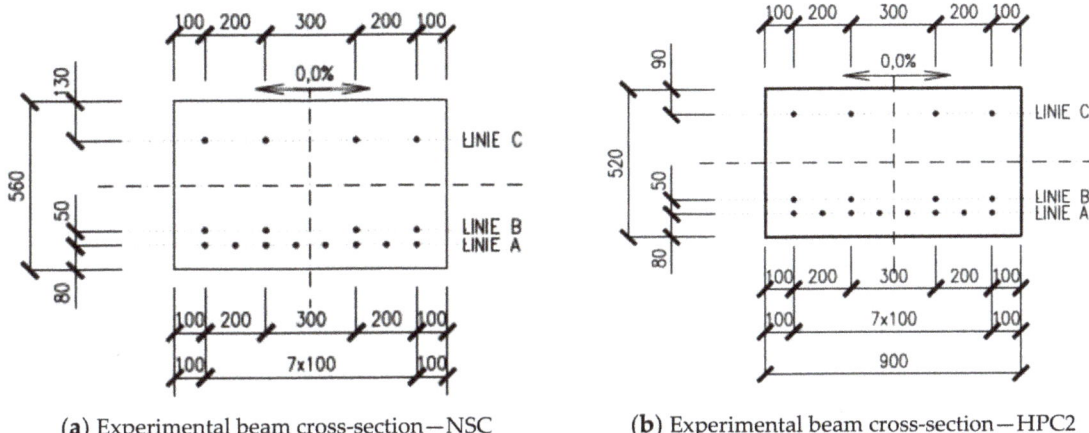

(a) Experimental beam cross-section—NSC   (b) Experimental beam cross-section—HPC2

Figure 6. Dimensions of experimental beams' cross-section, and layout of pretensioned reinforcement: cross-section of NSC beam (a), and cross-section of HPC2 beam (b).

The pretensioned reinforcement with a diameter of 15.7 mm and a strength of 1860 MPa was designed in three rows with a similar layout in both NSC and HPC cross-sections, Figure 6. The upper layer of the pretensioned reinforcement was designed to reduce the negative deformation. Together with the bottom pretensioned reinforcement, which is partially provided with separation, it eliminates tensile stress on the element's upper surface. Initial stress was 1400 MPa in all pretensioned reinforcements. Stress after releasing was calculated with a value of 1312 MPa and stress after long-term changes was 1129 MPa in the bottom layer of the pretensioned reinforcement.

There was also non-pretensioned reinforcement designed with a yield stress of 500 MPa. Four-legged stirrups with a profile of 8 mm and a distance in the longitudinal direction of 150 mm were designed to bear the shear force and to ensure accessories for manipulation. Stirrups were complemented with longitudinal reinforcement, with a profile of 16 mm.

Experimental beams were designed with approximately the same load capacity and cracking moment. Bending moment load capacities, cracking moments and appropriate loading forces together with testing peak load force are compared in Table 6. Load capacities were calculated according to the design code [17], based on the limit strain method, using the software IDEA statica [38]. Mean values of compressive and tensile strengths based on the test results were considered. Probabilistic assessment of the time-dependent load capacity is discussed in [39]. Cracking moment was calculated using the value of the pretensioned force in the appropriate time with respect to long-term changes. The designed shear reinforcement allows presupposing the bending failure of experimental beams.

Table 6. Load capacity of specimen beams.

	NSC		HPC2	
	Load Capacity	Crack Capacity	Load Capacity	Crack Capacity
Calculation—bending moment capacity [kNm]	1643	781	1581	779
Calculation—ultimate load [kN]	970	480	935	479
Experiment—peak load [kN]	1148	586	1062	546

Experimental beams were produced in the manufacturing plant of the industrial partner, Figure 7. Together with the experimental beams, specimens for laboratory testing were also concreted. Results of laboratory testing are listed in Chapter 2–Chapter 3 of this paper.

Figure 7. Prepared pretensioned and non-pretensioned reinforcements of experimental beams (a), and casting of concrete (b).

4.2. Testing of Experimental Beams

Experimental beams were exposed to a three-point loading test, Figure 8. Three beams were tested, one made of NSC and two made of HPC2. The testing frame equipped with a system of hydraulic cylinders allowed achieving a loading force up to 2000 kN. The loading force was recorded with a hydraulic electronic control system, and the deflection in the mid-span was measured with a precision of 0.01 mm. Initial negative deformation due to pretension was not measured. The concentrated load of the hydraulic cylinder was spread to the width of the experimental beam with the steel welded element, Figure 8. Experimental beams were supported with cylinder bearings to draw the supports close to the hinge support. The application of the load was controlled with deformation, and the loading step was 5 mm. After applying the load in one step, there was a 5–10 min time delay for the deformation setting.

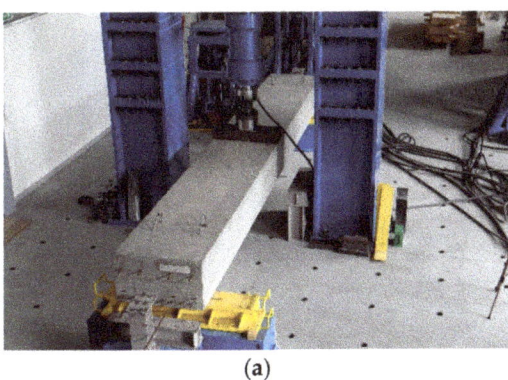

(a) (b)

Figure 8. Three-point loading test in Centre of Building Experiments (**a**), and spreading of concentrated force with steel element (**b**).

The tested peak load of the NSC experimental beam was 1148 kN and there is about 18% reserve when compared with the calculated value of the load capacity. The load x mid-span deflection diagram (LD diagram) is shown in Figure 9. Yielding of the reinforcement is apparent in the diagram, the rupture of pretensioned reinforcement was reached and there was also significant damage of the concrete in the compressed part of the cross-section, where also buckling of the compressed non-pretensioned reinforcement emerged, Figure 10. The first visible cracks, with a width of 0.05/0.1 mm, appeared for a loading force of about 586 kN. This value is also distinct in the diagram, and with this value of the loading, the linear part of the LD function finishes. The calculated value of the cracking force is lower.

The tested peak load of HPC experimental beams was 1080 kN and 1043 kN. There is about 14% reserve when comparing the mean value of the tested load capacity and the calculated load capacity. The load x mid-span deflection diagram is shown in Figure 9. Yielding of the reinforcement is apparent in the diagram, and rupture of the pretensioned reinforcement was not reached. There was significant damage of the concrete in the compressed part of the cross-section, and in the final stage of testing, the compressed part of the cross-section was detached, Figure 11. Buckling of the compressed non-pretensioned reinforcement emerged. The first visible crack, with a width of 0.05/0.1 mm, appeared for the loading force of about 546 kN. This value is also distinct in the diagram, and, approximately, for this value of the loading force, the linear part of the loading function finishes. The calculated value of the cracking force is 480 kN.

LD diagrams are very similar both for experimental HPC2 beams and the NSC beam. In the initial phase of loading, the LD function is linear. The first visible cracks appeared in the same loading step for the loading force of about 550 kN. As it had been supposed, the first cracks initiated in the mid-span of the tested beam from the bottom surface roughly perpendicular to the lower edge. Under further loading, there was a gradual development of cracks at nearly regular distances from the mid-span to the support area. When reaching the load of approximately 800 kN, the width of the crack increased up to 0.3 mm and the cracks progressed gradually to the upper surface of the beams. The decrease in bending rigidity is significant also in the load x displacement diagram.

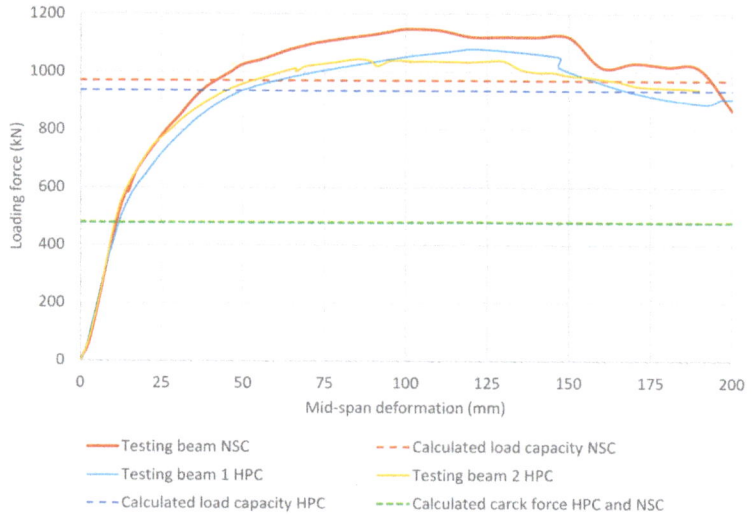

Figure 9. Load–deformation diagram testing of NSC beam compared with testing of HPC beams.

Figure 10. Failure of NSC beam: crack pattern (**a**), and rupture of pretensioned reinforcement (**b**).

A subsequent meaningful decrease in bending rigidity is in the load step of approximately 1000 kN, and the width of some cracks exceeds 1 mm. The increment of the loading force became smaller with the applied deformation. When the peak load was reached, the experimental beams did not collapse to downfall and the beams were gradually unloaded. In the final stage of loading, when testing the NSC beam, the rupture of the pretensioned reinforcement was reached and significant damage of concrete in the compressed part appeared together with buckling of the non-pretensioned reinforcement. In the final stage of testing of the HPC beam, the rupture of the pretensioned reinforcement was not reached; however, when applying the deformation with a nearly zero increment of force, the damaged area under the load-applying steel element was connected with a progressive crack, and this led to the detachment of the compressed upper part of the cross-section.

Figure 11. Failure of HPC beam: crack pattern (**a**), and rupture of the compressed part of the cross-section (**b**).

The load capacity was reached in bending as expected with the yielding of the pretensioned reinforcement. The reserve in the calculated load capacity is 18% for NSC and 14% for HPC. The reserve in the calculated crack load is 22% for NSC and 14% for HPC.

5. Tentative Price and Expenses

5.1. Tentative Price of Materials

The incorporation of precast load-bearing elements made of HPC in the portfolio of an industrial company is strongly connected with economical aspects. In Figure 12, a graphical comparison of tentative initial expenses on 1 m³ of concrete based on the price of raw materials is shown, according to data provided from the producer. Both variants of HPC are about 1.8 times more expensive than NSC, which corresponds to data from the professional literature, where the price of HPC is about 1.5–2.0 times more expensive than NSC. Prices of reinforcement in parametrical calculations are 0.9 E/m for pretensioned reinforcement, and 0.8 E/kg for non-pretensioned reinforcement.

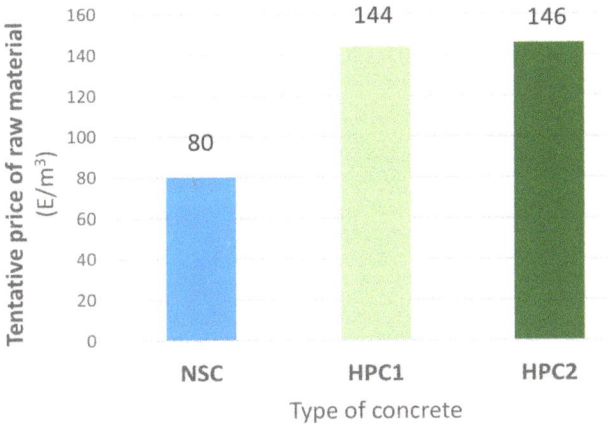

Figure 12. Comparison of tentative raw material price of NSC and variants of HPC.

5.2. Comparison of Testing Slab Beams

In Table 7, the initial expenses for testing slab beams are compared. Consumption of pretensioned and non-pretensioned reinforcements is nearly the same for both types of testing beams. The reduction in concrete consumption is small when retaining the same load and crack capacity. The increase in initial expenses for 1 kN of tested load capacity is 40% for the HPC testing beam compared with the NSC beam. It has to be emphasized that the comparison of expenses is only a basic rough assessment of prices of used materials.

Table 7. Tentative expenses of testing beams.

	NSC Testing Beam		HPC Testing Beam	
	Amount	Price [E]	Amount	Price [E]
Concrete	3.53 m^3	283	3.28 m^3	479
Pretensioned reinforcement	144 m	130	144 m	130
Non-pretensioned reinforcement	190 kg	152	185 kg	148
Total price		565		757
Expenses for 1 kN of tested load capacity		0.50		0.7

5.3. Study of T- and I-Shaped Beams

A deeper study of HPC utilization was conducted on a model road bridge made of six beams with a clear width of 9.5 m, Figure 13. The span of the parametric bridge is 15 m, 20 m, 25 m and 30 m. Utilization of the higher compressive strength of HPC enables an increase in the initial prestressing force. In each variant of the used material and span, it is possible to design a few variants of beams. In this study, beams with a minimum height/cross-sectional area are mentioned. In Table 8, heights and cross-sectional areas of T- and I-shaped bridge beams, designed for a parametrical bridge, are stated. The decrease in the cross-sectional area of T beams made of HPC is about 12% compared with NSC. In a slightly optimized variant of the bridge beam with an enlarged bottom flange, the I-shaped beam, the decrease in the cross-sectional area is about 30% compared with the T-shaped NSC beam. In Figure 14, cross-sections of bridge beams for a span of 30 m are compared together with the number of pretensioned strands.

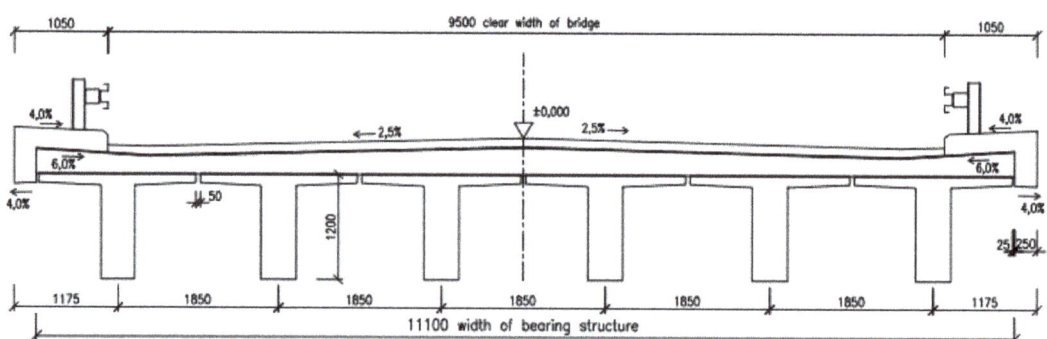

Figure 13. Section of model road bridge.

In Table 9, a comparison of raw material expenses for T- and I-shaped parametrical bridge beams made of HPC and NSC is presented. A considerable part of the price is represented by the pretensioned and non-pretensioned reinforcements, and this affects the final expenses balance, which is a 25% increase in expenses for the HPC T-shaped beam and only a 5% increase for the HPC I-shaped beam compared with T beam made of NSC; however, this variant is inconsiderably more laborious due to the more complicated formwork.

Table 8. Height of T beams designed for the study bridge.

Span [m]	T Beam—NSC Height [mm]	T Beam—NSC Area [m²]	T Beam—HPC Height [mm]	T Beam—HPC Area [m²]	I Beam—HPC Height [mm]	I Beam—HPC Area [m²]
15	900	0.509	750	0.449		
20	1150	0.629	950	0.529	850	0.431
25	1400	0.709	1200	0.629		
30	1650	0.809	1450	0.729	1350	0.583

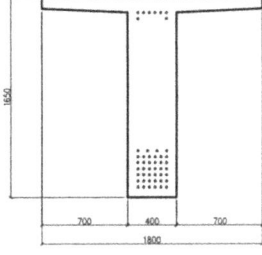
NSC: 40 + 8 strands

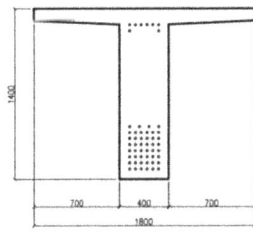
HPC—T beam: 46 + 8 strands

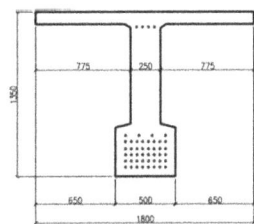
HPC—I beam: 44 + 4 strands

Figure 14. Cross-section of bridge beams with layout of pretensioned reinforcement, bridge span 30.

Table 9. Tentative expenses for parametrical bridge beams.

Span [m]		NSC—T Beam Amount	NSC—T Beam Expenses [E]	HPC—T Beam Amount	HPC—T Beam Expenses [E]	HPC—I Beam Amount	HPC—I Beam Expenses [E]
20	Concrete [m³]	13.2	1056	11.1	1622	9.1	1322
	Pretens. st. [m]	702	632	936	843	820	738
	Non-preten. st. [kg]	974	779	780	624	694	555
	Total		2467		3089		2615
30	Concrete [m³]	25.1	2006	22.6	3300	18.1	2639
	Pretens. st. [m]	1636	1472	1840	1656	1636	1472
	Non-preten. st. [kg]	2147	1717	1843	1475	1653	1323
	Total		5195		6431		5434

It has to be emphasized that the comparison of expenses is only a tentative rough assessment of the prices of the used material without taking into consideration costs concerning the used technology, build-in reinforcement and labor, e.g., due to the more complicated formwork of I-shaped beams. On the other hand, the decrease in structural elements' weight brings forth savings in transport expenses.

6. Discussion

Extensive material research focused on new compositions of HPC is carried out worldwide, with the aim to increase the strength characteristics together with improvements in other utility properties. The limiting factor of this research is, among other things, the context of laboratory small specimen testing and the industrial conditions of production.

Within the presented research, it was proved that within the optimization of the ordinary concrete composition, it is possible to achieve a compressive strength of about

100 MPa under the industrial conditions of the research partner. HPC was designed with a ternary binder with an increased cement portion, utilization of metakaolin and a superplasticizer and a change in the aggregate size ratio. The subject of different laboratory and industrial conditions is also discussed in [40], where the testing of precast bridge segments is described.

Based on previous testing, the second variant of HPC was preferred. In [41], long-term testing of the fracture energy and toughness of concrete with a lower water–cement ratio is described. A significant decrease in the fracture parameters of specimens made of concrete with an aggregate size of 22 mm was observed after one year, whilst the decrease in concrete with an aggregate size of 16 mm was smaller and no decrease in the fracture parameters was observed in concrete with an aggregate size of 8 mm. Strength characteristics stayed unaffected. The explanation is probably in the autogenous shrinkage of the cement paste and microcracks that arose on the aggregate's surface. The interfacial transition zone, which is developed especially around the higher-size aggregate, promotes this microcracking.

The HPC concrete mixture brings forth the change in microstructure, the increase in compactness and, consequently, increased durability. The case of bridge structure degradation due to exposure to chloride aerosols is the key factor for material durability and structural life span. Within the tests of the chloride diffusion coefficient, it was proved that the resistance to chloride penetration is about three times higher for both variants of HPC than for NSC.

When designing a new concrete mixture, it is valuable to complement laboratory testing of small specimens with the loading test of structural elements. The three pretensioned elements which were tested within the presented research showed very similar behavior. The positive influence of pretension resulted in a higher cracking load and slow crack development. Failure due to the pretensioned reinforcement yielding enabled high deformability, which was in the final stage of loading in 1/30 of the span. The calculated value of the load-bearing capacity and the value based on testing correspond well both for experimental beams made of NSC and HPC2. The disadvantage of HPC with the compressive strength on the border or outreaching the limits of validity of the current design code is the higher danger of brittle failure. Testing of the load capacity indicates that the ultimate bending moment calculation model reflected the danger of brittle failure in the value of the concrete ultimate limit strain and particular coefficients for the decrease in the compressed cross-sectional area [17,19]. The similar course of the loading test and failure mode allows assuming analogical computational methods both for NSC and HPC beams. An analogical discussion of HPC column design is presented in [42].

Conditions for wider utilization of HPC and UHPC in the precast industry are concisely and objectively discussed in [43]. Locally mixed and produced HPC/UHPC mixtures and optimized structural members are mainly cost-effective for buildings and bridges.

When taking into consideration the price of pretensioned and non-pretensioned reinforcements, the increase in the structural element total price is 40% for HPC in the case of experimental slab beams. However, in a deeper study of a parametrical bridge, it was quantified that HPC T-shaped beams are about 25% more expensive than NSC T beams and in a slightly optimized variation of HPC I-shaped beams, the increase in price is only 5%. Though all complex and hardly quantified relevant expenses are not taken into consideration, the information about the price of used materials is meaningful for primary comparison.

The decrease in the structural element's height could be decisive in the case of replacing an insufficient existing structure [8] and also from the perspective of the design value of the water level.

Nevertheless, the expenses should be inspected not only with respect to the manufacturing cost but also in the context of the whole life span, which is, in the case of bridges, 100 years. Convincing quantification of all relevant expenses, together with environmental impact, i.e., life cycle analysis, is the objective of many research tasks [44,45]. Their application in the case of bridge beams is the object of future research.

7. Conclusions

Complex research of HPC was presented which comprises the design of concrete mixtures, testing of mechanical properties of concrete cast in industrial conditions, testing of load capacity of pretensioned experimental beams made of commonly used and newly developed concrete, testing of durability and a tentative primary comparison of expenses.

In the paper, two variants of HPC were presented. Under industrial conditions, a progressive and perspective compressive strength of HPC of about 100 MPa was achieved.

Testing of pretensioned experimental beams indicates the conveniency of the calculation model of the ultimate bending moment capacity for structural elements made of concrete with compressive strength on the border or outreaching the limits of validity of the current design code.

An increase in resistance to chloride penetration of the new HPC variant compared to ordinarily used concrete was proved with AASTO and NORD laboratory testing.

The designed HPC is 1.8 times more expensive than NSC. Nevertheless, the inconsiderable item of the price is the cost of the reinforcement. The tentative raw comparison on parametrical bridge beams indicates that when taking into consideration the total cost, the increase in expenses of HPC bridge beams is acceptable at 25%, which is balanced with increased durability. In the slightly optimized but more laborious variant of the cross-section, the increase in expenses is only 5%.

Application of HPC in the construction industry leads to its further development and, in a wider context, to the construction of economical and sustainable structures with an increased life span.

Author Contributions: V.B., design of concrete mixture; O.S., supervisor of material characteristics tests, supervisor of loading tests of experimental beams in Centre of Building Experiments, VSB-TU Ostrava, evaluation of test results; P.M., design of precast bridge beams, preparing manuscript. All authors have read and agreed to the published version of the manuscript.

Funding: Czech Science Foundation (GACR) and conceptual development of science, research and innovation assigned to VŠB-TUO by the Ministry of Education, Youth and Sports of the Czech Republic.

Acknowledgments: The work was supported by project GACR 21-08772S—Influence of Self-Healing effects on structural fatigue life extension of structures made from high performance concrete (InShe), and The Operational Programme Enterprise and Innovation for Competitiveness (OP EIC), project CZ.01.1.02/0.0/0.0/15_019/0004505 Complex design of beams made of new advanced types of concrete. The work was also supported by means of the conceptual development of science, research and innovation assigned to VŠB-TUO by the Ministry of Education, Youth and Sports of the Czech Republic. The authors would like to appreciate the considerable support of the industrial partners, namely, Cevora, Rudolecka and Andres from divisions in Litice and Orlici.

Conflicts of Interest: The authors declare no conflict of interest.

References

1. Aitcin, P.C. *High Performance Concrete*; CRC Press: Boca Raton, FL, USA, 2018; ISBN 978036786598.
2. Park, S.; Wu, S.; Liu, Z.; Pyo, S. The Role of Supplementary Cementitious Materials (SCMs) in Ultra High Performance Concrete (UHPC): A Review. *Materials* **2021**, *14*, 1472. [CrossRef]
3. Thomas, M. The effect of supplementary cementing materials on alkali-silica reaction: A review. *Cem. Concr. Res.* **2011**, *41*, 1224–1231. [CrossRef]
4. Long, G.; Gao, Y.; Xie, Y. Designing more sustainable and greener self-compacting concrete. *Constr. Build. Mater.* **2015**, *84*, 301–306. [CrossRef]
5. Chiaia, B.; Fantilli, A.P.; Guerini, A.; Volpatti, G.; Zampini, D. Eco-mechanical index for structural concrete. *Constr. Build. Mater.* **2014**, *67*, 386–392. [CrossRef]
6. Kuruşcu, A.O.; Girgin, Z.C. Efficiency of Structural Materials in Sustainable Design. *J. Civ. Eng. Arch.* **2014**, *8*, 8. [CrossRef]
7. Kioumarsi, M.; Benenato, A.; Ferracuti, B.; Imperatore, S. Residual Flexural Capacity of Corroded Prestressed Reinforced Concrete Beams. *Metals* **2021**, *11*, 442. [CrossRef]
8. Moravcik, M.; Bujnakova, P.; Bahleda, F. Failure and damage of a first-generation precast prestressed bridge in Slovakia. *Struct. Concr.* **2020**, *21*, 2353–2362. [CrossRef]

9. Chu, H.; Wang, F.; Wang, L.; Feng, T.; Wang, D. Mechanical properties and environmental evaluation of ultra-high-performance concrete with aeolian sand. *Materials* **2020**, *13*, 3148. [CrossRef]
10. Vázquez-Rodríguez, F.J.; Elizondo-Villareal, N.; Verástegui, L.H.; Tovar, A.M.A.; López-Perales, J.F.; de León, J.E.C.; Gómez-Rodríguez, C.; Fernández-González, D.; Verdeja, L.F.; García-Quiñonez, L.V.; et al. Effect of Mineral Aggregates and Chemical Admixtures as Internal Curing Agents on the Mechanical Properties and Durability of High-Performance Concrete. *Materials* **2020**, *13*, 2090. [CrossRef]
11. Feo, L.; Ascione, F.; Penna, R.; Lau, D.; Lamberti, M. An experimental investigation on freezing and thawing durability of high performance fiber reinforced concrete (HPFRC). *Compos. Struct.* **2020**, *234*, 111673. [CrossRef]
12. Singh, S.P.; Singh, N. Reviewing the carbonation resistance of concrete. *J. Mater. Eng. Struct.* **2016**, *3*, 35–57.
13. Czarnecki, L.; Woyciechowski, P.; Adamczewski, G. Risk of concrete carbonation with mineral industrial by-products. *KSCE J. Civ. Eng.* **2017**, *22*, 755–764. [CrossRef]
14. Yoon, Y.S.; Kwon, S.J. Evaluation of time-dependent chloride resistance in HPC containing fly ash cured for 1 year. *J. Korea Inst. Struct. Maint. Insp.* **2018**, *22*, 52–59.
15. Li, L.; Zheng, J.; Ng, P.; Zhu, J.; Kwan, A. Cementing efficiencies and synergistic roles of silica fume and nano-silica in sulphate and chloride resistance of concrete. *Constr. Build. Mater.* **2019**, *223*, 965–975. [CrossRef]
16. Bilek, V., Sr.; Bilek, V., Jr.; Krutil, K.; Krutilova, K. Some aspect of durability of concrete with ternary binders. In Proceedings of the 8th CCC Durability of Concrete, Plitvice Lakes, Croatia, 4–6 October 2012; pp. 359–364, ISBN 978-953-7621-14-8.
17. *EN 1992-1-1: Eurocode 2: Design of Concrete Structures—Part 1–1: General Rules and Rules for Buildings*; British Standard Institution: London, UK, 2004.
18. *EN 1992-2: Eurocode 2: Design of Concrete Structures: Part 2: Concrete Bridges*; British Standard Institution: London, UK, 2005.
19. Cairns, J. *Model Code 2010 First Complete Draft*; fib Bulletin: Lausanne, Switzerland, 2010; Volume 1, ISBN 978-2-88394-095-6.
20. Siwiński, J.; Szcześniak, A.; Stolarski, A. Modified Formula for Designing Ultra-High-Performance Concrete with Experimental Verification. *Materials* **2020**, *13*, 4518. [CrossRef] [PubMed]
21. He, Z.-H.; Du, S.-G.; Chen, D. Microstructure of ultra high performance concrete containing lithium slag. *J. Hazard. Mater.* **2018**, *353*, 35–43. [CrossRef] [PubMed]
22. Park, G.-K.; Park, G.-J.; Park, J.-J.; Lee, N.; Kim, S.-W. Residual Tensile Properties and Explosive Spalling of High-Performance Fiber-Reinforced Cementitious Composites Exposed to Thermal Damage. *Materials* **2021**, *14*, 1608. [CrossRef] [PubMed]
23. Li, C.; Feng, Z.; Ke, L.; Pan, R.; Nie, J. Experimental Study on Shear Performance of Cast-In-Place Ultra-High Performance Concrete Structures. *Materials* **2019**, *12*, 3254. [CrossRef]
24. Karimipour, A.; Edalati, M. Shear and flexural performance of low, normal and high-strength concrete beams reinforced with longitudinal SMA, GFRP and steel rebars. *Eng. Struct.* **2020**, *221*, 111086. [CrossRef]
25. Tej, P.; Kolísko, J.; Kněž, P.; Čech, J. The overall research results of prestressed i-beams made of ultra-high performance concrete. In Proceedings of the IOP Conference Series: Materials Science and Engineering, Prague, Czech Republic, 13–16 September 2017; IOP Publishing: Bristol, UK, 2017; Volume 246, p. 12051.
26. Bilek, V.; Fiala, C.; Hajek, P. High Performance concrete for sustainable building elements and structures, In Proceedings of the 3rd R. N. Raikar Memorial International Conference Gettu-Kodur International Symposium on Advances in Science Technology of Concrete, Mumbai, India, 14–15 December 2018; pp. 213–218, ISBN 978-93-88237-28-4.
27. Sucharda, O.; Bilek, V.; Mateckova, P. Testing and mechanical properties of high strength concrete. In Proceedings of the IOP Conference Series: Materials Science and Engineering, Zuberec, Slovakia, 29–31 May 2019; IOP Publishing: Bristol, UK, 2019; Volume 549, p. 012012.
28. *CSN EN 12390-3:2020: Testing Hardened Concrete—Part 3: Compressive Strength of Test Specimens*; Czech Standardization Agency: Prague, Czech Republic, 2020.
29. *CSN EN 12390-6:2010: Testing Hardened Concrete—Part 6: Tensile Splitting Strength of Test Specimens*; Czech Standardization Agency: Prague, Czech Republic, 2010.
30. *CSN ISO 1920-10:2016: Testing of Concrete—Part 10: Determination of Static Modulus of Elasticity in Compression*; Czech Standardization Agency: Prague, Czech Republic, 2016.
31. Seitl, S.; Miarka, P.; Bílek, V. The mixed-mode fracture resistance of C 50/60 and its suitability for use in precast elements as determined by the Brazilian disc test and three-point bending specimens. *Theor. Appl. Fract. Mech.* **2018**, *97*, 108–119. [CrossRef]
32. Miarka, P.; Seitl, S.; Bílek, V. Mixed-mode fracture analysis in high-performance concrete using a Brazilian disc test. *Mater. Teh.* **2019**, *53*, 233–238. [CrossRef]
33. *AASHTO TP 95—AASHTO Method of Test for Surface Resistivity Indication of Concrete's Ability to Resist Chloride Ion Penetration*; American Association of State Highway and Transportation Officials: Washington, DC, USA, 2011.
34. *NT BUILD 443—Concrete, Hardened: Accelerated Chloride Penetration*; Nordtest: Espoo, Finland, 1995.
35. Tran, Q.; Ghosh, P.; Lehner, P.; Konečný, P. Determination of Time Dependent Diffusion Coefficient Aging Factor of HPC Mixtures. *Key Eng. Mater.* **2020**, *832*, 11–20. [CrossRef]
36. Konečný, P.; Lehner, P.; Ghosh, P.; Morávková, Z.; Tran, Q. Comparison of procedures for the evaluation of time dependent concrete diffusion coefficient model. *Constr. Build. Mater.* **2020**, *258*, 119535. [CrossRef]
37. Lehner, P.; Konečný, P.; Le, D.T.; Bílek, V. Contribution to comparison of methods for the investigation of chloride ingress related resistance. *Res. Model. Civ. Eng.* **2019**, *1*, 31–40.

38. IDEA, Idea Statica [Software]. Available online: https://www.ideastatica.com/concrete (accessed on 10 January 2021).
39. Le, T.D.; Konecny, P.; Mateckova, P. Time dependent variation of carrying capacity of prestressed precast beam. *IOP Conf. Ser. Earth Environ. Sci.* **2018**, *143*, 012013. [CrossRef]
40. Vitek, J.L.; Coufal, R.; Čítek, D. UHPC—Development and Testing on Structural Elements. *Procedia Eng.* **2013**, *65*, 218–223. [CrossRef]
41. Bilek, V. Possibility of explanation of interesting development of mechanical characteristics of concrete. In *Life Prediction and Aging Management of Concrete Structures*; RILEM Publications SARL: Bratislava, Slovakia, 1999; pp. 213–218.
42. Fiala, C.; Hejl, J.; Tomalová, V.; Bílek, V.; Pavlů, T.; Vlach, T.; Volf, M.; Novotná, M.; Hajek, P. Structural Design and Experimental Verification of Precast Columns from High Performance Concrete. *Adv. Mater. Res.* **2015**, *1106*, 110–113. [CrossRef]
43. Tadros, M.K.; Gee, D.; Asaad, M.; Lawler, J. Ultra-High-Performance Concrete: A Game Changer in the Precast Concrete Industry. *PCI J.* **2020**, *65*, 33–36. [CrossRef]
44. Hrabová, K.; Teplý, B.; Hájek, P. Concrete, Sustainability and Limit States. *IOP Conf. Ser. Earth Environ. Sci.* **2019**, *290*, 012049. [CrossRef]
45. Geiker, M.R.; Michel, A.; Stang, H.; Vikan, H.; Lepech, M.D. Design and maintenance of concrete structures requires both engineering and sustainability limit states. In Proceedings of the 6th International Symposium on Life-Cycle Civil Engineering, Ghent, Belgium, 28–31 October 2018; CRC Press: Boca Raton, FL, USA, 2019; pp. 987–991.

Article

The Effect of Lateral Load Type on Shear Lag of Concrete Tubular Structures with Different Plan Geometries

Mostafa Moghadasi [1,*], Soheil Taeepoor [2], Seyed Saeid Rahimian Koloor [3] and Michal Petrů [3]

1. Department of Civil Engineering, Faculty of Engineering, Bu-Ali Sina University, Hamedan 6516738695, Iran
2. M.Sc. Graduated, Department of Civil Engineering, Faculty of Engineering, Islamic Azad University, South Tehran, Tehran 1584743311, Iran; soheiltaeepoor@ymail.com
3. Institute for Nanomaterials, Advanced Technologies and Innovation (CXI), Technical University of Liberec (TUL), Studentska 2, 461 17 Liberec, Czech Republic; s.s.r.koloor@gmail.com (S.S.R.K.); michalpetru@tul.cz (M.P.)
* Correspondence: m.moghadasi@basu.ac.ir

Received: 3 September 2020; Accepted: 30 September 2020; Published: 3 October 2020

Abstract: Tubular structures are extensively recognized as a high efficiency and economically reasonable structural system for the design and construction of skyscrapers. The periphery of the building plan in a tubular system consists of closely spaced columns connected by circumferential deep spandrels. When a cantilever tube is subjected to a lateral load, it is expected that the axial stress in each column located in the flange frame of the tube is the same, but because of the flexibility of peripheral beams, the axial stress in the corner columns and middle columns is distributed unequally. This anomaly is called "shear lag", and it is a leading cause of the reduction in efficiency of the structure. In this paper, the possible relation between shear lag and the type of lateral load subjected to these systems is investigated. The above relation is not yet considered in previous literatures. Three various plan shapes including rectangular, triangular and hexagon were modeled, analyzed, designed and subjected to the earthquake and wind load, separately. Further work is carried out to compare the shear lag factor of these structures with distinct plan shapes against different types of lateral load. It is observed that all types of structures with various plan geometry subjected to the wind load had a greater amount of shear lag factor in comparison with structures subjected to the static and dynamic earthquake loads. In addition, shear lag in structures with the hexagon shaped plan was at the minimum.

Keywords: tall building; framed tube; lateral load type; plan geometry; shear lag; structural behavior

1. Introduction

The framed tube idea is an effective framing system for high-rise buildings. This type of structural system is mainly comprises closely spaced circumferential columns, which are connected by deep spandrel beams. The whole system works as a giant vertical cantilever, and its high efficiency is due to the large distance between windward and leeward columns. Among the most important specifications of tubular systems is their high economic efficiency. A case in point is that the material consumed in this kind of system is reduced by half in comparison with other systems [1]. In a rigid frame the "strong" bending direction of columns is aligned perpendicular to the face, while this factor is typically aligned along the face of the building in a framed tube system. In a framed tube system, the tube form resists overturning produced by lateral load—a leading cause of compression and tension in columns. Bending in columns and beams or rotation of the beam-column joint in the web section resists the shear force produced by lateral load. Gravity loads are resisted partly by exterior frames and partly by interior columns [2].

In an ideal tubular structure, circumferential columns and beams are assumed to be completely rigid, so both web and flange panels act separately and bend against lateral loads like a true cantilever. While the above system has a tubular form, it also has a more complicated behavior than a solid tube. To be more specific, the components in a framed tube cannot be completely rigid due to technical and economic constraints [3]. This will be a leading cause of nonuniform distribution of loads in columns.

Consequently, in a framed-tube structure under lateral load, the stress distribution in the flange wall panels is nonuniform and is nonlinear in the web wall panels. This anomaly which reduces the efficiency of the structures is referred to as "shear lag" [4]. To obtain a better understanding of this phenomenon, a factor called the shear lag factor is defined. This factor is a ratio of the corner column axial force to the middle column axial force. When the stresses in the corner columns of the flange frame panels exceed those in the middle columns, the shear lag is positive. Nevertheless, in some cases, it is vice-versa where the stress in the middle columns exceeds those in the corner columns, and this is referred to as the negative shear lag.

This paper studied the effect of lateral load type on the shear lag phenomenon in framed-tube reinforced concrete tall buildings with different plan geometries.

2. Review of Literature

With regard to the great importance of framed tube systems in the construction of high-rise buildings, it is no wonder that multifarious research has been carried out on this structural system and its shortcomings in order to make framed tube systems more effective. However, previous studies did not pay enough attention to the subjected load type and its relation to the shear lag factor and the shape of the structures, which play an important role in the amount of this factor.

In 1969, Fazlur Khan proposed a chart named "structural systems for height" which has classified the different types of tubular systems with regard to their efficiency for high-rise buildings of different heights [5]. In the same year, Chang and Zheng tried to find out more about negative shear lag and its influential factors on a cantilever box girder. In this research, they found that negative shear lag will change with the different boundary conditions of displacement and external force applied to the girder [6]. In 1988, Shiraishi et al. studied the aerodynamic stability effects on rectangular cylinders by altering their shapes subjected to the lateral loads such as wind and water flow. They cut some squares with different sizes in each edge in a rectangular cross section and observed that it has a controlling effect on the separated shear layer generating from the leading edge. These sections with various sizes of corner-cuts had totally different behaviors against wind force and water flume [7]. In 1990, Hayashida and Iwasa investigated the effects of the geometry of structures on aerodynamic forces and displacement response for tall buildings. They tested four different plan shapes, with and without corner-cuts in a wind tunnel and identified the aerodynamic damping effects produced by changing some parts of the basic cross section and also the aerodynamic character of basic shapes [8]. In 1991, Connor and Pouangare presented a simple model for the design of framed tube structures. They modeled the structures as a series of stringers which resisted axial forces without bending rigidity and shear panels which resisted shear forces without bending or axial rigidity. They proposed a model that gives accurate results for the preliminary design and analysis of tubular structures with different geometry and material properties [9]. Kwan, in 1994, proposed a hand calculation method for approximate analysis in framed tube structures by considering the shear lag factor. This method could be useful for preliminary design and quick evaluation and could provide a better perception of the effect of multifarious parameters on the structure's behavior [10]. In 2000, Han et al. investigated the shear lag factor in the web panels of shear-core walls [11]. Lee et al., in 2002, looked into the behavior of the shear lag of framed tube structures with and without internal tube(s) for the behavioral characteristics of the structures and also the relation between their performance and various structural parameters. They also proposed a simple numerical method for the prediction of the shear lag effect in framed tube structures. It has been found that the stiffness factor has an effective role in producing shear lag in tubular structures [12]. Haji-Kazemi and Company proposed a new method to analyze shear lag in framed tube structures using an analogy between the shear lag behavior of a cantilever box which represents a uniform framed tube building. This method is able to accurately analyze positive as well as negative shear lag effects in tubular structures accurately [13]. Furthermore, Moghadasi and Keramati, in 2009, studied the effects of internal tubes in shear lag reduction. They reduced the lateral displacement and shear lag amount in high-rise buildings by adding internal tubes to the framed tube structures [14].

In 2012, Shin et al. investigated different parameters such as depth and width of beams and columns on the behavior of shear lag in a frame-wall tube building. The results showed that the effect of column depth on the shear lag behavior of framed tube was more outstanding than other parameters [15]. In 2014, Mazinani et al. compared the shear lag amount of pure tube structural systems with braced tube systems and different types of X-diagonal bracing. It was observed that these braces started from corner-to-corner, increased the stiffness of the structure and consequently reduced the story drift and shear lag factor in the tubular system [16]. In addition, Nagvekar and Hampali studied the shear lag phenomenon in both the web and wing panel of a hollow structure and measured it in various heights of a structure [17]. The plan geometry, building's body form, the ratio of height to width and three-dimensional stiffness for the transfer of wind and seismic loads are the most important structural system properties affect the behavior of tall buildings, as reported by Szolomicki and Golasz-Szolomicka [18]. Alaghmandan et al. in 2016 inquired about the architectural strategies on wind effects in tall buildings. These tactics included altering the geometry of the whole building scale such as tapering and setbacks, and attenuated the wind effects in some models by architectural strategies [19]. In 2019, Shi and Zhang proposed a simplified method for calculation of shear lag in diagrid framed tube structures. In their study, the diagrid tube structure is assumed to be equivalent to continuous orthogonal elastic membrane. They tried to solve two key problems of finding the optimized angle of the diagonal column and the shear lag assessment in the preliminary design of the above structures [20].

In terms of microstructural view of concrete subjected to dynamic loading at high strain rates, Hentz et al. used a 3D discrete element method and verified it [21]. Furthermore, the evaluation of concrete cracks occurring in complex states of stress was studied by Golewski and Sadowski [22]. In their study, crack development at shear was investigated through experimental tests using two types of aggregates.

As seen from the review above of previous studies, the possible relation between shear lag and the type of lateral load subjected to these systems is not yet considered.

3. Structural Models and Analyses Specifications

3.1. Characteristics and Final Dimensions

In this paper, 12 reinforced concrete framed tube buildings with different heights and different shapes are modeled and analyzed. From a shape point of view, they were divided into three various groups of structures with (a) rectangular, (b) triangular and (c) hexagonal plan shapes (See Figure 1). Each structural plan shape consists of four different heights: 20-story, 40-story, 60-story and 80-story buildings. Figure 2 shows example pictures of each type of buildings illustrated in Figure 1. Table 1 shows the terminology of the models used in this paper. Figure 3 displays the 40-story models in ETABS software version 18.1.1 as examples. This software is chosen because it is developed specifically for the analysis and design of building structures. Furthermore, it has a high ability for static and dynamic analyses regardless of the number of nodes and stories.

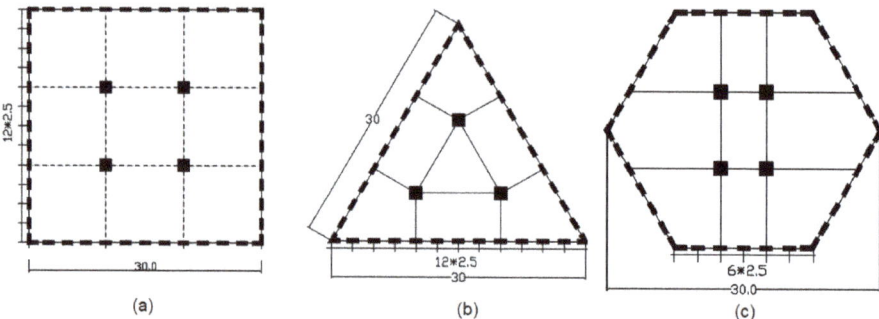

Figure 1. Three different plan shapes: (**a**) rectangular, (**b**) triangular and (**c**) hexagonal (all dimensions are in meters).

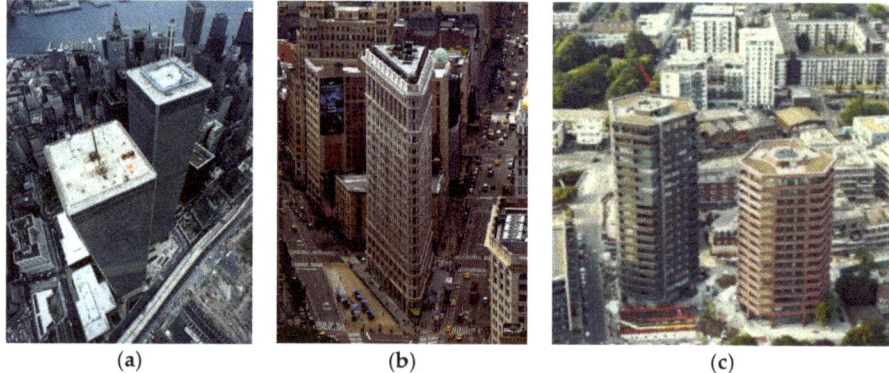

Figure 2. Three example pictures of: (**a**) rectangular plan building (WTC buildings, New York, USA) [23], (**b**) triangular plan building (Flatiron Building, New York, USA) [24] and (**c**) hexagonal plan (Hoxton Press towers, London, UK) [25].

Table 1. Terminology of models.

Number of Stories	Rectangular Plan	Triangular Plan	Hexagonal Plan
20	20R	20T	20H
40	40R	40T	40H
60	60R	60T	60H
80	80R	80T	80H

The spaces between columns are normal (10 m in rectangular and triangular plans and 5 m in hexagonal plans) on the first floor, due to the existence of entrances, and the circumference of the second floor is also isolated by deep beams. As a result, the tubular behavior of framed tube structures begins from the third floor. To obtain an accurate analogy, the equivalent length of 30 m for each side of the rectangular and triangular plans is assumed which is equal to the hexagonal plan diameters. Specifically, each side of the hexagonal plan is obtained as 15 m. The distances between columns in the first floor are 10 m for rectangular and triangular plans and five meters for hexagonal plans. The spaces between peripheral columns, which create a tubular form in the structures, are 2.5 m in all stories. The dimensions of all circumferential and gravitational beams and columns are lessened in each 10 stories from the bottom to top of the structures.

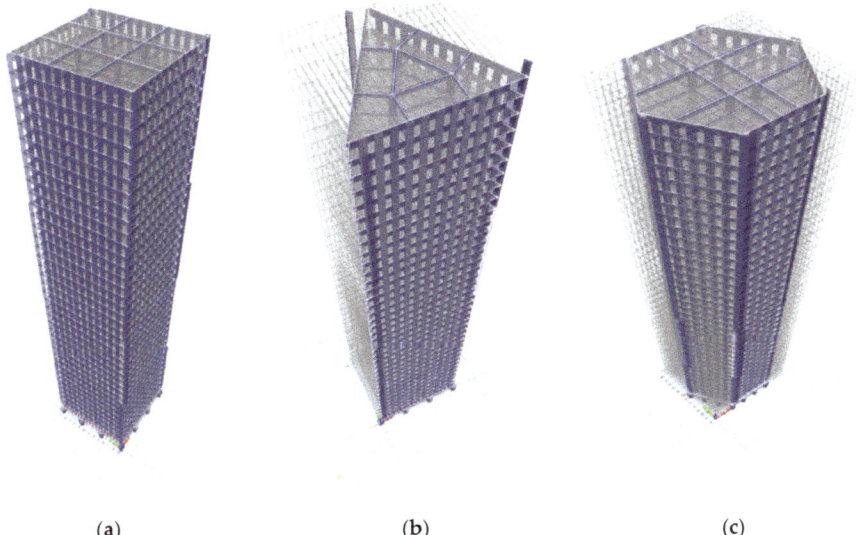

Figure 3. Three-dimensional view of 40-story models in ETABS Software: (**a**) rectangular, (**b**) triangular and (**c**) hexagonal.

3.2. Loading, Structural Analyses and Final Design Specifications

The national Iranian building code was used in order to calculate the loading of structures [26]. The total dead load for each model was 308 kg/m² with regard to this point that ETABS software automatically calculates result loads from beams, columns and slabs weights. The dead load consisted of two parts—63 kg/m² for floor weight and 245 kg/m² for partition weight. Moreover, 250 kg/m² was assumed as a live load.

The structures were analyzed by both static and pseudo-spectral dynamic methods, and to achieve a real and accurate structural analysis and design, earthquake and wind load were also applied separately on every 12 framed tube structures. The earthquake load was applied based on the Iranian seismic code [27], and the wind load specifications were according to the ASCE 7-16 code [28].

3.2.1. Earthquake Load

In this study, all the models were analyzed against the earthquake load in two manners—static equivalent and pseudo-spectral dynamic analyses—and the Iranian seismic code was used for both methods.

Static Equivalent Earthquake Load

Like every static equivalent analysis, the equivalent static base shear in this research was determined in accordance with the following equations [27]:

$$V = CW \tag{1}$$

where V is base shear, C is the seismic response coefficient and W is the effective weight of building (all dead load + a percentage of live load)
And:

$$C = \frac{ABI}{R} \tag{2}$$

where A is the design basis acceleration over the bedrock dependent on the seismic zone of the building's location; B is the reflection coefficient of the building related to seismic zone, soil type and vibration period of the structure (T) (See Figure 4); I is the importance factor; and R is the response modification factor which determines the nonlinear performance of the building during earthquakes and affected directly from the type of the structural system. It should be noticed that the structural system of the modeled structures is reinforced concrete special moment frames. A, I and R were considered as 0.3, 1.0 and 10, respectively, due to the Iranian seismic code. The minimum value of V was $V_{min} = 0.12AIW$ [27].

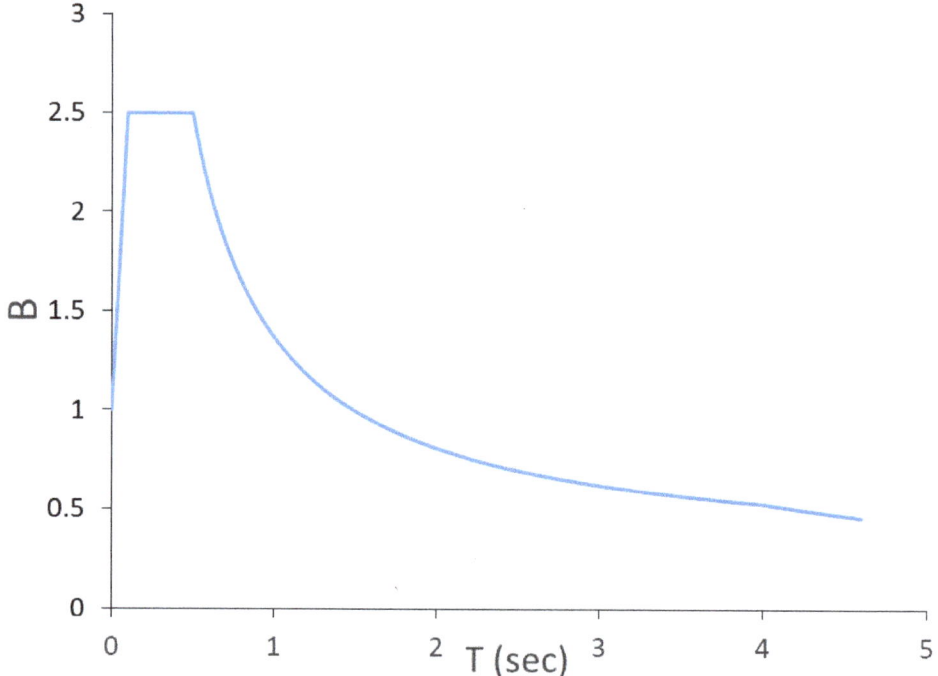

Figure 4. Response spectra intended for dynamic analysis (Iranian seismic code, 2015).

Dynamic Earthquake Load

In the dynamic analysis procedure, the lateral seismic load is determined from the dynamic response of a building subjected to an appropriate ground motion, and the pseudo-spectral dynamic analysis method was used in this paper according to the Iranian seismic code. The dynamic analysis spectrum was also obtained from the same code reflecting the effects of ground motion for design earthquake level. The spectrum shape is based on the type of soil, seismic zone and period of the structures. The information about the spectrum used in the dynamic analysis is shown in Figure 4 [27].

Wind Load

Wind load was applied to the aforementioned structures to obtain more accurate results and a comparison of framed tube behavior against different load types. Table 2 shows the specifications of the wind load obtained from the ASCE7-16 code [28].

Table 2. Specification of the wind load.

Parameters and Descriptions	Values
Basic wind speed, V	60 (m/s)
Exposure category	B
Importance factor	1
Directionality factor, K_d	0.85
Topographic factor, K_{zt}	1
Gust factor, G_f	0.85
Windward coefficient, C_p	0.8
Leeward coefficient, C_p	0.5

3.2.2. Specifications of the Models

To evaluate and analyze all 12 framed tube structures against earthquake and wind load, ETABS software version 9.7.3 was used. This choice was not made because the above software is developed specifically for analysis and design of building structures, but it also has a high ability for static and dynamic analyses regardless of the number of nodes and stories. The models were checked to have the minimum requirements of CSA 2019 code [29]. Circumferential columns and beams dimensions were lessened due to economic targets after every tenth story as the structures increased in height.

4. Results and Discussions

In this section, the results obtained from the static and dynamic analyses are discussed, and possible relationships between shear lag and the three different factors are studied in three sections. As was mentioned in the previous sections, the shear lag factor is the ratio of the axial force in a corner column to the middle column axial force in each story. It is worth mentioning that shear lag factors above one are referred to as positive shear lag and below one are referred to as negative. In the first part, the effect of lateral load types on the shear lag of framed tube structures is investigated. For this purpose, three types of lateral load including wind load, static earthquake load and dynamic earthquake load were considered to be subjected to the models. In the second section, the effects of plan geometry on the shear lag of tubular structures were studied and three different plan geometries with different heights were analyzed against different lateral loads. In the last section, the effects of height in framed tube structures on shear lag phenomena were discussed without considering any other factors.

4.1. The Effect of Lateral Load Type on Shear Lag of Framed Tube Structures

There was ample evidence from analysis results, which indicated that wind load could distribute forces among the columns more unequally. In most cases, shear lag phenomenon observed in structures subjected to the wind load were more severe than those for structures subjected to the dynamic or static earthquake load. Wind load in comparison with dynamic earthquake load caused a greater positive shear lag factor in almost all models. The positive Shear Lag factor obtained from the wind load for the 20R model was almost 1.5 times greater than the same factor for the same model against the dynamic earthquake load. Negative shear lag intensity of the 20R model against wind load was 1.22 times more than the same model subjected to the dynamic earthquake load. Although positive shear lag factors calculated from structures subjected to the wind load and dynamic earthquake load for 20T and 20H models were almost equal, negative shear lag factors for these structures subjected to the wind load were 39.08% and 21.18% more than the same models subjected to the dynamic earthquake load. The intensity of positive and negative shear lag factors observed from the 40R model subjected to the wind load was higher than the factors calculated from this model against dynamic earthquake load by 19.59% and 33.85%, respectively. The above percentages for 40T model were 3.3% and 25.71% and, for the 40H model, were 0.0% and 5.59%, respectively. This trend was observed for 60- and 80-story structures as well, and shear lag factors in structures subjected to wind load had greater amounts. Table 3 shows the percentage differences between shear lag obtained from wind load and dynamic earthquake load. The nature of wind load and its distribution and application on the surface of the structure in comparison with the seismic load applied on the center mas of the rigid diagram floors by ETABS could be the reason for the above differences illustrated in Table 3.

Table 3. The differences between shear lag obtained from the wind load and dynamic earthquake load.

Story	20R	20T	20H
Pos. Shear Lag differences	31.14%	0%	~0%
Neg. Shear Lag differences	18.39%	39.08%	21.18%
Story	40R	40T	40H
Pos. Shear Lag differences	19.59%	3.31%	~0%
Neg. Shear Lag differences	33.85%	25.71%	5.95%
Story	60R	60T	60H
Pos. Shear Lag differences	6.03%	6.84%	0.98%
Neg. Shear Lag differences	6.85%	10.26%	~0%
Story	80R	80T	80H
Pos. Shear Lag differences	5.22%	2.78%	0%
Neg. Shear Lag differences	4.295	6.41%	2.20%

Moreover, although the dynamic analysis shows a more accurate result in comparison with equal static analysis, the shear lag factors for structures subjected to the static earthquake load were also less than those factors for structures subjected to the wind load in more than 80% of the models.

4.2. The Effect of Geometry of Plan on Shear Lag of Framed Tube Structures

Analysis results indicated that the amount of shear lag in framed tube structures could be highly dependent on the plan geometry of the structure. Shear lag diagrams for three types of 80-story framed tube structures subjected to earthquake loads are shown in Figures 5 and 6. Although these diagrams are more similar to a straight line in a particular story, the framed tube structure has less shear lag in that story. Diagrams with a minimum in the middle represent the positive shear lag, and those with a maximum in the middle illustrate negative shear lag.

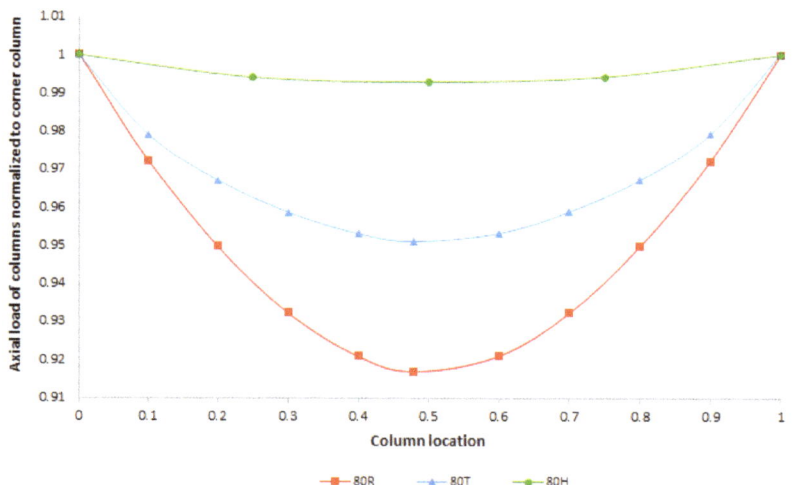

Figure 5. Shear lag diagram of 20th story of 80 story models subjected to the dynamic earthquake load.

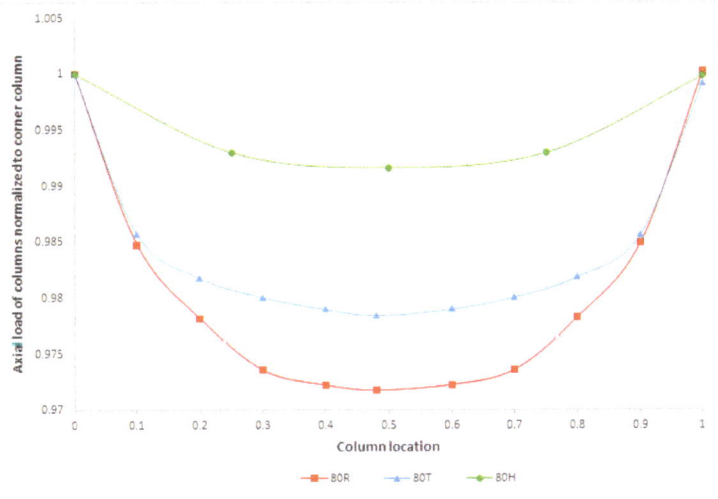

Figure 6. Shear lag diagram of 60th story of 80 story models subjected to the dynamic earthquake load.

According to these diagrams, it is observed that hexagon shaped plan structures have a better performance in terms of shear lag in comparison with the other two shapes. This structural behavior is in line with a previous study conducted by Awida, who stated that the octagon shape as a plan geometry can be the best in the structural response against wind load in comparison with other possible plan geometries [30]. Triangular-shaped structures act much better against lateral loads than rectangular ones which have the most inequality of load distribution in their flange columns. In addition, shear lag factors for every 10 stories in all structures were obtained, and the average of positive and negative shear lag factors for each structure was calculated as shown in Table 4. This table shows a similar trend in the case of shear lag in all structures. For instance, in terms of positive shear lag of a dynamic earthquake load, the 80H model acted 8.2% better than the 80R model. Moreover, the 80T model had 3.6% less shear lag in comparison with 80R. Furthermore, the negative shear lag factor of the 80H and 80T models—23% and 10.2%, respectively—performed better than the 80R model.

Table 4. The average of positive and negative shear lag factor in all models analyzed with three different load types.

Story	Dynamic Earthquake		Static Earthquake		Wind	
	Pos	Neg	Pos	Neg	Pos	Neg
20R	1.99	0.87	1.99	1.99	2.89	0.71
20T	1.74	0.87	1.73	1.73	1.74	0.53
20H	1.13	0.85	1.13	1.13	1.14	0.67
40R	1.19	0.65	1.24	1.24	1.48	0.43
40T	1.17	0.7	1.21	1.21	1.21	0.52
40H	1.02	0.84	1.01	1.01	1.01	0.79
60R	1.09	0.73	1.1	1.1	1.16	0.68
60T	1.09	0.78	1.1	1.1	1.17	0.7
60H	1.01	0.85	1.01	1.01	1.02	0.86
80R	1.09	0.7	1.15	1.15	1.15	0.67
80T	1.05	0.78	1.09	1.09	1.08	0.73
80H	1	0.91	1	1	1	0.89

To obtain a better understanding of shear lag fluctuation in the models, shear lag factors for odd stories were investigated, and Figures 7–9 shows the shear lag factor diagram for 80-story structures for three types of loads. In regard to these Figures, the shear lag factor diagrams for hexagonal-shaped plan structures in most of the stories are close to one, which means shear lag is at a minimum in these types. The shear lag phenomenon that showed up in rectangular-shaped plan structures was the maximum, especially in the first and last 10 stories. Furthermore, triangular-shaped plan structures exhibited a better behavior, in general, in terms of shear lag in comparison with rectangular ones mostly in the top half of the buildings, but still, shear lag in these structures was far higher than framed tube structures with hexagonal plan shape.

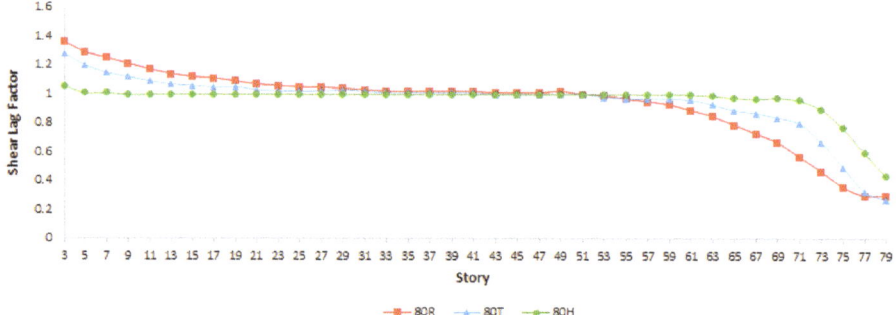

Figure 7. Shear lag factor diagram for 80 story structures subjected to the dynamic earthquake load.

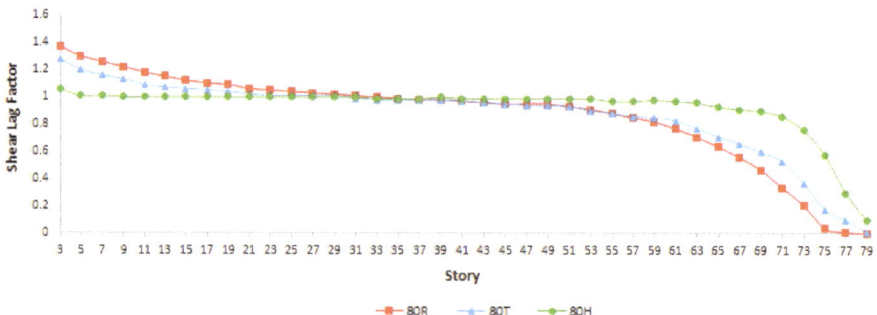

Figure 8. Shear lag factor diagram for 80 story structures subjected to the static earthquake load.

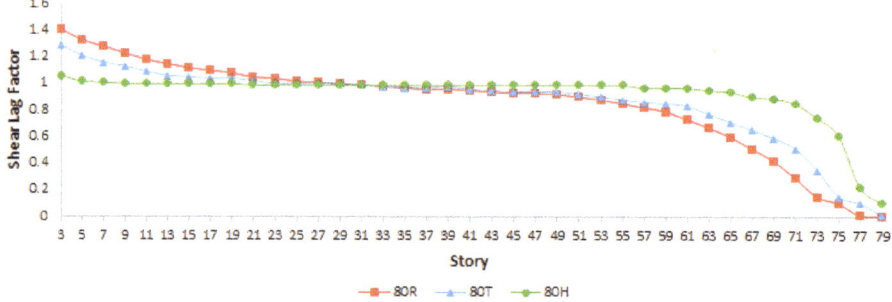

Figure 9. Shear lag factor diagram for 80 story structures subjected to the wind load.

Shear lag factors for static earthquake load are also available in Table 4. A lower positive shear lag up to 13% in the 80H model in comparison with 80R could be observed, and the 80T model had 5.21% less positive shear lag in comparison with 80R. This amount of reduction in negative shear lag was 21.3% for 80H model and 5.4% for 80T model as compared to the 80R model. This tendency could be observed in all the other models.

In addition to the earthquake load, lateral wind load was also applied to the structures in the static analysis method. The behavior in shear lag reduction in this part of analysis is also like previous sections (See Table 4). For positive shear lag, 80H and 80T specimens—13% and 6%, respectively—behaved better than the 80R model. Likewise, for negative shear lag, the 80H model responded at 24.7%, and the 80T model responded almost 8.2% better than 80R structures.

It is evident that from Figure 7, the shear lag effect is more intense in the top half of the structures, and this trend was observed in all other models. The shear lag switch-level from positive to negative in 20-story structures is between the 10th to 15th floors, for 40-story structures is between the 25th to 30th floors, for 60-story structures is between the 35th to 40th floors, and for 80-story structures is between the 45th to 50th floors.

4.3. The Effect of Structural Height on Shear Lag of Framed Tube Structures

Structural height in tubular buildings has a direct effect on shear lag phenomenon. As the height of a framed tube structure increases, the positive shear lag in each story will decrease as well.

Table 5 indicates that 20 story structures had the highest positive shear lag factor in all types of plan geometry. This factor decreased as the number of stories increased almost in all the subsequent models. For example, the positive shear lag factor for 20R model against the dynamic earthquake load was 1.99, and this number was 1.19, 1.09 and 1.09 for 40R, 60R and 80R models, respectively. Furthermore, the average of positive shear lag factors for all models subjected to various lateral loads, without considering their plan geometry, is shown in Table 5. It is observed that positive shear lag factors in taller structures are fewer than this factor in shorter models and has nothing to do with the factor of geometry. In addition, the average of positive shear lag factors without considering any other factor such as load type and plan geometry was calculated. The positive shear lag factor in 20-story structures was 31.91% fewer than in 40-story structures, 37.02% fewer than in 60-story structures and 37.92% fewer than in 80-story structures. It is concluded that the positive shear lag phenomenon has a negative correlation with the height in framed tube structures subjected to any type of lateral load and with any plan geometry.

Table 5. Average of positive shear lag factor among three types of plan geometry for each height.

Story/Load Type	20 Story	40 Story	60 Story	80 Story
Dynamic Earthquake	1.62	1.13	1.06	1.05
Static Earthquake	1.62	1.15	1.07	1.08
Wind	1.92	1.23	1.12	1.08

4.4. Comparison and Verification of the Results

In this study, a 40-story reinforced concrete framed tube building was chosen to compare the Matrix method [31] and Haji-Kazemi and Company [13] analyses results. Beams and columns dimensions in this example are 0.8 × 0.8 m. Each story is 3 m in height, and center to center spacing between columns is 2.5 m. The modulus of elasticity and shear modulus of concrete are 20 and 8.0 GPa, respectively. The external load is 120 kN/m and uniformly distributed along the height of the structure. Figures 10 and 11 show the axial forces in columns of the web and flange of the structure at the base and 10th floor of the framed tube, respectively.

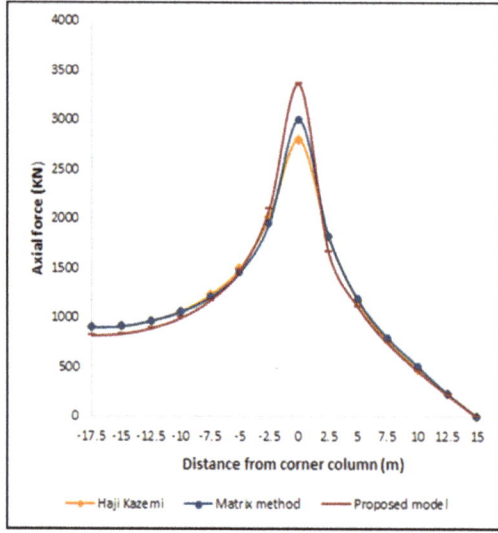

Figure 10. Axial force distribution in the flange and web columns at the base of the building.

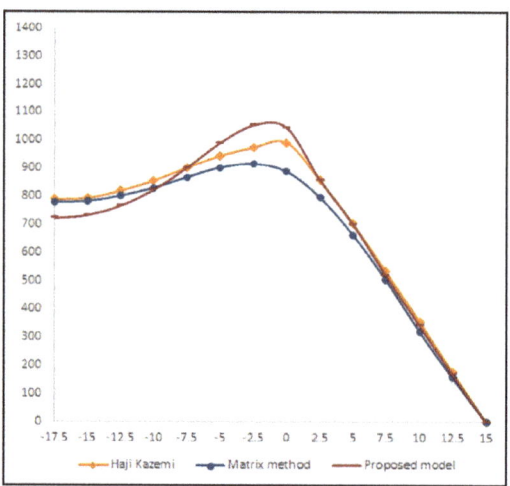

Figure 11. Axial force distribution in the flange and web columns at the 10th floor of the building.

Due to the symmetry of the structure, only half of the web and flange was considered. These diagrams illustrate that the axial forces in corner columns obtained from the proposed model have 9% and 10% difference in comparison with Haji-Kazemi and Company [13] and the Matrix method [31], respectively. In all models and analyses, the shear lag phenomenon was positive at lower heights and negative in the upper stories. Specifically, this anomaly was at a minimum in the middle of each structure.

This factor will decrease gradually in upper stories so that shear lag factor in the middle of the structures decreases to 1 which is considered no shear lag. This trend had no change, and the shear lag factor reduced till the last story. In sharp contrast with lower floors, minimum and negative shear lag factors were obvious in the upper stories. The above results are consistent with the previous studies [6].

5. Conclusions

This study investigated the effect of lateral load type, plan geometry and height on shear lag behavior of framed tube structures. For this goal, 12 models in four different heights and three different plan geometry against three different load types were considered. From the structural analyses performed, it could be concluded that:

(1) Type of the lateral load could affect the distribution of forces in peripheral columns in tubular structures. Wind load caused a greater amount of positive shear lag in comparison with the dynamic earthquake load and the static earthquake load by 9% and 7.5%, respectively. These numbers for negative shear lag were 14% and 1.5%, respectively. In regard to the importance of wind load in the design of high-rise structures and the severity of shear lag in framed tube structures designed based on it, the above results should be seriously considered by structural designers.

(2) Shear lag phenomenon could be affected significantly by the geometry plan in framed tube structures. Hexagon shaped plan structures had a reasonable behavior against lateral loads. Specifically, the average of positive and negative shear lag factors in the three types of analyses were 28.76%, and 25% less in hexagon shaped plan structures, respectively, in comparison with the control model (rectangular-shaped plan). This superiority may lead the structures towards being more laterally load resistant, of lighter weight and more economical due to its equal load distribution in the whole frame.

(3) Rectangular-shaped plan structures had the most inequality of axial force distribution in the flange frame columns. In the above mentioned structures, the average of positive shear lag rose to near two and even more and the average of negative shear lag fell down to below 0.5 in some cases. These amounts of shear lag are not evident in any other shaped plans.

(4) The structures with triangular-shaped plan had almost the same amount of shear lag with rectangular-shaped plan structures in shorter buildings, but the triangular plan had a better behavior in terms of shear lag than the rectangular plan in taller structures. The triangular performed almost 5% better, in the case of positive, and 8.4% in the case of negative shear lag on average in comparison with the rectangular-shaped plan.

(5) Shear lag of framed tube structures is highly affected by the height of the structure. Column axial forces were distributed more unequally in shorter structures, and taller buildings had smaller amount of shear lag factors. It can be concluded that in taller buildings, the structural behavior of the box-shaped cantilever beam that represents the whole building is more similar to Euler–Bernoulli beam than the shorter building and, as it is known from theory of structures, the effect of shear lag in Euler–Bernoulli cantilever beams (taller buildings) is lower than the shorter ones.

Author Contributions: Formal analysis, S.T.; Funding acquisition, S.S.R.K. and M.P.; Investigation, S.T.; Project administration, M.M., S.T., S.S.R.K. and M.P.; Resources, M.M. and S.S.R.K.; Software, S.T.; Supervision, M.M.; Writing—review & editing, M.M. and S.T.; and All authors have read and agreed to the published version of the manuscript.

Funding: This work was funded by the Ministry of Education, Youth, and Sports of the Czech Republic and the European Union (European Structural and Investment Funds Operational Program Research, Development, and Education) in the framework of the project "Modular platform for autonomous chassis of specialized electric vehicles for freight and equipment transportation", Reg. No. CZ.02.1.01/0.0/0.0/16_025/0007293, as well as financial support from internal grants in the Institute for Nanomaterials, Advanced Technologies and Innovations (CXI), Technical University of Liberec (TUL).

Acknowledgments: The authors would like to acknowledge the financial support by Ministry of Education, Youth and Sports of the Czech Republic and the European Union (European Structural and Investment Funds—Operational Program Research, Development and Education), Reg. No. CZ.02.1.01/0.0/0.0/16_025/0007293, as well as the financial support from internal grants in the Institute for Nanomaterials, Advanced Technologies and Innovations (CXI), Technical University of Liberec (TUL).

Conflicts of Interest: The authors declare no conflict of interest.

References

1. Taranath, B.S. *Tall Building Design: Steel, Concrete, and Composite Systems*; CRC Press: New York, NY, USA, 2016.
2. Taranath, B.S. *Wind and Earthquake Resistant Buildings: Structural Analysis and Design*; CRC Press: New York, NY, USA, 2004; pp. 298–300.
3. Smith, B.S.; Coull, A. *Tall Building Structures Analysis and Design*; Wiley-Interscience: New York, NY, USA, 1991.
4. Lee, K.-K.; Guan, H.; Loo, Y.C. Simplified analysis of shear-lag in framed-tube structures with multiple internal tubes. *Comput. Mech.* **2000**, *26*, 447–458.
5. Ali, M.M.; Moon, K.S. Structural developments in tall buildings: Current trends and future prospects. *Archit. Sci. Rev.* **2007**, *50*, 205–223. [CrossRef]
6. Chang, S.T.; Zheng, F.Z. Negative shear lag in cantilever box girder with constant depth. *J. Struct. Eng.* **1987**, *113*, 20–35. [CrossRef]
7. Shiraishi, N.; Matsumoto, M.; Shirato, H.; Ishizaki, H. On aerodynamic stability effects for bluff rectangular cylinders by their corner-cut. *J. Wind Eng. Ind. Aerodyn.* **1988**, *28*, 371–380. [CrossRef]
8. Hayashida, H.; Iwasa, Y. Aerodynamic shape effects of tall building for vortex induced vibration. *J. Wind Eng. Ind. Aerodyn.* **1990**, *33*, 237–242. [CrossRef]
9. Connor, J.J.; Pouangare, C.C. Simple model for design of framed-tube structures. *J. Struct. Eng.* **1991**, *117*, 3623–3644. [CrossRef]
10. Kwan, A.K.H. Simple method for approximate analysis of framed tube structures. *J. Struct. Eng.* **1994**, *120*, 1221–1239. [CrossRef]
11. Han, S.W.; Oh, Y.H.; Lee, L.H. Structural performance of shear wall with sectional shape in wall-type apartment building. *J. Korea Concr. Inst.* **2000**, *12*, 3–14.
12. Lee, K.K.; Lee, L.H.; Lee, E.J. Prediction of shear-lag effects in framed-tube structures with internal tube (s). *Struct. Des. Tall Spec. Build.* **2002**, *11*, 73–92. [CrossRef]
13. Haji-Kazemi, H.; Company, M. Exact method of analysis of shear lag in framed tube structures. *Struct. Des. Tall Spec. Build.* **2002**, *11*, 375–388. [CrossRef]
14. Moghadasi, M.; Keramati, A. Effects of adding internal tube(s) on behavior of framed tube tall buildings subjected to lateral load. In Proceedings of the 7th Asia Pacific Structural Engineering and Construction Conference (APSEC) and 2nd European Asian Civil Engineering Forum (EACEF), Langkawi, Malaysia, 25 July 2009; Volume 1.
15. Shin, M.; Kang, T.H.K.; Pimentel, B. Towards optimal design of high-rise building tube systems. *Struct. Des. Tall Spec. Build.* **2012**, *21*, 447–464. [CrossRef]
16. Mazinani, I.; Jumaat, M.Z.; Ismail, Z.; Chao, O.Z. Comparison of shear lag in structural steel building with framed tube and braced tube. *Struct. Eng. Mech.* **2014**, *49*, 297–309. [CrossRef]
17. Nagvekar, Y.D.; Hampali, M.P. Analysis of shear lag effect in hollow structures. *Int. J. Eng. Res. Technol.* **2014**, *3*, 936–937.
18. Szolomicki, J.; Golasz-Szolomicka, H. Technological Advances and Trends in Modern High-Rise Buildings. *Buildings* **2019**, *9*, 193. [CrossRef]
19. Alaghmandan, M.; Elnimeiri, M.; Krawczyk, R.J.; Buelow, P.V. Modifying Tall Building Form to Reduce the Along-Wind Effect. *CTBUH J.* **2016**, *1*, 34–39.
20. Shi, Q.; Zhang, F. Simplified calculation of shear lag effect for high-rise diagrid tube structure. *J. Build Eng.* **2019**, *22*, 486–495. [CrossRef]
21. Hentz, S.; Donzé, F.V.; Daudeville, L. Discrete Element modeling of concrete submitted to dynamics loading at high strain rates. *Comput. Struct.* **2004**, *82*, 2509–2524. [CrossRef]
22. Golewski, G.; Sadowski, T. Fracture toughness at shear (Mode II) of concretes made of natural and broken aggregates. *Brittle Matrix Compos. 8* **2006**, 537–546. [CrossRef]
23. Dec. 23, 1970: World Trade Center Tops Out. Available online: https://www.wired.com/2008/12/dayintech-1223/ (accessed on 2 October 2020).
24. Hoxton Press. N1. Available online: https://www.homeviews.com/development/hoxton-press-n1/ (accessed on 2 October 2020).
25. The Skyscaper Center. Available online: http://www.skyscrapercenter.com/building/flatiron-building/9014 (accessed on 2 October 2020).

26. *National Iranian Buildings Code, Part 6, Loads on Structures*; BHRC Publication: Tehran, Iran, 2009.
27. *Iranian Code of Practice for Seismic Resistant Design of Buildings*, 4th ed.; BHRC Publication: Tehran, Iran, 2015.
28. *Minimum Design for Buildings and Other Structures*; American Society of Civil Engineering (ASCE7-16 Standard): Fairfax, VA, USA, 2016.
29. *Design of Concrete Structures for Buildings*; Canadian Concrete Code (CSA A.23.3-19): Toronto, ON, Canada, 2019.
30. Awida, T.A. Impact of Building Plan Geometry on the Wind Response of Concrete Tall Buildings. In Proceedings of the Modern Methods and Advances in Structural Engineering and Construction (ISEC-6), Zurich, Switzerland, 21–26 June 2011.
31. Ha, K.H.; Moselhi, O.; Fazio, P. Orthotropic membrane for tall building analysis. *J. Struct. Div. ASCE* **1978**, *104*, 1495–1505.

© 2020 by the authors. Licensee MDPI, Basel, Switzerland. This article is an open access article distributed under the terms and conditions of the Creative Commons Attribution (CC BY) license (http://creativecommons.org/licenses/by/4.0/).

MDPI
St. Alban-Anlage 66
4052 Basel
Switzerland
Tel. +41 61 683 77 34
Fax +41 61 302 89 18
www.mdpi.com

Crystals Editorial Office
E-mail: crystals@mdpi.com
www.mdpi.com/journal/crystals